AF411410

Occurrence and Functions of Endophytic Fungi in Crop Species

Occurrence and Functions of Endophytic Fungi in Crop Species

Editor

Rosario Nicoletti

MDPI • Basel • Beijing • Wuhan • Barcelona • Belgrade • Manchester • Tokyo • Cluj • Tianjin

Editor
Rosario Nicoletti
Council for Agricultural
Research and Economics and
University of Naples Federico II
Italy

Editorial Office
MDPI
St. Alban-Anlage 66
4052 Basel, Switzerland

This is a reprint of articles from the Special Issue published online in the open access journal *Agriculture* (ISSN 2077-0472) (available at: https://www.mdpi.com/journal/agriculture/special_issues/endophytic_fungi).

For citation purposes, cite each article independently as indicated on the article page online and as indicated below:

LastName, A.A.; LastName, B.B.; LastName, C.C. Article Title. *Journal Name* **Year**, *Volume Number*, Page Range.

ISBN 978-3-0365-3272-1 (Hbk)
ISBN 978-3-0365-3273-8 (PDF)

Contents

About the Editor

Rosario Nicoletti was born in Napoli (Italy) on 2 May 1961. Nicoletti is married to Luisa and is the father of Simone. In 1985, he graduated in Agricultural Sciences from the Faculty of Agriculture, University of Naples Federico II. On 1 July 1988, Nicoletti entered the research staff of the Italian Ministry of Agriculture and Forestry, working at the Tobacco Experiment Institute, the Flower Crops Experiment Institute, and the Olive-Growing Experiment Institute. He is currently employed by the Council for Agricultural Research and Economics (CREA) and, since 2014, he has worked as a seconded senior researcher at the Department of Agricultural Sciences, University of Naples Federico II. He is the author/coauthor of approximately 200 papers published in scientific journals or presented at national and international congresses, of which 74 are registered in Scopus. He has served as a reviewer for approximately 100 scientific journals, and he is a member of the Editorial Boards of the journals *Agriculture* and *Folia Horticulturae.*

 *agriculture*

Editorial

Occurrence and Functions of Endophytic Fungi in Crop Species

Rosario Nicoletti [1,2]

[1] Council for Agricultural Research and Economics, Research Centre for Olive, Fruit and Citrus Crops, 81100 Caserta, Italy; rosario.nicoletti@crea.gov.it
[2] Department of Agricultural Sciences, University of Naples 'Federico II', 80055 Portici, Italy

Citation: Nicoletti, R. Occurrence and Functions of Endophytic Fungi in Crop Species. *Agriculture* **2021**, *11*, 18. https://doi.org/10.3390/agriculture11010018

Received: 16 December 2020
Accepted: 28 December 2020
Published: 30 December 2020

Publisher's Note: MDPI stays neutral with regard to jurisdictional claims in published maps and institutional affiliations.

After having been initially boosted by the general aim to exploit biodiversity in natural environments, research on endophytic microorganisms has recently started considering their occurrence in crop species. Many studies have shown that these microbial associates may improve plant fitness through various biological mechanisms of interaction, and have a major impact on plant growth and productive parameters. Besides the relevance of interesting case studies disclosing the effects/properties of single strains/species, a more comprehensive ecological approach should consider that endophytes more effectively play these functional roles in the form of interacting consortia. For this reason, it is important to organize, analyse, and implement the available information on the occurrence and functions of microbes that are part of the crop biocoenosis as a fundamental condition to define possible translational applications in view of enhancing crop performances.

A series of reviews have been recently delivered in literature dealing with the occurrence of endophytic fungi in cultivated plants, considering general aspects [1–4], specific crops [5–7], or implications in crop management [8–12]. This Special Issue is aimed at providing a contribution through making available a collection of papers reviewing the state of the art concerning the occurrence and properties of endophytic fungi associated with crop species or other plants of economic importance. It includes reviews concerning citrus [13], tomato [14], the Amaryllidaceae family [15], and medicinal plants, such as sages [16] and species in the Asteraceae family [17]. Another paper considers aspects pertaining to the trade of ornamentals, following concerns advanced by the European Food Safety Authority for the circulation of pathogens in asymptomatic plant materials [18]. This topic is also the thread of a review dedicated to one of such pathogens, *Lasiodiplodia theobromae*, which, besides concerns of its potential spread via the plant trade, has recently been spreading to temperate areas as a consequence of global warming [19]. More directly considering fungi whose ecological role is exploited in the integrated pest management of crops, the endophytic occurrence of species in the closely related genera *Lecanicillium* and *Akanthomyces* is examined in [20]. Finally, two case studies are proposed touching aspects related to the possible relevance of endophytic fungi in crops, such as mycotoxin production [21] and nutritional interactions concerning fertilizers [22].

Funding: This research received no external funding.

Data Availability Statement: No new data were created or analyzed in this study. Data sharing is not applicable to this article.

Conflicts of Interest: The author declares no conflict of interest.

References

1. Suryanarayanan, T.S.; Devarajan, P.T.; Girivasan, K.P.; Govindarajulu, M.B.; Kumaresan, V.; Murali, T.S.; Rajamani, T.; Thirunavukkarasu, N.; Venkatesan, G. The host range of multi-host endophytic fungi. *Curr. Sci.* **2018**, *115*, 1963–1969. [CrossRef]
2. Nicoletti, R.; Fiorentino, A. Plant bioactive metabolites and drugs produced by endophytic fungi of Spermatophyta. *Agriculture* **2015**, *5*, 918–970. [CrossRef]
3. Lugtenberg, B.J.; Caradus, J.R.; Johnson, L.J. Fungal endophytes for sustainable crop production. *FEMS Microbiol. Ecol.* **2016**, *92*, fiw194. [CrossRef] [PubMed]
4. Yan, L.; Zhu, J.; Zhao, X.; Shi, J.; Jiang, C.; Shao, D. Beneficial effects of endophytic fungi colonization on plants. *Appl. Microbiol. Biotechnol.* **2019**, *103*, 3327–3340. [CrossRef]
5. Nicoletti, R.; Di Vaio, C.; Cirillo, C. Endophytic fungi of olive tree. *Microorganisms* **2020**, *8*, 1321. [CrossRef]
6. Card, S.D.; Hume, D.E.; Roodi, D.; McGill, C.R.; Millner, J.P.; Johnson, R.D. Beneficial endophytic microorganisms of *Brassica*—A review. *Biol. Control* **2015**, *90*, 102–112. [CrossRef]
7. Jia, M.; Chen, L.; Xin, H.L.; Zheng, C.J.; Rahman, K.; Han, T.; Qin, L.P. A friendly relationship between endophytic fungi and medicinal plants: A systematic review. *Front. Microbiol.* **2016**, *7*, 906. [CrossRef]
8. Eberl, F.; Uhe, C.; Unsicker, S.B. Friend or foe? The role of leaf-inhabiting fungal pathogens and endophytes in tree-insect interactions. *Fungal Ecol.* **2019**, *38*, 104–112. [CrossRef]
9. Busby, P.E.; Ridout, M.; Newcombe, G. Fungal endophytes: Modifiers of plant disease. *Plant Mol. Biol.* **2016**, *90*, 645–655. [CrossRef]
10. Barelli, L.; Moonjely, S.; Behie, S.W.; Bidochka, M.J. Fungi with multifunctional lifestyles: Endophytic insect pathogenic fungi. *Plant Mol. Biol.* **2016**, *90*, 657–664. [CrossRef]
11. Mehta, P.; Sharma, R.; Putatunda, C.; Walia, A. Endophytic fungi: Role in phosphate solubilization. In *Advances in Endophytic Fungal Research*; Singh, B.P., Ed.; Springer: Cham, Switzerland, 2019; pp. 183–209.
12. Vega, F.E. The use of fungal entomopathogens as endophytes in biological control: A review. *Mycologia* **2018**, *110*, 4–30. [CrossRef] [PubMed]
13. Nicoletti, R. Endophytic fungi of citrus plants. *Agriculture* **2019**, *9*, 247. [CrossRef]
14. Sinno, M.; Ranesi, M.; Gioia, L.; d'Errico, G.; Woo, S.L. Endophytic fungi of tomato and their potential applications for crop improvement. *Agriculture* **2020**, *10*, 587. [CrossRef]
15. Caruso, G.; Golubkina, N.; Tallarita, A.; Abdelhamid, M.T.; Sekara, A. Biodiversity, ecology, and secondary metabolites production of endophytic fungi associated with Amaryllidaceae crops. *Agriculture* **2020**, *10*, 533. [CrossRef]
16. Zimowska, B.; Bielecka, M.; Abramczyk, B.; Nicoletti, R. Bioactive products from endophytic fungi of sages (*Salvia* spp.). *Agriculture* **2020**, *10*, 543. [CrossRef]
17. Caruso, G.; Abdelhamid, M.T.; Kalisz, A.; Sekara, A. Linking endophytic fungi to medicinal plants therapeutic activity. A case study on Asteraceae. *Agriculture* **2020**, *10*, 286. [CrossRef]
18. Gioia, L.; d'Errico, G.; Sinno, M.; Ranesi, M.; Woo, S.L.; Vinale, F. A survey of endophytic fungi associated with high risk plants imported for ornamental purposes. *Agriculture* **2020**, *10*, 643. [CrossRef]
19. Salvatore, M.M.; Andolfi, A.; Nicoletti, R. The thin line between pathogenicity and endophytism: The case of *Lasiodiplodia theobromae*. *Agriculture* **2020**, *10*, 488. [CrossRef]
20. Nicoletti, R.; Becchimanzi, A. Endophytism of *Lecanicillium* and *Akanthomyces*. *Agriculture* **2020**, *10*, 205. [CrossRef]
21. Manganiello, G.; Marra, R.; Staropoli, A.; Lombardi, N.; Vinale, F.; Nicoletti, R. The shifting mycotoxin profiles of endophytic *Fusarium* strains: A case study. *Agriculture* **2019**, *9*, 143. [CrossRef]
22. Jastrzębska, M.; Wachowska, U.; Kostrzewska, M.K. Pathogenic and non-pathogenic fungal communities in wheat grain as influenced by recycled phosphorus fertilizers: A case study. *Agriculture* **2020**, *10*, 239. [CrossRef]

 *agriculture*

Article

The Shifting Mycotoxin Profiles of Endophytic *Fusarium* Strains: A Case Study

Gelsomina Manganiello [1], Roberta Marra [1], Alessia Staropoli [1,2], Nadia Lombardi [1], Francesco Vinale [1,2] and Rosario Nicoletti [1,3,*]

[1] Department of Agricultural Sciences, University of Naples Federico II, 80055 Portici, Italy
[2] Institute for Sustainable Plant Protection (IPSP), National Research Council (CNR), 80055 Portici, Italy
[3] Council for Agricultural Research and Economics, Research Centre for Olive, Citrus and Tree Fruit, 81100 Caserta, Italy
*   Correspondence: rosario.nicoletti@crea.gov.it

Received: 1 June 2019; Accepted: 3 July 2019; Published: 5 July 2019

Abstract: *Fusarium* species are known to establish manifold interactions with wild and crop plants ranging from pathogenicity to endophytism. One of the key factors involved in the regulation of such relationships is represented by the production of secondary metabolites. These include several mycotoxins, which can accumulate in foodstuffs causing severe health problems to humans and animals. In the present study, an endophytic isolate (A1021B), preliminarily ascribed to the *Fusarium incarnatum-equiseti* species complex (FIESC), was subjected to biochemical and molecular characterization. The metabolomic analysis of axenic cultures of A1021B detected up to 206 compounds, whose production was significantly affected by the medium composition. Among the most representative products, fusaric acid (FA), its derivatives fusarinol and 9,10-dehydro-FA, culmorin and bikaverin were detected. These results were in contrast with previous assessments reporting FIESC members as trichothecene rather than FA producers. However, molecular analysis provided a conclusive indication that A1021B actually belongs to the species *Fusarium babinda*. These findings highlight the importance of phylogenetic analyses of *Fusarium* species to avoid misleading identifications, and the opportunity to extend databases with the outcome of metabolomic investigations of strains from natural contexts. The possible contribution of endophytic strains in the differentiation of lineages with an uneven mycotoxin assortment is discussed in view of its ensuing impact on crop productions.

Keywords: endophytic fungi; *Fusarium*; species complexes; mycotoxins; fusaric acid; trichothecenes; biosynthetic gene clusters

---

1. Introduction

Fusarium species have been commonly reported in the majority of bioclimatic regions and ecosystems, where they occur as endophytes, latent plant pathogens, or soil saprobes, thus showing a considerable ecological plasticity [1–3]. Some *Fusarium* species may cause severe plant diseases, and contaminate crop productions with mycotoxins, which are secondary metabolites (SMs) of major concern to food and feed safety worldwide [4–7]. This aptitude may not only involve pre- and post-harvest plant pathogens, but also strains which develop endophytically without causing disease symptoms [8,9]. On the other hand, the release of bioactive SMs in plant tissues by endophytic strains may induce defensive responses against pests and pathogens [10], with positive implications on plant growth [11].

The genus *Fusarium* is also well known for a controversial taxonomic history, where species descriptions were basically founded on key morphological characters [3]. In the last decades several studies considered data on SM production as a possible support in *Fusarium* taxonomy [12–14]. However, such a sound approach has been impaired by the finding that synapomorphy, i.e., the occurrence of certain common characters in distantly related organisms, has notably affected the classification of *Fusarium* strains in the past [15]. More recently, the advances in the DNA-sequencing technology allowed the identification of *Fusarium* spp. based on multi-gene genealogies [16–19], thus improving the phylogenetic accuracy and the taxonomic resolution [20–22]. Nevertheless, the ongoing deposition of DNA sequences in database resulting from the manifold surveys of natural populations of *Fusarium*, together with the characterization of novel species, may result in incorrect matches and, subsequently, in misleading identifications [23].

In this work, we report a case study describing a *Fusarium* strain (A1021B) that was recovered as an endophyte of common spindle (*Euonymus europaeus*) at the Astroni Natural Reserve near Naples, Italy. The fungus was provisionally defined to belong to the *Fusarium incarnatum-equiseti* species complex (FIESC) [24]. Fusarinol, a derivative of fusaric acid (FA), was the main extrolite purified from cultures grown in Czapek-Dox broth (CDB) [25]. Afterwards, we found FA to be the major SM produced by A1021B in potato dextrose broth (PDB), thus confirming that FA production in vitro is influenced by the culture medium composition [26–28]. Previous studies reported that FA production is strain-dependent even in species known as common producers, and it can be stimulated in some reluctant *Fusarium* strains by co-cultivation with other fungi [29]. Factors regulating gene expression are fundamental in explaining variation in SM production. Gene clusters for FA synthesis have been detected in many *Fusarium* spp. [15,30], and the deletion of specific genes has been reported to affect the production of FA and related compounds [31,32].

Our results appeared in contrast with the mycotoxin profile commonly associated with FIESC members. In fact, previous investigations failed to detect the production of FA in species/strains ascribed to this species complex [26,33–35], which are mainly known as trichothecene producers [35,36]. FA was listed among the mycotoxins of *F. equiseti* in a couple of recent papers [15,28], but no specific references were provided supporting this inference. Therefore, further investigations were undertaken concerning both the authentic taxonomic identity of strain A1021B and its biochemical and molecular characterization.

2. Materials and Methods

2.1. Fungal Strain and Culture Conditions

The *Fusarium* strain A1021B was maintained on potato dextrose agar (PDA, HI-Media, Mumbai, India) at 4 °C, and subcultured bimonthly. For metabolomic investigation CDB and PDB (both from HI-Media) cultures were prepared in 250-mL flasks containing 100 mL of broth. Twelve flasks were prepared for each medium and inoculated with 10 plugs (5 mm diameter) from 6-day old PDA cultures. Six flasks were incubated at 25 °C in a growth chamber with 16:8 h photoperiod, while the remaining 6 were incubated at 25 °C in darkness. These batches were further divided into two groups (each including three replicates) which were grown for one or two weeks, respectively. Then fungal debris were filtered through three layers of cheesecloth, and the filtrates were stored at −20 °C.

Solid fermentation of strain A1021B was carried out on maize kernels (MK). After rinsing three times in sterile water, 30 g of kernels were placed in each of twelve 250-mL flasks and sterilized (121 °C, 20 min). Five milliliters of sterile water were added to each flask, that were subsequently inoculated as described for liquid cultures. Similarly, the flasks were grouped in 4 batches, each consisting of three replicates, and incubated at 25 °C as described above. After one or two weeks, a 10 g sample was taken from each MK flask and separately ground to be further processed.

2.2. Culture Extraction and LC-MS Analysis

MK samples were extracted in 8 mL of 50% methanol in water. Samples were centrifuged (10 min at 16,100 g, 4 °C), and the supernatants were collected. These, as well as samples from liquid cultures, were filtered through 0.2 μm polyvinylidene fluoride filters (Chromacol, Welwyn Garden City, UK).

SM profiling was carried out through a 6540 Ultra High Definition (UHD) Accurate Quadrupole Time-of-Flight (Q-TOF) Liquid Chromatography tandem Mass-Spectrometry (LC-MS/MS) mass spectrometer (Agilent Technologies, Santa Clara, CA, USA) with a Dual Electrospray Ionization (ESI) source, coupled to a 1200 series Rapid Resolution High Performance Liquid Chromatography (HPLC) with a Diode Array Detector (DAD) system (all from Agilent Technologies). Samples (7 μL) were injected onto a Poroshell 120EC-C18 1.8 pm, 2.1 × 5 mm reverse phase analytical column (Agilent Technologies) at a constant temperature (35 °C). Mobile phases consisted of (A) water (Cromasolv® Plus, LC-MS-Sigma) and (B) acetonitrile (Cromasolv® Plus, LC-MS-Sigma) both acidified with 0.1% LC-MS grade formic acid. The analyses were carried out at a flow rate of 0.6 mL min^{-1} with the following gradient: 0 min—5% B; 12 min—100% B; 15 min—100% B; 17 min—95% B; 20 min—95% B; 2 min post-time. The UV spectra were collected by DAD every 0.4 s from 190 to 750 nm with a resolution of 2 nm. The source conditions for electrospray ionization were the following: nitrogen gas temperature was 350 °C with a drying gas flow rate of 11 L min^{-1} and a nebulizer pressure of 45 psig. The fragmentor voltage was 180 V and skimmer voltage 45 V. The range acquisition of TOF spectra was from 50 to 1600 m/z with an acquisition rate value of 3 spectra s^{-1}. Data were collected in positive ion mode. The real-time lock mass correction was performed by using two reference mass solutions including purine ($C_5H_4N_4$ at m/z 121.050873, 10 μmol L^{-1}) and hexakis (1H,1H,3H-tetrafluoropentoxy)-phosphazene ($C_{18}H_{18}O_6N_3P_3F_{24}$ at m/z 922.009798, 2 μmol L^{-1}). These solutions were purchased from Agilent Technologies and injected into MS by an isocratic pump at a constant flow rate (0.06 mL min^{-1}). Solvents were LC–MS grade, and all other chemicals were analytical grade. All were from Sigma-Aldrich (Steinheim, Germany) unless otherwise stated.

Mass spectra were analyzed through the MassHunter Qualitative Analysis Software B.06.00 (Agilent Technologies), and then through the MassProfile Professional Software (Agilent Technologies) to compute the annotation and statistical analyses. LC-MS data were compared to known compounds included in an in-house database, as previously described [37,38].

Graphical representations were performed using ClustVis, a web-based multivariate data analysis tool. The principal component analysis (PCA) was performed using the Singular Value Decomposition (SVD) with imputation algorithm in ClustVis online tool. Data on SMs were summarized using the heatmap function in ClustVis tool with row centered and unit variance scaling applied. The hierarchical clustering was obtained using correlation method. Compounds with normalized intensity values >2 were used to analyze common and unique entities in the different treatments by Venn diagrams with the online tool jvenn (http://jvenn.toulouse.inra.fr/app/index.html).

2.3. DNA Extraction and PCR Conditions

Isolate A1021B was grown in PDB on a rotary shaker at 120 rpm for 96 h at 25 °C. Fresh mycelium was collected after vacuum filtration through No. 4 Whatman filter paper (Whatman Biosystems Ltd., Maidstone, UK), then frozen in liquid nitrogen, ground to a fine powder and stored at −80 °C until further processing. Total genomic DNA was extracted from 10 mg of ground mycelium by using the NucleoSpin® Soil kit (Macherey-Nagel, Düren, Germany) according to the manufacturer's protocol. Sequences of the housekeeping genes calmodulin (*CAL1*), translation elongation factor (*TEF1*), β-tubulin (*TUB2*) and internal transcribed spacer 1–4 (ITS) were amplified using the following PCR program: denaturation at 96 °C for 2 min; 35 cycles of denaturation at 94 °C for 30 s, annealing at 55 °C for 30 s; extension at 68 °C for 75 s; and final extension at 68 °C for 10 min. Before sequencing, PCR products were purified using PureLink PCR purification kit (Invitrogen, Paisley, UK) following the manufacturer's instructions. Furthermore, the presence of amplicons related to the trichothecene biosynthetic genes *TRI1*, *TRI4*, *TRI5*, *TRI8*, *TRI11* was investigated. All primers used in this work are reported in Table 1.

Table 1. Primers used in this study for DNA sequence amplifications.

Target Gene	5′–3′ Sequence	References
TRI1	GCGTCTCAGCTTCATCAAGGCAKCKAMTGAWTCG	[39]
	CTTGACTTSMTTGGCKGCAAAGAARCGACCA	
TRI4	CCAATCAGYCAYGCTRTTGGGATACTG	[39]
	ACCCGGATTTCRCCAACATGCT	
TRI5	GGCATGGTCGTGTACTCTTGGGTCAAGGT	[39]
	GCCTGMYCAWAGAAYTTGCRGAACTT	
TRI8	GACCAGNAYCACSGYCAACAGTTCAG	[35]
	GAACAGCCRCTCCRWAACTATTGTC	
TRI11	TWCCCCACAAGRAACAYCTYGARCT	[35]
	TCCCASACTGTYCTSGCMAGCATCAT	
CAL	GARTWCAAGGAGGCCTTCTC	[17]
	TTTTGCATCATGAGTTGGAC	
TEF1	ATGGGTAAGGARGACAAGAC	[16]
	GGARGTACCAGTSATCATGTT	
TUB2	GGTAACCAAATCGGTGCT	[40]
	ACCCTCAGTGTAGTGACCCTYTGGC	
ITS 1–4	CTTGGTCATTTAGAGGAAGTAA	
	TCCTCCGCTTATTGATATGC	

2.4. Species Identification and Phylogenetic Analysis

Phylogenetic relationships of strain A1021B were investigated on account of *CAL1*, *TEF1* and *TUB2* sequences as reported in [15]. DNA sequences were blasted against the NCBI GenBank database using default parameters and then aligned with isolates belonging to the FIESC [15,41] and *Fusarium babinda* by the Clustal W algorithm [42] with MEGA7 software [43]. Phylogenetic trees were inferred using the maximum likelihood method based on Tamura-Nei model applied to the whole set of manually edited aligned sequences. The confidence of the branching was estimated by bootstrap (BP) analysis (1000 BP). A strain of *Fusarium concolor* was used as outgroup for rooting the phylogenetic tree. DNA sequences of the three loci were submitted to GenBank, with the following accession numbers: MK968883 (ITS), MK984207 (*CAL1*), MK984206 (*TEF1*), and MK984208 (*TUB2*).

3. Results

3.1. Metabolome Analysis

The investigation on SM production in liquid (CDB, PDB) or solid (MK) media, the latter representing a commonly used substrate to evaluate mycotoxin production in *Fusarium* spp. [28], revealed that up to 206 compounds are synthesized by strain A1021B in axenic cultures. The PCA score plot demonstrated a differential and significant effect of the medium composition on SM production (Figure 1A). Moreover, the assortment of SMs produced in CDB or PDB was less affected by light exposure than by the culturing time (1 vs. 2 weeks). On the other hand, on MK the SM profile was particularly influenced by the former factor, i.e., multiple compounds after one week of growth in darkness were produced. To obtain a simplified representation of the different assortments, a heatmap clustering compounds was generated (Figure 1B) by selecting 29 entities which made it possible to discriminate among the different treatments, selected on PCA. Among them, FA (179.0974 Da) and its derivatives fusarinol (165.1181 Da) and 9,10-dehydro-FA (177.0799 Da) were detected. Other putatively identified SMs were bikaverin (382.1126 Da), a tetracyclic benzoxanthone whose genetic base is reported to be clustered to FA [44,45], and culmorin (238.1446 Da), a sesquiterpenoid which is often associated with trichothecene production [46,47].

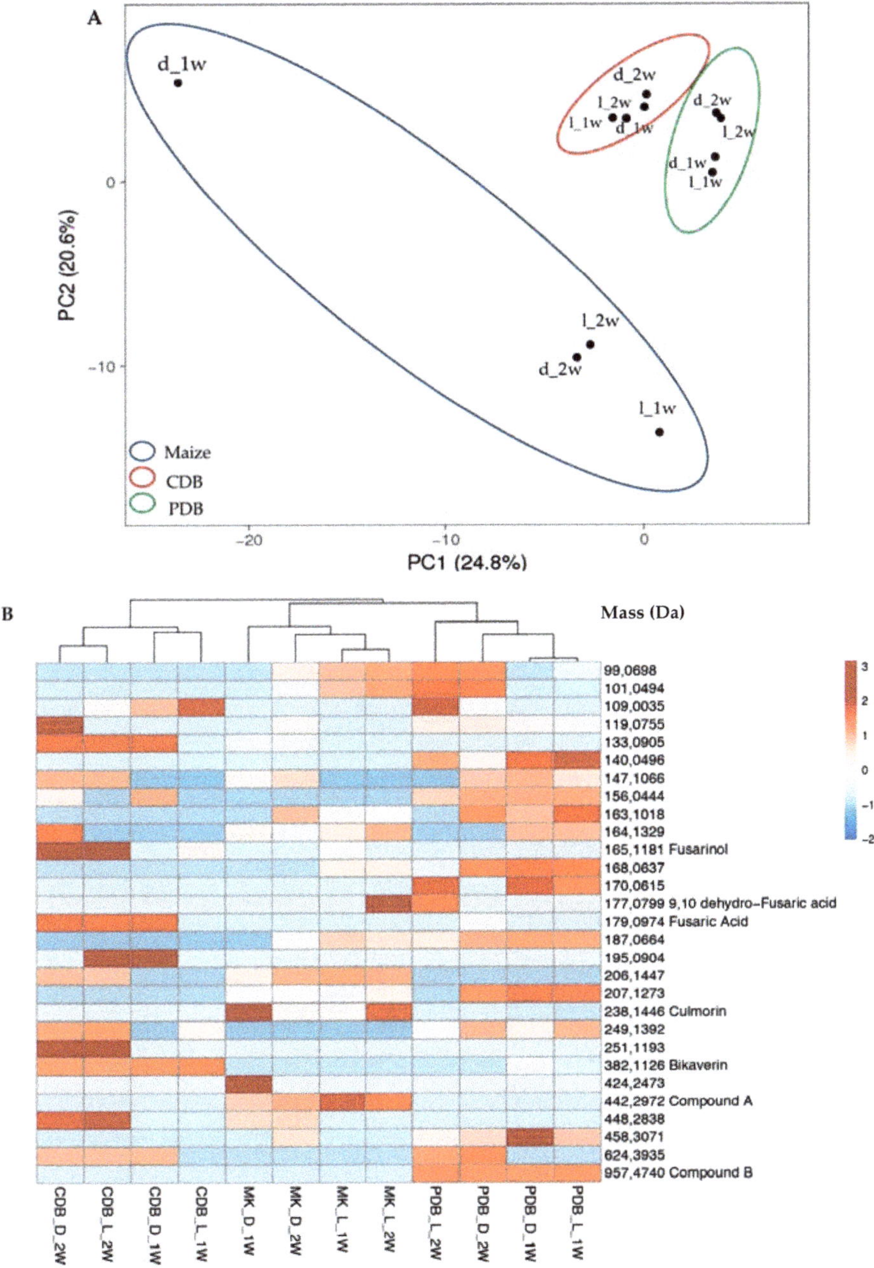

Figure 1. (**A**) Principal component analysis (PCA) score plot of secondary metabolites (SMs) produced by A1021B under different growth conditions. (**B**) Heat map illustrating the abundance of the main SMs in A1021B cultures, visualized through the color scale reported on the right. Each row represents differentially abundant products ordered by their mass (Da), while columns correspond to the different culturing conditions. MK = maize kernels; CDB = Czapek-Dox broth; PDB = potato dextrose broth; d = darkness; l = light; 1 w = 1 week; 2 w = 2 weeks.

Among the identified compounds, the LC-MS Q-TOF analysis revealed that fusarinol was predominantly produced in CDB, in dark as well as in light conditions, while 9,10-dehydro-FA accumulated in PDB and MK. However, both compounds were found only after two weeks of growth. The production of FA was mostly observed in CDB maintained in darkness, or in light exposure after two weeks only. A similar biosynthetic course is displayed by culmorin in PDB, while the production of bikaverin was mainly detected in CDB cultures regardless to the presence/absence of light. Furthermore, among the unidentified molecules, compound A and compound B (Figure 1B) were particularly affected by medium composition. In fact, compound A was detected exclusively in MK while compound B was produced only in PDB, and their production was not related to specific growth condition (1–2 w; light/darkness).

Our analysis did not show the production by A1021B of 8-O-methylbostrycoidin, a polyketide pigment which has been reported in association with FA [48]. Furthermore, no trichothecenes were detected in any cultivation condition.

Venn diagrams showed that A1021B was able to synthesize specific compounds in the different media, and that only a small part of them was in common among the three conditions (Figure 2). In darkness, the growth on MK enhanced the production of specific SMs at both time points considered (81 and 75 compounds, respectively, after 1 or 2 weeks of growth), while CDB was the least inductive medium (6 and 11 compounds, respectively). Moreover, very few compounds (3 and 4, respectively, after 1 or 2 weeks of growth) accumulated constitutively in darkness regardless to the medium. A similar distribution was observed when A1021B was cultivated under light exposure. In fact, MK represented the most inductive substrate, while in CDB few compounds accumulated. No specific SMs were detected in CDB at the first time point. Overall, SM production was higher in darkness (Figure 2), and MK was more effective in enhancing the production of certain compounds.

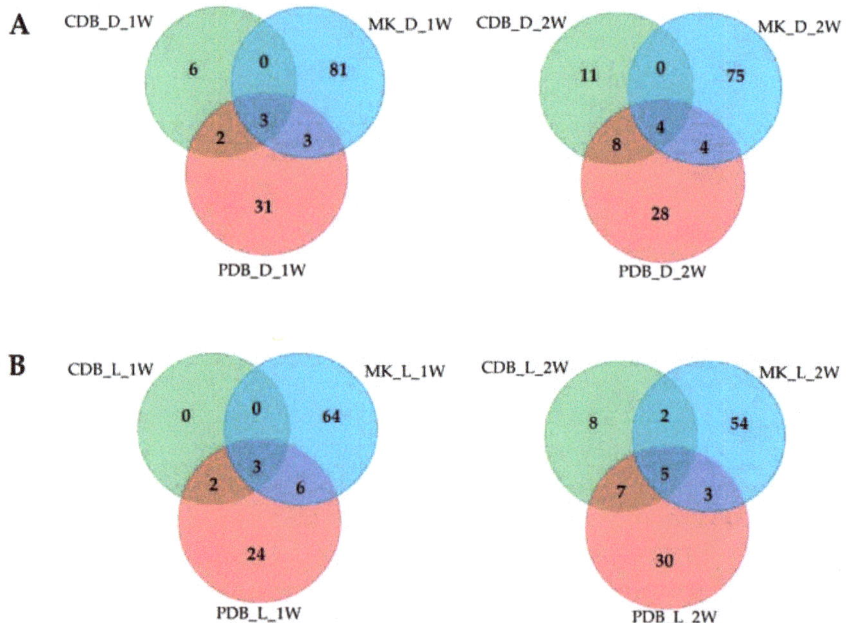

Figure 2. Venn diagrams showing the number of unique and overlapping products in A1021B cultures under the different growth conditions. (**A**) Secondary metabolites (SMs) produced in darkness (d) after one (1 w) or two weeks (2 w). (**B**) SMs produced under light exposure (l) after one (1 w) or two weeks (2 w). MK = maize kernels; CDB = Czapek-Dox broth; PDB = potato dextrose broth.

3.2. Genetic and Phylogenetic Analysis, and Species Identification

As trichothecenes may well characterize the mycotoxin profile of FIESC members, the presence in A1021B of genes involved in biosynthesis of these compounds was investigated by PCR as previously described [15]. Amplicons of all the selected regions (*TRI-1, TRI-4, TRI-5, TRI-8, TRI-11*) were detected (data not shown), indicating that strain A1021B actually holds the genetic features to produce these mycotoxins. Nevertheless, the related SMs were not detected in any of the culture conditions used in this study.

Even if the genetic data matched with the hypothesis that A1021B might belong to the FIESC, a different indication resulted from the phylogenetic analysis, conducted using concatamers of ITS, *TEF1* and *CAL1* sequences previously employed in the characterization of this species complex. In this experiment, a strain of *F. concolor* was used as the outgroup [41]. Interestingly, A1021B clustered with the latter instead of any of the several identified or unidentified FIESC members (Figure 3). Nevertheless, a new BLAST search in the NCBI database based on *TEF1* sequences did not yield a consistent homology with the available strains of *F. concolor*.

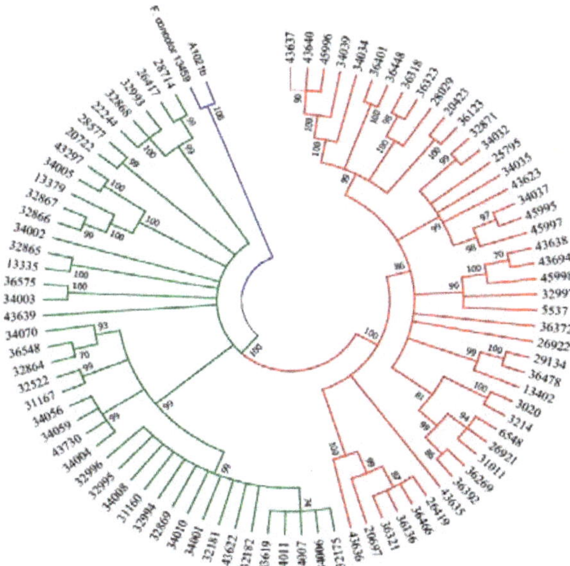

Figure 3. Maximum likelihood tree inferred from ITS-TEF1-CAL1 concatamers. Phylogenetic analysis including A1021B, FIESC members and *F. concolor* as outgroup inferred using the maximum likelihood method (MEGA7). The bootstrap consensus tree inferred from 1000 replicates is taken to represent the evolutionary history of the taxa analyzed.

The hypothesis that strain A1021B represented a novel taxon could explain such discrepancy. However, a subsequent BLAST search carried out in December 2018 revealed an unexpected 100% homology with a series of *TEF1* sequences from the species *F. babinda* [49], which were made available in October 2018 after another notable taxonomic revision [45]. Following this finding, another phylogenetic tree including isolates of *F. concolor*, *F. babinda* and FIESC was generated where A1021B clearly clustered with the strains of *F. babinda* (Figure 4).

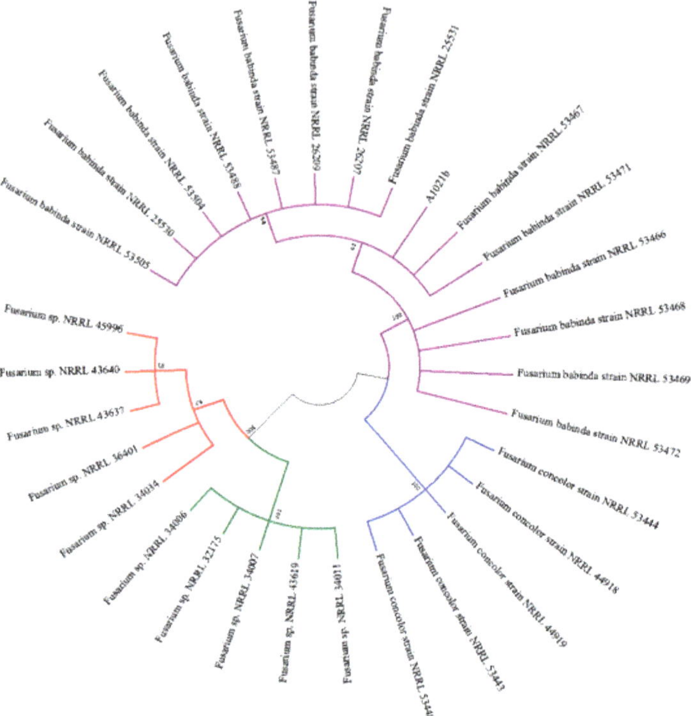

Figure 4. Maximum likelihood tree inferred from TEF1. Phylogenetic analysis including A1021B, strains of FIESC, *F. babinda* and *F. concolor* inferred using the maximum likelihood method (MEGA7). The bootstrap consensus tree inferred from 1000 replicates is taken to represent the evolutionary history of the taxa analyzed.

4. Discussion

A recent study [45] demonstrated that *F. babinda* represents the correct identification for many strains previously ascribed to *F. polyphialidicum*. This species, on the other hand, has now been reported as a synonym of *F. concolor*, deserving priority in taxonomy as an older accepted species name [50]. While *F. concolor/polyphialidicum* is known as a typical trichothecene producer [28,45], the mycotoxin profile of *F. babinda* seems to be centered on FA, and no clues of trichothecene biosynthetic abilities were detected in the limited assessments carried out so far. An analysis concerning the genetic basis for trichothecene synthesis in a single strain of this species (NRRL 25539) also provided negative results [45]. The same study reported that strain NRRL 25539 has the gene clusters for the production of some compounds (enniatins, fusarin, fusarubin), which, however, were absent in our cultures.

In this work, metabolomic analysis confirmed that in axenic cultures strain A1021B basically produced FA and some known compounds. Interestingly, bikaverin was found to accumulate mainly in CDB cultures, where the carbon source is represented by sucrose, in consistency with a previous report that the availability of this sugar stimulates bikaverin production in vitro [51]. Molecular data indicated the presence of trichothecene biosynthetic gene clusters, but they were not expressed under the culture conditions we tested, thus making A1021B divergent from strain NRRL 25539 [45]. Considering that *F. babinda*, which formerly had been reported only from Australia, turned out to have a worldwide diffusion [45], and that *F. polyphialidicum* was described as a typical producer of type-A trichothecenes [28], our finding highlights the need for more exhaustive investigations on the mycotoxin profile of this emerging species. In this respect, an assessment concerning occurrence of

TRI-5 in *F. equiseti* detected this gene sequence in 50% of the examined strains only [36], confirming previous evidence of uneven production of trichothecenes in this species [52].

Recent evaluations of the mycotoxin-producing ability indicate that *Fusarium* phylogenetic relationship may vary, and non-conforming strains, new species or lineages often result after the exploration of new ecological contexts, particularly those involving endophytic fungi [53–55]. In fact, an intriguing ability to synthesize unexpected SMs can be ascribed to endophytes, which are able to establish physical contacts and eventually interact through horizontal gene transfer (HGT) with both plants and other microorganisms living in this particular ecological niche [56,57]. Indeed, ecological proximity has been considered to favor HGT [57].

In fungi, genes coding for the synthesis of SMs are typically adjacent to one another in clusters of co-expressed genes, including a core gene responsible for the synthesis of a basic structure, and side genes which control chemical modifications, transport, and regulation [58]. Biosynthesis of FA, bikaverin, culmorin and trichothecenes is governed by polyketide synthases, large multi-domain enzymes that catalyze sequential condensation of simple carboxylic acids. A few hundreds of gene sequences involved in the biosynthesis of polyketides have been detected in *Fusarium* spp., which corresponded to 67 clades in a phylogenetic analysis, where each clade refers to distinct products. This analysis also pointed out a genetic potential to synthesize compounds which are the same or similar to those known to be produced from other fungi, but not reported in *Fusarium* so far [59].

From an evolutionary viewpoint, HGT of gene clusters regulating mycotoxin biosynthesis is theoretically supported by the reasonable inference that clustering confers a selective advantage to the cluster itself [56,60]. In addition, the hypothesis that the *TRI-5* gene cluster may have spread among unrelated fungal species through HGT has already been advanced in the past [61]. HGT was also indicated as the means of transmission of a 5-gene cluster presiding over the synthesis of bikaverin from *Fusarium* to *Botrytis cinerea* [51], and as a more general evolutionary mechanism in *Fusarium* [62]. Moreover, it has been demonstrated that transfer of lineage-specific genomic regions occurred in *Fusarium*, including even entire chromosomes up to more than one-quarter of the genome, and involving genes related to pathogenicity. These were effective in converting pathogenic strains into non-pathogenic ones, and were possibly responsible for the emergence of new pathogenic lineages [63]. Therefore, natural ecosystems are recognized to play a role as reservoirs of novel crop pathogens with a meaningful impact on disease management and biosecurity [64].

5. Conclusions

In this work, we reported a case-study investigating the taxonomy and SM production in the endophytic *Fusarium* strain A1021B. As a consequence of the ongoing updates in the phylogenetic relationships of *Fusarium* species, the analyses of mycotoxin profile and selected gene sequences lead us to identify this isolate as *F. babinda*. Our findings support previous observations that SM production in axenic cultures by *Fusarium* strains does not necessarily conform to genetically based analyses, and that this limitation could be overcome in vivo where interaction with the host plant or other endophytic microorganisms may result in the activation of silent genes.

Besides sequences deposited in GenBank, strain A1021B is available on request for inclusion in phylogenetic and metabolomic studies.

Author Contributions: Conceptualization, R.N. and F.V.; methodology, G.M., R.M., A.S., N.L. and F.V.; writing—review and editing, G.M., R.M., F.V. and R.N.

Funding: The research activity of F.V., A.S., R.M. and G.M. was funded by MIURPON [grant number Linfa 03PE_00026_1; grant number Marea 03PE_00106]; MIUR-GPS [grant number Sicura DM29156]; POR FESR CAMPANIA 2014/2020- O.S. 1.1 [grant number Bioagro 559]; MISE [grant number Protection F/050421/01-03/X32].

Conflicts of Interest: The authors declare no conflict of interest.

References

1.   Gordon, T.R.; Martyn, R.D. The evolutionary biology of *Fusarium oxysporum*. *Ann. Rev. Phytopathol.* **1997**, *35*, 111–128. [CrossRef] [PubMed]
2.   Backhouse, D.; Burgess, L.W.; Summerell, B.A. Biogeography of *Fusarium*. In *Fusarium*; Summerell, B.A., Leslie, J.F., Backhouse, D., Bryden, W., Burgess, L.W., Eds.; The American Phytopathology Society: Saint Paul, MN, USA, 2001; pp. 122–137.
3.   Summerell, B.A.; Laurence, M.H.; Liew, E.C.; Leslie, J.F. Biogeography and phylogeography of *Fusarium*: A review. *Fungal Divers.* **2010**, *44*, 3–13. [CrossRef]
4.   Desjardins, A.E.; Hohn, T.M. Mycotoxins in plant pathogenesis. *Mol. Plant Microbe Interact.* **1997**, *10*, 147–152. [CrossRef]
5.   Bryden, W.L. Mycotoxin contamination of the feed supply chain: Implications for animal productivity and feed security. *Anim. Feed Sci. Technol.* **2012**, *173*, 134–158. [CrossRef]
6.   Vendl, O.; Crews, C.; MacDonald, S.; Krska, R.; Berthiller, F. Occurrence of free and conjugated *Fusarium* mycotoxins in cereal-based food. *Food Addit. Contamin.* **2010**, *27*, 1148–1152. [CrossRef] [PubMed]
7.   Marin, S.; Ramos, A.J.; Cano-Sancho, G.; Sanchis, V. Mycotoxins: Occurrence, toxicology, and exposure assessment. *Food Chem. Toxicol.* **2013**, *60*, 218–237. [CrossRef] [PubMed]
8.   Bacon, C.W.; Glenn, A.E.; Yates, I.E. *Fusarium verticillioides*: Managing the endophytic association with maize for reduced fumonisins accumulation. *Toxin Rev.* **2008**, *27*, 411–446. [CrossRef]
9.   Lofgren, L.A.; LeBlanc, N.R.; Certano, A.K.; Nachtigall, J.; LaBine, K.M.; Riddle, J.; Broz, K.; Dong, Y.; Bethan, B.; Kafer, C.W.; et al. *Fusarium graminearum*: Pathogen or endophyte of North American grasses? *New Phytol.* **2018**, *217*, 1203–1212. [CrossRef]
10. Bacon, C.W.; White, J.F. Functions, mechanisms and regulation of endophytic and epiphytic microbial communities of plants. *Symbiosis* **2016**, *68*, 87–98. [CrossRef]
11. Bamisile, B.S.; Dash, C.K.; Akutse, K.S.; Keppanan, R.; Wang, L. Fungal endophytes: Beyond herbivore management. *Front. Microbiol.* **2018**, *9*, 544. [CrossRef]
12. Logrieco, A.; Bottalico, A.; Altomare, C. Chemotaxonomic observations on zearalenone and trichothecene production by *Gibberella zeae* from cereals in southern Italy. *Mycologia* **1988**, *80*, 892–895. [CrossRef]
13. Thrane, U. Grouping *Fusarium* section Discolor isolates by statistical analysis of quantitative high performance liquid chromatographic data on secondary metabolite production. *J. Microbiol. Methods* **1990**, *12*, 23–39. [CrossRef]
14. Frisvad, J.C.; Andersen, B.; Thrane, U. The use of secondary metabolite profiling in chemotaxonomy of filamentous fungi. *Mycol. Res.* **2008**, *112*, 231–240. [CrossRef] [PubMed]
15. Stępień, Ł. The use of *Fusarium* secondary metabolite biosynthetic genes in chemotypic and phylogenetic studies. *Crit. Rev. Microbiol.* **2014**, *40*, 176–185. [CrossRef]
16. O'Donnell, K.; Kistler, H.C.; Cigelnik, E.; Ploetz, R.C. Multiple evolutionary origins of the fungus causing Panama disease of banana: Concordant evidence from nuclear and mitochondrial gene genealogies. *Proc. Nat. Acad. Sci. USA* **1998**, *95*, 2044–2049. [CrossRef] [PubMed]
17. O'Donnell, K.; Nirenberg, H.I.; Aoki, T.; Cigelnik, E. A multigene phylogeny of the *Gibberella fujikuroi* species complex: Detection of additional phylogenetically distinct species. *Mycoscience* **2000**, *41*, 61–78. [CrossRef]
18. O'Donnell, K.; Ward, T.J.; Geiser, D.M.; Kistler, H.C.; Aoki, T. Genealogical concordance between the mating type locus and seven other nuclear genes supports formal recognition of nine phylogenetically distinct species within the *Fusarium graminearum* clade. *Fungal Genet. Biol.* **2004**, *41*, 600–623. [CrossRef]
19. Laurence, M.H.; Summerell, B.A.; Burgess, L.W.; Liew, E.C.Y. Genealogical concordance phylogenetic species recognition in the *Fusarium oxysporum* species complex. *Fungal Biol.* **2014**, *118*, 374–384. [CrossRef]
20. Zwickl, D.J.; Hillis, D.M. Increased taxon sampling greatly reduces phylogenetic error. *Syst. Biol.* **2002**, *51*, 588–598. [CrossRef]
21. Pollock, D.D.; Zwickl, D.J.M.; McGuire, J.A.; Hillis, D.M. Increased taxon sampling is advantageous for phylogenetic inference. *Syst. Biol.* **2002**, *51*, 664–671. [CrossRef]
22. Hillis, D.M.; Pollock, D.D.; McGuire, J.A.; Zwickl, D.J. Is sparse taxon sampling a problem for phylogenetic inference? *Syst. Biol.* **2003**, *52*, 124–126. [CrossRef]

23. Geiser, D.M.; Del Mar Jiménez-Gasco, M.; Kang, S.; Makalowska, I.; Veeraraghavan, N.; Ward, T.J.; Zang, N.; Kuldau, G.A.; O'Donnell, K. FUSARIUM-ID v. 1.0: A DNA sequence database for identifying *Fusarium*. *Eur. J. Plant Pathol.* **2004**, *110*, 473–479. [CrossRef]

24. Nicoletti, R.; De Filippis, A.; Buommino, E. Antagonistic aptitude and antiproliferative properties on tumor cells of fungal endophytes from the Astroni Nature Reserve, Italy. *Afr. J. Microbiol. Res.* **2013**, *7*, 4073–4083.

25. Scognamiglio, M.; Nicoletti, R.; Pacifico, S.; D'Abrosca, B.; Fiorentino, A. Spectroscopic characterization of a pyridine alkaloid from an endophytic strain of the *Fusarium incarnatum-equiseti* species complex. *Curr. Bioact. Comp.* **2014**, *10*, 196–200. [CrossRef]

26. Bacon, C.W.; Porter, J.K.; Norred, W.P.; Leslie, J.F. Production of fusaric acid by *Fusarium* species. *Appl. Environ. Microbiol.* **1996**, *62*, 4039–4043. [PubMed]

27. Wu, H.; Shen, S.; Han, J.; Liu, Y.; Liu, S. The effect in vitro of exogenously applied *p*-hydroxybenzoic acid on *Fusarium oxysporum* f. sp. *niveum*. *Phytopathol. Mediterr.* **2010**, *48*, 439–446.

28. Shi, W.; Tan, Y.; Wang, S.; Gardiner, D.M.; De Saeger, S.; Liao, Y.; Wang, C.; Fan, Y.; Wang, Z.; Wu, A. Mycotoxigenic potentials of *Fusarium* species in various culture matrices revealed by mycotoxin profiling. *Toxins* **2017**, *9*, 6. [CrossRef] [PubMed]

29. Bohni, N.; Hofstetter, V.; Gindro, K.; Buyck, B.; Schumpp, O.; Bertrand, S.; Monod, M.; Wolfender, J.L. Production of fusaric acid by *Fusarium* spp. in pure culture and in solid medium co-cultures. *Molecules* **2016**, *21*, 370. [CrossRef] [PubMed]

30. Crutcher, F.K.; Liu, J.; Puckhaber, L.S.; Stipanovic, R.D.; Bell, A.A.; Nichols, R.L. FUBT, a putative MFS transporter, promotes secretion of fusaric acid in the cotton pathogen *Fusarium oxysporum* f. sp. *vasinfectum*. *Microbiology* **2015**, *161*, 875–883. [CrossRef]

31. López-Díaz, C.; Rahjoo, V.; Sulyok, M.; Ghionna, V.; Martín-Vicente, A.; Capilla, J.; Di Pietro, A.; López-Berges, M.S. Fusaric acid contributes to virulence of *Fusarium oxysporum* on plant and mammalian hosts. *Mol. Plant Pathol.* **2018**, *19*, 440–453. [CrossRef]

32. Ding, Z.; Yang, L.; Wang, G.; Guo, L.; Liu, L.; Wang, J.; Huang, J. Fusaric acid is a virulence factor of *Fusarium oxysporum* f. sp. *cubense on banana plantlets*. *Trop. Plant Pathol.* **2018**, *43*, 297–305. [CrossRef]

33. Kosiak, E.B.; Holst-Jensen, A.; Rundberget, T.; Gonzalez Jaen, M.T.; Torp, M. Morphological, chemical and molecular differentiation of *Fusarium equiseti* isolated from Norwegian cereals. *Int. J. Food Microbiol.* **2005**, *99*, 195–206. [CrossRef]

34. Marín, P.; Moretti, A.; Ritieni, A.; Jurado, M.; Vazquez, C.; Gonzalez-Jaen, M.T. Phylogenetic analyses and toxigenic profiles of *Fusarium equiseti* and *Fusarium acuminatum* isolated from cereals from Southern Europe. *Food Microbiol.* **2012**, *31*, 229–237. [CrossRef]

35. Villani, A.; Moretti, A.; De Saeger, S.; Han, Z.; Di Mavungu, J.D.; Soares, C.M.G.; Proctor, R.H.; Venâncio, A.; Lima, N.; Stea, G.; et al. A polyphasic approach for characterization of a collection of cereal isolates of the *Fusarium incarnatum-equiseti* species complex. *Int. J. Food Microbiol.* **2016**, *234*, 24–35. [CrossRef]

36. Stępień, Ł.; Gromadzka, K.; Chełkowski, J. Polymorphism of mycotoxin biosynthetic genes among *Fusarium equiseti* isolates from Italy and Poland. *J. Appl. Genet.* **2012**, *53*, 227–236. [CrossRef]

37. Lacatena, F.; Marra, R.; Mazzei, P.; Piccolo, A.; Digilio, M.C.; Giorgini, M.; Woo, S.L.; Cavallo, P.; Lorito, M.; Vinale, F. Chlamyphilone, a novel *Pochonia chlamydosporia* metabolite with insecticidal activity. *Molecules* **2019**, *24*, 750. [CrossRef]

38. Vinale, F.; Nicoletti, R.; Borrelli, F.; Mangoni, A.; Parisi, O.A.; Marra, R.; Lombardi, N.; Lacatena, F.; Grauso, L.; Finizio, S.; et al. Co-culture of plant beneficial microbes as source of bioactive metabolites. *Sci. Rep.* **2017**, *7*, 14330. [CrossRef] [PubMed]

39. Proctor, R.H.; McCormick, S.P.; Alexander, N.J.; Desjardins, A.E. Evidence that a secondary metabolic biosynthetic gene cluster has grown by gene relocation during evolution of the filamentous fungus Fusarium. *Mol. Microbiol.* **2009**, *74*, 1128–1142. [CrossRef] [PubMed]

40. Glass, N.L.; Donaldson, G.C. Development of primer sets designed for use with the PCR to amplify conserved genes from filamentous ascomycetes. *Appl. Environ. Microbiol.* **1995**, *61*, 1323–1330. [PubMed]

41. O'Donnell, K.; Sutton, D.A.; Rinaldi, M.G.; Gueidan, C.; Crous, P.W.; Geiser, D.M. Novel multilocus sequence typing scheme reveals high genetic diversity of human pathogenic members of the *Fusarium incarnatum–F. equiseti* and *F. chlamydosporum* species complexes within the United States. *J. Clin. Microbiol.* **2009**, *47*, 3851–3861. [CrossRef] [PubMed]

42. Thompson, J.D.; Higgins, D.G.; Gibson, T.J. CLUSTAL W: Improving the sensitivity of progressive multiple sequence alignment through sequence weighting, position-specific gap penalties and weight matrix choice. *Nucleic Acids Res.* **1994**, *22*, 4673–4680. [CrossRef] [PubMed]

43. Kumar, S.; Stecher, G.; Tamura, K. MEGA7: Molecular Evolutionary Genetics Analysis Version 7.0 for bigger datasets. *Mol. Biol. Evol.* **2016**, *33*, 1870–1874. [CrossRef] [PubMed]

44. Busman, M.; Butchko, R.A.; Proctor, R.H. LC-MS/MS method for the determination of the fungal pigment bikaverin in maize kernels as an indicator of ear rot. *Food Addit. Contam. Part A* **2012**, *29*, 1736–1742. [CrossRef] [PubMed]

45. Jacobs-Venter, A.; Laraba, I.; Geiser, D.M.; Busman, M.; Vaughan, M.M.; Proctor, R.H.; McCormick, S.P.; O'Donnell, K. Molecular systematics of two sister clades, the *Fusarium concolor* and *F. babinda* species complexes, and the discovery of a novel microcycle macroconidium–producing species from South Africa. *Mycologia* **2018**, *110*, 1189–1204. [CrossRef] [PubMed]

46. Lauren, D.R.; Sayer, S.T.; Di Menna, M.E. Trichothecene production by *Fusarium* species isolated from grain and pasture throughout New Zealand. *Mycopathologia* **1992**, *120*, 167–176. [CrossRef]

47. Weber, J.; Vaclavikova, M.; Wiesenberger, G.; Haider, M.; Hametner, C.; Fröhlich, J.; Berthiller, F.; Adam, G.; Mikula, H.; Fruhmann, P. Chemical synthesis of culmorin metabolites and their biologic role in culmorin and acetyl-culmorin treated wheat cells. *Org. Biomol. Chem.* **2018**, *16*, 2043–2048. [CrossRef]

48. Busman, M. Utilization of high performance liquid chromatography coupled to tandem mass spectrometry for characterization of 8-O-methylbostrycoidin production by species of the fungus *Fusarium*. *J. Fungi* **2017**, *3*, 43. [CrossRef]

49. Summerell, B.A.; Rugg, C.A.; Burgess, L.W. Characterization of *Fusarium babinda* sp. nov. *Mycol. Res.* **1995**, *99*, 1345–1348. [CrossRef]

50. Balmas, V.; Migheli, Q.; Scherm, B.; Garau, P.; O'Donnell, K.; Ceccherelli, G.; Kang, S.; Geiser, D.M. Multilocus phylogenetics show high levels of endemic fusaria inhabiting Sardinian soils (Tyrrhenian Islands). *Mycologia* **2010**, *102*, 803–812. [CrossRef]

51. Campbell, M.A.; Rokas, A.; Slot, J.C. Horizontal transfer and death of a fungal secondary metabolic gene cluster. *Genome Biol. Evol.* **2012**, *4*, 289–293. [CrossRef]

52. Hestbjerg, H.; Nielsen, K.F.; Thrane, U.; Elmholt, S. Production of trichothecenes and other secondary metabolites by *Fusarium culmorum* and *Fusarium equiseti* on common laboratory media and a soil organic matter agar: An ecological interpretation. *J. Agric. Food Chem.* **2002**, *50*, 7593–7599. [CrossRef] [PubMed]

53. Walsh, J.L.; Laurence, M.H.; Liew, E.C.; Sangalang, A.E.; Burgess, L.W.; Summerell, B.A.; Petrovic, T. *Fusarium*: Two endophytic novel species from tropical grasses of northern Australia. *Fungal Divers.* **2010**, *44*, 149–159. [CrossRef]

54. Laurence, M.H.; Summerell, B.A.; Burgess, L.W.; Liew, E.C.Y. *Fusarium burgessii* sp. nov. representing a novel lineage in the genus *Fusarium*. *Fungal Divers.* **2011**, *49*, 101–112. [CrossRef]

55. Laurence, M.H.; Walsh, J.L.; Shuttleworth, L.A.; Robinson, D.M.; Johansen, R.M.; Petrovic, T.; Vu, T.T.H.; Burgess, L.W.; Summerell, B.A.; Liew, E.C.Y. Six novel species of *Fusarium* from natural ecosystems in Australia. *Fungal Divers.* **2016**, *77*, 349–366.

56. Walton, J.D. Horizontal gene transfer and the evolution of secondary metabolite gene clusters in fungi: An hypothesis. *Fungal Genet. Biol.* **2000**, *30*, 167–171. [CrossRef] [PubMed]

57. Fitzpatrick, D.A. Horizontal gene transfer in fungi. *FEMS Microbiol. Lett.* **2012**, *329*, 1–8. [CrossRef] [PubMed]

58. Brown, D.W.; Butchko, R.A.; Busman, M.; Proctor, R.H. Identification of gene clusters associated with fusaric acid, fusarin, and perithecial pigment production in *Fusarium verticillioides*. *Fungal Genet. Biol.* **2012**, *49*, 521–532. [PubMed]

59. Brown, D.W.; Proctor, R.H. Insights into natural products biosynthesis from analysis of 490 polyketide synthases from *Fusarium*. *Fungal Genet. Biol.* **2016**, *89*, 37–51. [CrossRef]

60. Soanes, D.; Richards, T.A. Horizontal gene transfer in eukaryotic plant pathogens. *Ann. Rev. Phytopathol.* **2014**, *52*, 583–614.

61. Kimura, M.; Tokai, T.; Takahashi-Ando, N.; Ohsato, S.; Fujimura, M. Molecular and genetic studies of *Fusarium* trichothecene biosynthesis: Pathways, genes, and evolution. *Biosci. Biotechnol. Biochem.* **2007**, *71*, 2105–2123. [CrossRef]

62. Sieber, C.M.; Lee, W.; Wong, P.; Münsterkötter, M.; Mewes, H.W.; Schmeitzl, C.; Varga, E.; Berthiller, F.; Adam, G.; Güldener, U. The *Fusarium graminearum* genome reveals more secondary metabolite gene clusters and hints of horizontal gene transfer. *PloS ONE* **2014**, *9*, e110311. [CrossRef] [PubMed]

63. Ma, L.J.; Van der Does, H.C.; Borkovich, K.A.; Coleman, J.J.; Daboussi, M.J.; Di Pietro, A.; Dufresne, M.; Freitag, M.; Grabherr, M.; Henrissat, B.; et al. Comparative genomics reveals mobile pathogenicity chromosomes in *Fusarium*. *Nature* **2010**, *464*, 367–373. [CrossRef] [PubMed]

64. Burgess, L.W. 2011 McAlpine Memorial Lecture-A love affair with *Fusarium*. *Australas. Plant Pathol.* **2014**, *43*, 359–368. [CrossRef]

Review

Endophytic Fungi of Citrus Plants

Rosario Nicoletti [1,2]

[1] Council for Agricultural Research and Economics, Research Centre for Olive, Citrus and Tree Fruit, 81100 Caserta, Italy; rosario.nicoletti@crea.gov.it

[2] Department of Agricultural Sciences, University of Naples Federico II, 80055 Portici, Italy

Received: 12 October 2019; Accepted: 19 November 2019; Published: 21 November 2019

Abstract: Besides a diffuse research activity on drug discovery and biodiversity carried out in natural contexts, more recently, investigations concerning endophytic fungi have started considering their occurrence in crops based on the major role that these microorganisms have been recognized to play in plant protection and growth promotion. Fruit growing is particularly involved in this new wave, by reason that the pluriannual crop cycle likely implies a higher impact of these symbiotic interactions. Aspects concerning occurrence and effects of endophytic fungi associated with citrus species are revised in the present paper.

Keywords: *Citrus* spp.; endophytes; antagonism; defensive mutualism; plant growth promotion; bioactive compounds

---

1. Introduction

Despite the early pioneering observations dating back to the nineteenth century [1], a settled prejudice that pathogens basically were the only microorganisms able to colonize living plant tissues has long delayed the awareness that endophytic fungi are constantly associated to plants, and remarkably influence their ecological fitness. Overcoming an apparent vagueness of the concept of 'endophyte', scientists working in the field have agreed on the opportunity of delimiting what belongs to this functional category. Thus, a series of definitions have been enunciated which are all based on the condition of not causing any immediate overt negative effect to the host [2].

Besides being prompted by the general theoretical intent that all components of biodiversity from natural contexts ought to be exploited for the benefit of humanity, investigations on the endophytic microbiota, or endosphere [3], have also been undertaken with reference to crops. In this respect, it can be said that endophytes are even more relevant in orchards, where the time factor confers higher impact to the establishment of an equilibrium among the species which are part of the tree biocoenosis, and to its possible disruption. Hence, all sorts of contributions have recently been proliferating in the literature, to such an extent that an organization of the available information is now appropriate in order to support the scientific community in achieving further progress. In view of this perspective, the present paper offers a review of the state-of-the-art of research concerning occurrence and effects of endophytic fungi associated with citrus species.

2. Endophytic Occurrence of Citrus Pathogens

The agent of citrus black spot (CBS) *Phyllosticta citricarpa*, also known under the teleomorph name *Guignardia citricarpa* (Dothideomycetes, Botryosphaeriaceae), is one of the most noxious pests of these crops in subtropical regions, and it is subject to phytosanitary restrictions by the European Union and the United States. The employment of biomolecular methods has provided substantial support to the distinction between pathogenic isolates, typically slow-growing in axenic cultures and producing a yellow halo on oatmeal agar, and non-pathogenic isolates, which are morphologically similar but fast-growing, and producing conidia embedded within a thicker mucoid sheath [4–8].

The latter, characterized as a different species (*Phyllosticta capitalensis*), are known to be ubiquitous as endophytes in woody plants, having been reported from at least 70 botanical families [6,9,10]. *Guignardia endophyllicola*, treated as a separate species in a work also emphasizing its widespread endophytic occurrence [11], is at present recognized as a synonym. Differences between the two sister species also concern their metagenetic cycle. In fact, it has been ascertained that *P. citricarpa* is heterothallic, while *P. capitalensis* is homothallic [8]. This consolidated taxonomic distinction supports the exclusion from quarantine measures of plant material harbouring *P. capitalensis*. To this purpose, several rapid PCR assays have been developed [12–20]. The applicative use of these assays has enabled to exclude the presence of the pathogen in New Zealand, unlike what was previously assumed [21], and has supported the hypothesis of the possible endophytic occurrence of *P. citricarpa* in asymptomatic *Citrus* spp., as pointed out by several investigations (Table 1). Moreover, the two species have been clearly differentiated on account of their enzymatic profiles, with a higher expression of amylases, endoglucanases, and pectinases in *P. citricarpa*, suggesting a likely involvement of these enzymes in the pathogenic aptitude of the CBS agent [22]. Differences in terms of pathogenesis-related proteins have been confirmed after the genome sequencing of the two species, disclosing a higher number of coding sequences in *P. citricarpa* (15,206 versus 14,797). Such a difference has been interpreted considering the presence of growth and developmental genes involved in the expression of pathogenicity [23].

The issue of detection of contaminated material imported from areas where the pathogen is endemic has also prompted investigations concerning the assortment of *Phyllosticta* spp. able to colonize citrus plants in either symptomatic or latent courses. Several revisions have been published [17,24], and novel species characterized, which consistently enlarge the citrus-associated consortium within this widespread genus. Particularly, the pathogenic *P. citriasiana* from south-east Asia [25], *P. citrichinaensis* from China [26], *P. citrimaxima* from Thailand [24], and *P. paracitricarpa* from Greece [27], and the non-pathogenic endophytic *P. citribraziliensis* from Brazil [28] and *P. paracapitalensis* from New Zealand, Italy, and Spain [27]. The isolation by the latter research group of *P. citricarpa* from specimen collected in citrus groves in Italy, Malta, and Portugal, following analogue findings in Florida [19,29], is expected to provide impulse for a more thorough assessment of distribution and pathogenicity of this species [30]. A very recent investigation carried out in Australia on several *Citrus* spp. and growing conditions, has disclosed *P. paracapitalensis* to be even more widespread than *P. capitalensis*. Strains of both species were confirmed to be non-pathogenic on fruits under field conditions, and displayed antagonistic effects against the CBS agent, introducing their possible exploitation in the integrated management of this disease [31]. In this respect, it has been speculated that, rather than depending on intrinsic genetic factors, resistance to CBS by *C. latifolia* could be due to its systematic colonization by *P. capitalensis*, as disclosed by a dedicated investigation carried out in Brazil [32].

Colletotrichum (Sordariomycetes, Glomerellaceae) is another important ascomycete genus in course of coherent taxonomic revision. Besides *Colletotrichum gloeosporioides*, the agent of citrus anthracnose, it includes many species known for their endophytic aptitude. A recent investigation carried out in China on several *Citrus* spp. has shown a high proportion of endophytic strains to belong to *C. gloeosporioides sensu stricto*, calling for further investigations concerning the asymptomatic occurrence of this pathogen in citrus orchards. Additional identified species are *Colletotrichum fructicola* from *Citrus reticulata* cv. Nanfengmiju and *Citrus japonica* (=*Fortunella margarita*), and *Colletotrichum karstii* [33]. A similar widespread occurrence of *C. gloeosporioides* has been more recently confirmed in Brazil, where just one out of 188 isolates was found to be able to induce post-bloom fruit drop. This syndrome is more frequently associated to the species *Colletotrichum abscissum*, which, however, does not display an endophytic habit [34]. Endophytic *C. gloeosporioides* were also previously reported from *Citrus limon* in Argentina [35] and Cameroon [36].

One more meaningful example of endophytic fungus converting to pathogenic when plants are exposed to stress factors is represented by another member of the Botryosphaeriaceae, Lasiodiplodia theobromae. Characterized by a widespread endophytic occurrence [37,38], this species has been reported to exacerbate pre-harvest fruit drop and post-harvest fruit decay in plants of Citrus sinensis hit by the huanglongbing syndrome [39].

A quite intricate case deserving further investigations with reference to the epidemiological impact by endophytic strains is represented by members of the genus Diaporthe (Sordariomycetes, Diaporthaceae), also known under the anamorph name Phomopsis [40,41], which are widespread in different ecological contexts [41,42]. Besides the longtime known D. citri, more species in this genus have been recently identified as the causal agents of melanose, stem-end rot, and gummosis on *Citrus* spp., particularly, D. citriasiana and *D. citrichinensis* in China [43], and *D. limonicola, D. melitensis, D. baccae, D. foeniculina*, and *D. novem* in Europe [44]. Even more species have been reported for their endophytic occurrence as a result of a phylogenetic reassessment carried out in China, with eight known (D. arecae species complex, *D. citri, D. citriasiana, D. citrichinensis, D. endophytica, D. eres, D. hongkongensis*, and *D. sojae*) and seven new species (*D. biconispora, D. biguttulata, D. discoidispora, D. multiguttulata, D. ovalispora, D. subclavata*, and *D. unshiuensis*) [45].

Table 1. Endophytic fungi reported from *Citrus* species.

Endophyte [1]	Plant Species	Country	Reference
Alternaria alternata	C. limon, C. tangelo	Florida	[46]
	Citrus spp.	Japan	[47]
	C. limon	Argentina	[35]
	C. reticulata	Iran	[48]
Alternaria brassicicola	C. reticulata	Iran	[48]
Alternaria carthami	C. reticulata	Iran	[48]
Alternaria citri	C. sinensis	Iran	[49]
Alternaria infectoria	C. sinensis	Iran	[49]
Alternaria rosae	C. sinensis	Iran	[49]
Alternaria sp.	C. kotokan	Taiwan	[52]
	C. sinensis	Iran	[49]
Annulohypoxylon stygium	C. sinensis	Iran	[49]
Arthrinium sp.	C. japonica	Taiwan	[52]
Ascochyta medicaginicola	C. reticulata	Iran	[48]
Aspergillus nidulans	C. sinensis	Iran	[49]
Aspergillus niger	C. reticulata	Iran	[48]
Aspergillus pallidofulvus	C. reticulata	Iran	[48]
Aspergillus terreus	C. sinensis	Iran	[49]
Aureobasidium iranianum	C. reticulata	Iran	[48]
Aureobasidium melanogenum	C. reticulata	Iran	[48]
Aureobasidium pullulans	C. sinensis	Brazil	[53]
	C. japonica	Uruguay	[54]
	C. reticulata	Iran	[48]
Beauveria bassiana	C. limon	China	[55]
Biscogniauxia mediterranea	C. sinensis	Iran	[49]
Biscogniauxia nummularia	C. sinensis	Iran	[49]
Bjerkandera adusta	C. sinensis	Iran	[49]
Botryosphaeria sp.	C. aurantium	Taiwan	[52]
Camarosporium sp.	C. aurantium, C. medica var. sarcodactylis	Taiwan	[52]
Candida parapsilosis	C. sinensis	Brazil	[53]
Cercospora sp.	C. limon	Cameroon	[36]
	C. sinensis	Iran	[49]
Chaetomium globosum	C. sinensis	Iran	[49]
Chaetomium sp.	C. sinensis	Taiwan	[52]
Cladosporium cladosporioides	C. reticulata	Iran	[48]
Cladosporium sp.	C. limon, C. reshni, C. sinensis, C. sunki, C. trifoliata, C. volkameriana	Brazil	[56]
Cladosporium xanthochromaticum	C. reticulata	Iran	[48]
Colletotrichum boninense	C. limon	Cameroon	[36]
	C. sinensis	Iran	[49]

Table 1. *Cont.*

Endophyte [1]	Plant Species	Country	Reference
Colletotrichum fructicola	*C. japonica, C. reticulata*	China	[43]
	C. limon, C. reshni, C. sinensis, C. sunki, C. trifoliata, C. volkameriana	Brazil	[56]
Colletotrichum gloeosporioides	*C. limon*	Argentina	[35]
		Cameroon	[36]
	C. grandis, C. reticulata, C. sinensis, C. unshiu	China	[43]
	C. sinensis	Iran	[49]
Colletotrichum karstii	*C. grandis, C. limon*	China	[43]
Colletotrichum sp.	*C. aurantium, C. medica* var. *sarcodactylis, C. sinensis*	Taiwan	[52]
	C. deliciosa, C. reticulata	Brazil	[57]
	C, aurantifolia	India	[58]
Coprinellus radians	*C. sinensis*	Iran	[49]
Coprinopsis sp.	*C. medica*	Taiwan	[52]
Cryptococcus flavescens	*C. sinensis*	Brazil	[53]
Cryptococcus laurentii	*C. sinensis*	Brazil	[53]
Cyanodermella sp.	*C. medica* var. *sarcodactylis, Citrus* sp.	Taiwan	[52]
Diaporthe arecae s.c. [2]	*C. grandis, C. limon, C. reticulata, C. sinensis, Citrus* sp., *C. unshiu*	China	[45]
Diaporthe biconispora [2,*]	*C. grandis, C. japonica, C. sinensis*	China	[45]
Diaporthe biguttulata [2,*]	*C. limon*	China	[45]
Diaporthe citri [2]	*C. reticulata, C. unshiu*	China	[43,45]
Diaporthe citriasiana [2]	*C. unshiu*	China	[43]
Diaporthe citrichinensis [2]	*C. grandis, C. japonica*	China	[45]
Diaporthe discoidispora [2,*]	*C. sinensis, C. unshiu*	China	[45]
Diaporthe endophytica [2]	*C. limon*	China	[45]
Diaporthe eres [2]	*C. japonica, Citrus* sp., *C. unshiu*	China	[45]
Diaporthe eucalyptorum [2]	*C. limon*	Cameroon	[36]
Diaporthe foeniculina [2]	*C. sinensis*	Iran	[49]
Diaporthe hongkongensis [2]	*C. grandis, C. reticulata, C. sinensis, C. unshiu*	China	[45]
Diaporthe multiguttulata [2,*]	*C. grandis*	China	[45]
Diaporthe ovalispora [2,*]	*C. limon*	China	[45]
Diaporthe phaseolorum [2]	*C. limon*	Cameroon	[36]
Diaporthe sojae [2]	*C. limon, C. reticulata, C. unshiu*	China	[45]
	C. limon	Cameroon	[36]
Diaporthe sp. [2]	*C. aurantium, C. medica, C. sinensis*	Taiwan	[52]
	C. japonica	China	[45]
	C. reticulata	Iran	[48]
Diaporthe unshiuensis [2,*]	*C. japonica*	China	[45]
Didymella microchlamydospora	*C. reticulata*	Iran	[48]
Discostroma sp.	*C. medica*	Taiwan	[52]
Epicoccum nigrum	*C. sinensis*	Iran	[49]
Eutypella sp.	*C. medica* var. *sarcodactylis*	Taiwan	[52]
Fusarium culmorum	*C. sinensis*	Iran	[49]
Fusarium incarnatum	*C. sinensis*	Iran	[49]
Fusarium oxysporum	*C. reticulata*	Iran	[48]
Fusarium proliferatum	*C. sinensis*	Iran	[49]
Fusarium sarcochroum	*C. limon, C. reticulata*	Italy, Spain	[50]
Fusarium sp.	*C. sinensis*	Taiwan	[52]
	C. reticulata	Iran	[48]
Hanseniaspora opuntiae	*C. reticulata*	China	[59]
Hypholoma fasciculare	*C. sinensis*	Iran	[49]
Hypoxylon investiens	*C. sinensis*	Iran	[49]
Lasiodiplodia theobromae	*C. sinensis*	China	[39]
Lasmenia sp.	*C. medica* var. *sarcodactylis*	Taiwan	[52]
Meira geulakonigae	*C. paradisi*	Israel	[60]
Meyerozyma caribbica	*C. reticulata*	Iran	[48]
Meyerozyma guilliermondii	*C. sinensis*	Brazil	[53]
	C. reticulata	China	[58]
Muscodor sp.	*C. sinensis*	Brazil	[61]
Mycoleptodiscus sp.	*C. aurantium*	Taiwan	[52]
Mycosphaerella sp.	*C. limon*	Cameroon	[36]
Myrothecium sp.	*C. reticulata*	Iran	[48]
Neocosmospora solani	*C. reticulata*	Iran	[48]

Table 1. *Cont.*

Endophyte [1]	Plant Species	Country	Reference
Neosetophoma sp.	*C. reticulata*	Iran	[48]
Nigrospora oryzae	*C. sinensis*	Iran	[49]
Nigrospora sphaerica	*C. limon*	Argentina	[35]
Nodulisporium sp.	*C. limon*	Argentina	[35]
Passalora loranthi	*C. limon*	Cameroon	[36]
Penicillium citrinum	*C. reticulata*	Iran	[48]
Pestalotiopsis mangiferae	*C. limon*	Cameroon	[36]
Pestalotiopsis microspora	*C. limon*	Cameroon	[36]
Pestalotiopsis sp.	*C. limon*	Cameroon	[36]
Phaeoacremonium parasiticum	*C. reticulata*	Iran	[48]
Phialophora sp.	*C. sinensis*	Brazil	[53]
Phoma sp.	*C. limon*	Cameroon	[36]
	Citrus spp.	South Africa	[4]
	C. deliciosa, C. reticulata	Brazil	[57]
	C. aurantium, C. natsudaidai, C. trifoliata	Japan	[11]
	C. aurantium	Brazil	[62]
	C. latifolia	Brazil	[17]
Phyllosticta capitalensis [2]	*C. limonia, C. sinensis, Citrus* sp.	Brazil	[28]
	C. aurantium, C. australasica	Australia	[63]
	C. limon	Cameroon	[36]
		Italy, Malta, Spain Greece, Portugal	[27]
	C. aurantifolia	Italy	
	C. sinensis	Iran	[49]
Phyllosticta citribraziliensis [2,*]	*Citrus* sp.	Brazil	[28]
	Citrus sp.	South Africa	[64]
	C. reshni, C. sinensis, C. sunki, C. trifoliata, C. volkameriana	Brazil	[56]
Phyllosticta citricarpa [2]	*C. deliciosa, C. reticulata*	Brazil	[65]
	C. limon	Argentina	[35]
	C. latifolia	Brazil	[17]
	C. sinensis	Florida	[29]
	C. aurantifolia	New Zealand	
Phyllosticta paracapitalensis [2,*]	*C. floridana*	Italy	[27]
	C. limon	Spain	
	C. aurantium, C. australasica, C. hystrix, C. japonica, C. maxima, C. reticulata, C. wintersii	Australia	[31]
Phyllosticta sp. [2]	*C. medica* var. *sarcodactylis*	Taiwan	[52]
Physoderma citri	*Citrus* spp.	Florida	[51]
Pichia kluyveri	*C. reticulata*	China	[59]
Pseudocercospora sp.	*C. japonica*	Taiwan	[52]
Pseudopestalotiopsis theae	*C. limon*	Cameroon	[36]
Pseudozyma flocculosa	*C. reticulata*	Iran	[48]
Rhodotorula dairenensis	*C. sinensis*	Brazil	[53]
Rhodotorula mucilaginosa	*C. sinensis*	Brazil	[53]
Rosellinia sp.	*C. sinensis*	Iran	[49]
Sarocladium subulatum	*C. sinensis*	Iran	[49]
Scedosporium apiospermum	*C. reticulata*	Iran	[48]
Sordaria fimicola	*C. sinensis*	Iran	[49]
Sporobolomyces sp.	*C. sinensis*	Brazil	[53]
Sporormiella minima	*C. limon*	Argentina	[35]
	C. sinensis	Iran	[49]
Stemphylium sp.	*C. aurantium, C. japonica*	Taiwan	[52]
Stenella sp.	*C. limon*	Cameroon	[36]
Talaromyces purpurogenus	*C. reticulata*	Iran	[48]
Talaromyces trachyspermus	*C. reticulata*	Iran	[48]
Xylaria cubensis	*C. sinensis*	Iran	[49]
Xylaria sp.	*C. limon*	Cameroon	[36]
	C. japonica	Taiwan	[52]
Zasmidium sp.	*C. limon*	Cameroon	[36]

[1] Species are reported according to the latest accepted name, which might not correspond to the one used in the corresponding reference. [2] Conforming to the principle 'one fungus—one name' [66], the older genus names *Diaporthe* and *Phyllosticta* have been considered to deserve priority over *Phomopsis* and *Guignardia*, respectively. * Novel species described for the first time from this plant source.

Endophytic occurrence has also been reported for other citrus pathogens, such as the leaf-spot agents *Alternaria alternata* [35,46–48] and *Alternaria citri* [49], *Fusarium oxysporum* [48], and *Fusarium sarcochroum*, known as a possible agent of dieback of twigs on mandarin and lemon [50]. The latter study also introduces new *Fusarium* spp. (*F. citricola*, *F. salinense*, *F. siculi*), causing cankers on several citrus species. Considering that pathogenic *Fusaria* often present an early latent stage, this finding claims for further assessments concerning their possible endophytic occurrence. Finally, it is worth mentioning *Physoderma citri*, a species ascribed to the *phylum* Blastocladiomycota reported to cause vessel occlusion, but also found in asymptomatic plants of several *Citrus* spp. [51].

3. Other Endophytic Fungi and Their Interactions with Pests and Pathogens of *Citrus*

Besides the above reports, essentially dedicated to pathogenic species/genera upon the aim to assess the epidemiological impact of latent endophytic stages, additional data have been recorded on the overall species assemblage in a few contexts (Table 1). A study carried out on *C. limon* in Cameroon [36] pointed out that yellowing of leaves affects foliar endophytic communities, and that interactions among endophytes may be a factor driving the yellowing process. In fact, yellow leaves presented a less varied species assortment dominated by *C. gloeosporioides* in the absence of species belonging to the Mycosphaerellaceae, otherwise common in healthy leaves. In vitro observations in dual cultures showed that the latter were inhibited and overgrown by *C. gloeosporioides*, even if capable to revert this inhibitory effect when pre-inoculated, which was interpreted as deriving from production of fungitoxic metabolites. This study also demonstrated a low occurrence of species in the Xylariaceae, which are usually quite widespread as tree endophytes [67,68].

The endophytic occurrence of a few yeast species was documented in an investigation carried out on *C. sinensis* in Brazil [53]. By means of scanning electron microscopy, it was observed that these microorganisms are mostly localized around stomata and in xylem vessels. Isolates of the species *Rhodotorula mucilaginosa*, *Meyerozyma* (*Pichia*) *guilliermondii*, and *Cryptococcus flavescens* were inoculated in healthy plants, and re-isolated, without causing any kind of disease symptoms. Quite interestingly, the authors noted that *M. guilliermondii* primarily occurred in plants colonized by *Xylella fastidiosa*, the causal agent of citrus variegated chlorosis (CVC), and that the bacterium could thrive on a supernatant separated from cultures of a strain of this species. This finding represents an indication that the presence of the yeast could stimulate the pathogen and could be responsible for more severe disease symptoms. More recently, strains of *M. guilliermondii* have been recovered, along with strains of *Hanseniaspora opuntiae* and *Pichia kluyveri*, from tangerine peel in China. However, it is questionable if this record can actually concern endophytic occurrence considering that authors refer that fruits were purchased on the market rather than being directly collected in the field [59].

Indeed, interactions between endophytic bacteria and fungi are complex, and the assortment of strains which can be recovered is largely influenced by the antagonistic interactions as mediated by the production of antibiotics. In this respect, strains of *P. citricarpa* isolated from *Citrus* spp. in Brazil were found to possess inhibitory properties toward several endophytic *Bacillus* spp. from the same source, while a stimulatory effect was assessed towards the gram-negative *Pantoea agglomerans*, which can be taken as an indication of the opportunity to investigate possible interference with the development of *X. fastidiosa* [56].

Antagonistic properties by an isolate of *Muscodor* sp. from *C. sinensis* were reported against *P. citrocarpa* as deriving from the production of volatile organic compounds (VOCs) [61]. Actually, such properties are known for endophytic isolates of *Muscodor* and other genera of xylariaceous fungi, such as *Hypoxylon* (=*Nodulisporium*) and *Xylaria*, reported from many plant species [69] and also occurring in citrus plants [35,36,49,52].

Endophytic strains belonging to two species of *Diaporthe*, *D. terebinthifolii* and the already-mentioned *D. endophytica*, displayed inhibitory properties against *P. citrocarpa* in vitro and on detached fruits. Moreover, their transformants expressing the fluorescent protein DsRed proved to be able to actively colonize citrus seedlings, and to remain viable in the plant tissues for one year at least. These evidences support their possible use in the biocontrol of this pathogen [70]. Antifungal properties have also been reported for a strain of another fungus belonging to the Diaporthales (*Lasmenia* sp.), which was recovered from *C. medica* var. *sarcodactylis* [52].

Rather than just concerning agents of cryptogamic diseases, protective effects by endophytic fungi may pertain several kinds of pests [71,72]. Actually, data available in the literature concerning citrus plants are limited but encourage further assessments. For instance, a ustilaginomycetous yeast endophytic in grapefruit (*Citrus paradisi*), *Meira geulakonigae*, was found to be able to reduce populations of the citrus rust mite (*Phyllocoptruta oleivora*) [60]. More recently, two strains of *Beauveria bassiana* were inoculated in seedlings of *C. limon* through foliar sprays and proved to be able to colonize the plants endophytically. Besides increasing plant growth, they caused 10%–15% mortality on adults of the Asian citrus psyllid (*Diaphorina citri*), and the females feeding on the treated plants laid significantly fewer eggs [55]. It is not unlikely that more evidence in this respect can be gathered from targeted investigations concerning naturally occurring endophytes, considering that protective effects have been documented for endophytic strains of *F. oxysporum* against aphids [73] and nematodes [74].

As a general ecological trait, endophytic fungi seem to be absent in seeds of citrus species [65]. This is to be taken as an indication that these microorganisms are not adapted to a vertical spread, and most likely colonize citrus plants coming from the surrounding environment.

4. Biotechnological Implications

The involvement of endophytic fungi in a tripartite relationship with the host plant and its pests and pathogens highlights their basic role in establishing an equilibrium in such a fragile biocoenosis. Indeed, a major biotechnological application of endophytic strains consists in the exploitation of their aptitude to defensive mutualism.

The endophytic habit is conducive for interactions with other microorganisms sharing the same micro-environment. There is strong evidence that these interactions entangle the genetic level, and that interspecific transfer of gene pools regularly occurs. Probably, the best example in this respect is represented by genes encoding for the synthesis of polyketide secondary metabolites, which are usually grouped in clusters and are influenced in their expression by several external factors [75,76]. Horizontal gene transfer from other endophytic microorganisms may eventually explain the ability by a strain of *P. citricarpa* [77] to produce the blockbuster drug taxol, first extracted from *Taxus* spp. and afterwards as a secondary metabolite of a high number of endophytic fungi [69,78].

P. citricarpa has been further characterized with reference to production of secondary metabolites. Particularly, it has been reported to produce the new dioxolanone phenguignardic acid butyl ester, along with four previously reported compounds: phenguignardic acid methyl ester, peniisocoumarin G, protocatechuic acid, and tyrosol [79]. *Phyllosticta* spp. have been reported to have a similar metabolomic profile, including the dioxolanone phytotoxins which are regarded as potential virulence factors. However, one of these products, guignardic acid, has also been reported from *P. capitalensis* [80]. Biosynthetic abilities by endophytic strains of the latter species also refer to meroterpenes, such as compounds in the guignardone series [81–84] and the manginoids [85]. Besides a likely implication in the relationships with other citrus-associated microbial species, the bioactive properties of the dioxolanones and the related meroterpene compounds deserve to be further investigated in view of possible pharmaceutical exploitation [79,86].

Protocatechuic acid was again reported from an unidentified fungal strain recovered from leaves of *Citrus jambhiri*, along with indole-3-acetic acid (IAA) and acropyrone [87]. The latter compound was shown to possess antibiotic properties against *Staphylococcus aureus*, while the finding of IAA is in line with the many reports of plant hormones produced by endophytic fungi [69], which at least in part unfold the growth-promoting effects exerted on their hosts [88,89]. Production of IAA was also reported from strains of the yeasts *Hanseniaspora opuntiae* and *Meyerozyma guilliermondii* from *Citrus reticulata*, which were able to induce growth-promoting effects on seedlings of *Triticum aestivum* [59].

The above-mentioned VOCs reported from an endophytic strain of *Muscodor* sp. from *C. sinensis* include several sesquiterpenes, namely azulene, cis/trans-α-bergamotene, cedrene, (Z)-β-farnesene, farnesene epoxide, α-himachalene, α-longipinene, thujopsene, 2,4,6-trimethyl-1,3,6-heptatriene, 2-methyl-5,7-dimethylene-1-8-nonadiene, and cis-Z-bisabolene epoxide [61]. Mixtures of these compounds have a possible biotechnological application for the mycofumigation of fruits, proposed for the control of CBS and various post-harvest pathogens [90–92]. Concerning VOCs, another possible investigational subject consists in assessing if any endophyte of citrus plants is able to produce compounds occurring in the typical aroma spread by flowers and fruits of these plants, which are exploited by the pharmaceutical and the perfume industries. In this respect, the production of bergapten, a psoralen compound known from bergamot (*Citrus bergamia*), has already been pointed out by endophytic strains of *Penicillium* sp. [93] and *L. theobromae* [94]. Although these findings concern plants other than citrus, it is worth considering that these fungi are also reported as citrus endophytes (Table 1).

Antimicrobial properties of fungi do not just depend on the production of bioactive compounds. In fact, a strain of *P. capitalensis* (Bios PTK 4) recovered from an unidentified citrus plant was found to be able to synthesize silver nanoparticles extracellularly. These nanoparticles, which were spherical, 5–30 nm in size, well-dispersed, and extremely stable, have been characterized for their antibacterial and antifungal properties [95].

5. Conclusions

Revision of literature in the field shows that a major part of the research activity carried out on endophytic fungi of citrus plants consists in investigations on the occurrence of pathogens, and their discrimination from other ecologically associated taxa. Such a limited approach has, anyway, turned to be useful to disclose an epidemiological relevance of these microorganisms, as related to a modulatory role in the spread of citrus diseases. On that account, possible interactions in conducive contexts with other important pathogens, such as the agent of mal secco *Phoma tracheiphila* and species of *Phytophthora* causing foot and root rot, should be attentively considered. Even when there is no apparent direct interaction with disease agents, such as in the cases of CVC incited by *X. fastidiosa*, tristeza, and other viroses, the possible effect by endophytic fungi in stimulating plant defense reaction, or, more in general, to act as plant disease modifiers [96], should not be disregarded. In this respect, data concerning occasional isolations might well disclose some relevance. Unfortunately, description of the endophytic assemblages in several papers is often approximate or incomplete, such as in a mentioned survey concerning sweet orange (*C. sinensis*), where just a single strain was characterized out of a sample of over 400 endophytes [61]. It is to be recommended that future investigations in the field be more circumstantial in their approach to describe this component of biodiversity, in the aim of increasing opportunities for its biotechnological exploitation.

Encouraging examples in this direction are represented by two very recent publications from Iran [48,49]. Indeed, the focus on endophytic fungi is gradually evolving from a basically descriptive phase to the analysis of factors influencing the structure and composition of microbiomes, in view of their manipulation for increasing plant protection and productivity. A better comprehension of the already introduced genetic interactions among members of the associated biota and the host tree is crucial for the success of any practical application of endophytic fungi in sustainable agriculture [97]. Moreover, the observed effects of the host genotype [98,99] could be adequately considered in breeding programs, in the aim to select suitable recipient genotypes for fungal inoculants.

Funding: This research received no external funding.

Conflicts of Interest: The author declares no conflict of interest.

References

1. De Bary, A. *Morphologie und Physiologie der Pilze, Flechten und Myxomyceten*; Engelmann: Leipzig, Germany, 1866.
2. Hyde, K.D.; Soytong, K. The fungal endophyte dilemma. *Fungal Divers.* **2008**, *33*, 163–173.
3. Turner, T.R.; James, E.K.; Poole, P.S. The plant microbiome. *Genome Biol.* **2013**, *14*, 209. [CrossRef] [PubMed]
4. McOnie, K.C. The latent occurrence in *Citrus* and other hosts of a *Guignardia* easily confused with *G. citricarpa*, the black spot pathogen. *Phytopathology* **1964**, *54*, 64–67.
5. Meyer, L.; Slippers, B.; Korsten, L.; Kotze, J.M.; Wingfield, M.J. Two distinct *Guignardia* species associated with citrus in South Africa. *S. Afr. J. Sci.* **2001**, *97*, 191–194.
6. Baayen, R.P.; Bonants, P.J.M.; Verkley, G.; Carroll, G.C.; van der Aa, H.A.; de Weerdt, M.; van Brouwershaven, I.R.; Schutte, G.C.; Maccheroni, W., Jr.; Glienke de Blanco, C.; et al. Nonpathogenic isolates of the citrus black spot fungus, *Guignardia citricarpa*, identified as a cosmopolitan endophyte of woody plants, *G. mangiferae* (*Phyllosticta capitalensis*). *Phytopathology* **2002**, *92*, 464–477. [CrossRef]
7. Zavala, M.G.M.; Er, H.L.; Goss, E.M.; Wang, N.Y.; Dewdney, M.; van Bruggen, A.H.C. Genetic variation among *Phyllosticta* strains isolated from citrus in Florida that are pathogenic or nonpathogenic to citrus. *Trop. Plant Pathol.* **2014**, *39*, 119–128. [CrossRef]
8. Guarnaccia, V.; Gehrmann, T.; Silva-Junior, G.J.; Fourie, P.H.; Haridas, S.; Vu, D.; Spatafora, J.; Martin, F.M.; Robert, V.; Grigoriev, I.V.; et al. *Phyllosticta citricarpa* and sister species of global importance to *Citrus*. *Mol. Plant Pathol.* **2019**. [CrossRef]
9. Wikee, S.; Udayanga, D.; Crous, P.W.; Chukeatirote, E.; McKenzie, E.H.; Bahkali, A.H.; Dai, D.Q.; Hyde, K.D. *Phyllosticta*—An overview of current status of species recognition. *Fungal Divers.* **2011**, *51*, 43–61. [CrossRef]
10. Wikee, S.; Lombard, L.; Crous, P.W.; Nakashima, C.; Motohashi, K.; Chukeatirote, E.; Alias, S.A.; McKenzie, E.H.C.; Hyde, K.D. *Phyllosticta capitalensis*, a widespread endophyte of plants. *Fungal Divers.* **2013**, *60*, 91–105. [CrossRef]
11. Okane, I.; Nakagiri, A.; Ito, T.; Lumyong, S. Extensive host range of an endophytic fungus, *Guignardia endophyllicola* (anamorph: *Phyllosticta capitalensis*). *Mycoscience* **2003**, *44*, 353–363. [CrossRef]
12. Bonants, P.J.; Carroll, G.C.; De Weerdt, M.; van Brouwershaven, I.R.; Baayen, R.P. Development and validation of a fast PCR-based detection method for pathogenic isolates of the citrus black spot fungus, *Guignardia citricarpa*. *Eur. J. Plant Pathol.* **2003**, *109*, 503–513. [CrossRef]
13. Meyer, L.; Sanders, G.M.; Jacobs, R.; Korsten, L. A one-day sensitive method to detect and distinguish between the citrus black spot pathogen *Guignardia citricarpa* and the endophyte *Guignardia mangiferae*. *Plant Dis.* **2006**, *90*, 97–101. [CrossRef]
14. Meyer, L.; Jacobs, R.; Kotzé, J.M.; Truter, M.; Korsten, L. Detection and molecular identification protocols for *Phyllosticta citricarpa* from citrus matter. *S. Afr. J. Sci.* **2012**, *108*, 53–59. [CrossRef]
15. Peres, N.A.; Harakava, R.; Carroll, G.C.; Adaskaveg, J.E.; Timmer, L.W. Comparison of molecular procedures for detection and identification of *Guignardia citricarpa* and *G. mangiferae*. *Plant Dis.* **2007**, *91*, 525–531. [CrossRef]
16. Van Gent-Pelzer, M.P.E.; Van Brouwershaven, I.R.; Kox, L.F.F.; Bonants, P.J.M. A TaqMan PCR method for routine diagnosis of the quarantine fungus *Guignardia citricarpa* on citrus fruit. *J. Phytopathol.* **2007**, *155*, 357–363. [CrossRef]
17. Baldassari, R.B.; Wickert, E.; de Goes, A. Pathogenicity, colony morphology and diversity of isolates of *Guignardia citricarpa* and *G. mangiferae* isolated from *Citrus* spp. *Eur. J. Plant Pathol.* **2008**, *120*, 103–110. [CrossRef]
18. Stringari, D.; Glienke, C.; Christo, D.D.; Maccheroni, W., Jr.; Azevedo, J.L.D. High molecular diversity of the fungus *Guignardia citricarpa* and *Guignardia mangiferae* and new primers for the diagnosis of the citrus black spot. *Braz. Arch. Biol. Technol.* **2009**, *52*, 1063–1073. [CrossRef]
19. Hu, J.; Johnson, E.G.; Wang, N.Y.; Davoglio, T.; Dewdney, M.M. qPCR quantification of pathogenic *Guignardia citricarpa* and nonpathogenic *G. mangiferae* in citrus. *Plant Dis.* **2014**, *98*, 112–120. [CrossRef]
20. Schirmacher, A.M.; Tomlinson, J.A.; Barnes, A.V.; Barton, V.C. Species-specific real-time PCR for diagnosis of *Phyllosticta citricarpa* on *Citrus* species. *EPPO Bull.* **2019**. [CrossRef]
21. Everett, K.R.; Rees-George, J. Reclassification of an isolate of *Guignardia citricarpa* from New Zealand as *Guignardia mangiferae* by sequence analysis. *Plant Pathol.* **2006**, *55*, 194–199. [CrossRef]

22. Romão, A.S.; Spósito, M.B.; Andreote, F.D.; Azevedo, J.L.D.; Araújo, W.L. Enzymatic differences between the endophyte *Guignardia mangiferae* (Botryosphaeriaceae) and the citrus pathogen *G. citricarpa*. *Genet. Mol. Res.* **2011**, *10*, 243–252. [CrossRef] [PubMed]

23. Munari Rodrigues, C.; Takita, M.A.; Silva, N.V.; Ribeiro-Alves, M.; Machado, M.A. Comparative genome analysis of *Phyllosticta citricarpa* and *Phyllosticta capitalensis*, two fungi species that share the same host. *BMC Genom.* **2019**, *20*, 554. [CrossRef]

24. Wikee, S.; Lombard, L.; Nakashima, C.; Motohashi, K.; Chukeatirote, E.; Cheewangkoon, R.; McKenzie, E.H.C.; Hyde, K.D.; Crous, P.W. A phylogenetic re-evaluation of *Phyllosticta* (Botryosphaeriales). *Stud. Mycol.* **2013**, *76*, 1–29. [CrossRef] [PubMed]

25. Wulandari, N.F.; Toanun, C.; Hyde, K.D.; Duong, L.M.; de Gruyter, J.; Meffert, J.P.; Groenewald, J.Z.; Crous, P.W. *Phyllosticta citriasiana* sp. nov., the cause of *Citrus* tan spot of *Citrus maxima* in Asia. *Fungal Divers.* **2009**, *34*, 23–39.

26. Wang, X.; Chen, G.; Huang, F.; Zhang, J.; Hyde, K.D.; Li, H. *Phyllosticta* species associated with citrus diseases in China. *Fungal Divers.* **2012**, *52*, 209–224. [CrossRef]

27. Guarnaccia, V.; Groenewald, J.Z.; Li, H.; Glienke, C.; Carstens, E.; Hattingh, V.; Fourie, P.H.; Crous, P.W. First report of *Phyllosticta citricarpa* and description of two new species, *P. paracapitalensis* and *P. paracitricarpa*, from citrus in Europe. *Stud. Mycol.* **2017**, *87*, 161–185. [CrossRef]

28. Glienke, C.; Pereira, O.L.; Stringari, D.; Fabris, J.; Kava-Cordeiro, V.; Galli-Terasawa, L.; Cunnington, J.; Shivas, R.G.; Groenewald, J.Z.; Crous, P.W. Endophytic and pathogenic *Phyllosticta* species, with reference to those associated with citrus black spot. *Persoonia* **2011**, *26*, 47–56. [CrossRef]

29. Schubert, T.S.; Dewdney, M.M.; Peres, N.A.; Palm, M.E.; Jeyaprakash, A.; Sutton, B.; Mondal, S.N.; Wang, N.Y.; Rascoe, J.; Picton, D.D. First report of *Guignardia citricarpa* associated with citrus black spot on sweet orange (*Citrus sinensis*) in North America. *Plant Dis.* **2012**, *96*, 1225. [CrossRef]

30. Jeger, M.; Bragard, C.; Caffier, D.; Candresse, T.; Chatzivassiliou, E.; Dehnen-Schmutz, K.; Gilioli, G.; Gregoire, J.C.; Jaques Miret, J.A.; MacLeod, A.; et al. Evaluation of a paper by Guarnaccia et al. (2017) on the first report of *Phyllosticta citricarpa* in Europe. *EFSA J.* **2018**, *16*, 5114.

31. Tran, N.T.; Miles, A.K.; Dietzgen, R.G.; Drenth, A. *Phyllosticta capitalensis* and *P. paracapitalensis* are endophytic fungi that show potential to inhibit pathogenic *P. citricarpa* on citrus. *Australas. Plant Pathol.* **2019**, *48*, 281–296. [CrossRef]

32. Wickert, E.; de Macedo Lemos, E.G.; Kishi, L.T.; de Souza, A.; de Goes, A. Genetic diversity and population differentiation of *Guignardia mangiferae* from "Tahiti" acid lime. *Sci. World J.* **2012**. [CrossRef]

33. Huang, F.; Chen, G.Q.; Hou, X.; Fu, Y.S.; Cai, L.; Hyde, K.D.; Li, H.Y. *Colletotrichum* species associated with cultivated citrus in China. *Fungal Divers.* **2013**, *61*, 61–74. [CrossRef]

34. Waculicz-Andrade, C.E.; Savi, D.C.; Bini, A.P.; Adamoski, D.; Goulin, E.H.; Silva, G.J., Jr.; Massola, N.S., Jr.; Terasawa, L.G.; Kava, V.; Glienke, C. *Colletotrichum gloeosporioides sensu stricto*: An endophytic species or citrus pathogen in Brazil? *Australas. Plant Pathol.* **2017**, *46*, 191–203. [CrossRef]

35. Durán, E.L.; Ploper, L.D.; Ramallo, J.C.; Piccolo Grandi, R.A.; Hupper Giancoli, Á.C.; Azevedo, J.L. The foliar fungal endophytes of *Citrus limon* in Argentina. *Can. J. Bot.* **2005**, *83*, 350–355. [CrossRef]

36. Douanla-Meli, C.; Langer, E.; Mouafo, F.T. Fungal endophyte diversity and community patterns in healthy and yellowing leaves of *Citrus limon*. *Fungal Ecol.* **2013**, *6*, 212–222. [CrossRef]

37. Mohali, S.; Burgess, T.I.; Wingfield, M.J. Diversity and host association of the tropical tree endophyte *Lasiodiplodia theobromae* revealed using simple sequence repeat markers. *For. Pathol.* **2005**, *35*, 385–396. [CrossRef]

38. Slippers, B.; Wingfield, M.J. Botryosphaeriaceae as endophytes and latent pathogens of woody plants: Diversity, ecology and impact. *Fungal Biol. Rev.* **2007**, *21*, 90–106. [CrossRef]

39. Zhao, W.; Bai, J.; McCollum, G.; Baldwin, E. High incidence of preharvest colonization of huanglongbing-symptomatic *Citrus sinensis* fruit by *Lasiodiplodia theobromae* (*Diplodia natalensis*) and exacerbation of postharvest fruit decay by that fungus. *Appl. Environ. Microbiol.* **2015**, *81*, 364–372. [CrossRef]

40. Udayanga, D.; Liu, X.; McKenzie, E.H.C.; Chukeatirote, E.; Bahkali, A.H.A.; Hyde, K.D. The genus *Phomopsis*: Biology, applications, species concepts and names of common phytopathogens. *Fungal Divers.* **2011**, *50*, 189–225. [CrossRef]

41. Gomes, R.R.; Glienke, C.; Videira, S.I.R.; Lombard, L.; Groenewald, J.Z.; Crous, P.W. *Diaporthe*: A genus of endophytic, saprobic and plant pathogenic fungi. *Persoonia* **2013**, *31*, 1–41. [CrossRef]

42. Savi, D.C.; Aluizio, R.; Glienke, C. Brazilian plants: An unexplored source of endophytes as producers of active metabolites. *Planta Med.* **2019**, *85*, 619–636.

43. Huang, F.; Hou, X.; Dewdney, M.M.; Fu, Y.; Chen, G.; Hyde, K.D.; Li, H. *Diaporthe* species occurring on citrus in China. *Fungal Divers.* **2013**, *61*, 237–250. [CrossRef]

44. Guarnaccia, V.; Crous, P.W. Emerging citrus diseases in Europe caused by species of *Diaporthe*. *IMA Fungus* **2017**, *8*, 317–334. [CrossRef]

45. Huang, F.; Udayanga, D.; Wang, X.; Hou, X.; Mei, X.; Fu, Y.; Hyde, K.D.; Li, H. Endophytic *Diaporthe* associated with *Citrus*: A phylogenetic reassessment with seven new species from China. *Fungal Biol.* **2015**, *119*, 331–347. [CrossRef]

46. Peever, T.L.; Canihos, Y.; Olsen, L.; Ibáñez, A.; Liu, Y.C.; Timmer, L.W. Population genetic structure and host specificity of *Alternaria* spp. causing brown spot of Minneola tangelo and rough lemon in Florida. *Phytopathology* **1999**, *89*, 851–860. [CrossRef]

47. Akimitsu, K.; Peever, T.L.; Timmer, L.W. Molecular, ecological and evolutionary approaches to understanding *Alternaria* diseases of citrus. *Mol. Plant Pathol.* **2003**, *4*, 435–446. [CrossRef]

48. Sadeghi, F.; Samsampour, D.; Seyahooei, M.A.; Bagheri, A.; Soltani, J. Diversity and spatiotemporal distribution of fungal endophytes associated with *Citrus reticulata* cv. Siyahoo. *Curr. Microbiol.* **2019**, *76*, 279–289. [CrossRef]

49. Juybari, H.Z.; Tajick Ghanbary, M.A.; Rahimian, H.; Karimi, K.; Arzanlou, M. Seasonal, tissue and age influences on frequency and biodiversity of endophytic fungi of *Citrus sinensis* in Iran. *For. Pathol.* **2019**, e12559. [CrossRef]

50. Sandoval-Denis, M.; Guarnaccia, V.; Polizzi, G.; Crous, P.W. Symptomatic *Citrus* trees reveal a new pathogenic lineage in *Fusarium* and two new *Neocosmospora* species. *Persoonia* **2018**, *40*, 1–25. [CrossRef]

51. Childs, J.F.L.; Kopp, L.E.; Johnson, R.E. A species of *Physoderma* present in *Citrus* and related species. *Phytopathology* **1965**, *55*, 681–687.

52. Ho, M.Y.; Chung, W.C.; Huang, H.C.; Chung, W.H.; Chung, W.H. Identification of endophytic fungi of medicinal herbs of Lauraceae and Rutaceae with antimicrobial property. *Taiwania* **2012**, *57*, 229–241.

53. Santos Gai, C.; Teixeira Lacava, P.; Maccheroni, W., Jr.; Glienke, C.; Araújo, W.L.; Miller, T.A.; Azevedo, J.L. Diversity of endophytic yeasts from sweet orange and their localization by scanning electron microscopy. *J. Basic Microbiol.* **2009**, *49*, 441–451.

54. Rodríguez, P.; Reyes, B.; Barton, M.; Coronel, C.; Menéndez, P.; Gonzalez, D.; Rodríguez, S. Stereoselective biotransformation of α-alkyl-β-keto esters by endophytic bacteria and yeast. *J. Mol. Catal. B Enzym.* **2011**, *71*, 90–94. [CrossRef]

55. Bamisile, B.S.; Dash, C.K.; Akutse, K.S.; Qasim, M.; Ramos Aguila, L.C.; Wang, F.; Keppanan, R.; Wang, L. Endophytic *Beauveria bassiana* in foliar-treated *Citrus limon* plants acting as a growth suppressor to three successive generations of *Diaphorina citri* Kuwayama (Hemiptera: Liviidae). *Insects* **2019**, *10*, 176. [CrossRef]

56. Araújo, W.L.; Maccheroni, W., Jr.; Aguilar-Vildoso, C.I.; Barroso, P.A.; Saridakis, H.O.; Azevedo, J.L. Variability and interactions between endophytic bacteria and fungi isolated from leaf tissues of citrus rootstocks. *Can. J. Microbiol.* **2001**, *47*, 229–236. [CrossRef]

57. Glienke-Blanco, C.; Aguilar-Vildoso, C.I.; Vieira, M.L.C.; Barroso, P.A.V.; Azevedo, J.L. Genetic variability in the endophytic fungus *Guignardia citricarpa* isolated from citrus plants. *Genet. Mol. Biol.* **2002**, *25*, 251–255. [CrossRef]

58. Manoharan, G.; Sairam, T.; Thangamani, R.; Ramakrishnan, D.; Tiwari, M.K.; Lee, J.K.; Marimuthu, J. Identification and characterization of type III polyketide synthase genes from culturable endophytes of ethnomedicinal plants. *Enzyme Microb. Technol.* **2019**, *131*, 109396. [CrossRef]

59. Ling, L.; Li, Z.; Jiao, Z.; Zhang, X.; Ma, W.; Feng, J.; Zhang, J.; Lu, L. Identification of novel endophytic yeast strains from tangerine peel. *Curr. Microbiol.* **2019**, *76*, 1066–1072. [CrossRef]

60. Paz, Z.; Burdman, S.; Gerson, U.; Sztejnberg, A. Antagonistic effects of the endophytic fungus *Meira geulakonigii* on the citrus rust mite *Phyllocoptruta oleivora*. *J. Appl. Microbiol.* **2007**, *103*, 2570–2579. [CrossRef]

61. Pena, L.C.; Jung, L.F.; Savi, D.C.; Servienski, A.; Aluizio, R.; Goulin, E.H.; Galli-Terasawa, L.V.; Lameiro de Noronha Sales Maia, B.H.; Annies, V.; Cavichiolo Franco, C.R.; et al. A *Muscodor* strain isolated from *Citrus sinensis* and its production of volatile organic compounds inhibiting *Phyllosticta citricarpa* growth. *J. Plant Dis. Prot.* **2017**, *124*, 349–360. [CrossRef]

62. Rodrigues, K.F.; Sieber, T.N.; Grünig, C.R.; Holdenrieder, O. Characterization of *Guignardia mangiferae* isolated from tropical plants based on morphology, ISSR-PCR amplifications and ITS1-5.8 S-ITS2 sequences. *Mycol. Res.* **2004**, *108*, 45–52. [CrossRef]

63. Miles, A.K.; Tan, Y.P.; Tan, M.K.; Donovan, N.J.; Ghalayini, A.; Drenth, A. *Phyllosticta* spp. on cultivated citrus in Australia. *Australas. Plant Pathol.* **2013**, *42*, 461–467. [CrossRef]

64. Schüepp, H. Untersuchungen über *Guignardia citricarpa* Kiely, den Erreger der Schwarzfleckenkrankheit auf *Citrus*. *J. Phytopathol.* **1960**, *40*, 258–271. [CrossRef]

65. Azevedo, J.L.; Maccheroni, W., Jr.; Pereira, J.O.; de Araújo, W.L. Endophytic microorganisms: A review on insect control and recent advances on tropical plants. *Electron. J. Biotechnol.* **2000**, *3*, 15–16. [CrossRef]

66. Hawksworth, D.L.; Crous, P.W.; Redhead, S.A.; Reynolds, D.R.; Samson, R.A.; Seifert, K.A.; Taylor, J.W.; Wingfield, M.J.; Abaci, O.; Aime, C.; et al. The Amsterdam declaration on fungal nomenclature. *IMA Fungus* **2011**, *2*, 105–112. [CrossRef]

67. Davis, E.C.; Franklin, J.B.; Shaw, A.J.; Vilgalys, R. Endophytic *Xylaria* (Xylariaceae) among liverworts and angiosperms: Phylogenetics, distribution, and symbiosis. *Am. J. Bot.* **2003**, *90*, 1661–1667. [CrossRef]

68. U'Ren, J.M.; Miadlikowska, J.; Zimmerman, N.B.; Lutzoni, F.; Stajich, J.E.; Arnold, A.E. Contributions of North American endophytes to the phylogeny, ecology, and taxonomy of Xylariaceae (Sordariomycetes, Ascomycota). *Mol. Phylogenet. Evol.* **2016**, *98*, 210–232. [CrossRef]

69. Nicoletti, R.; Fiorentino, A. Plant bioactive metabolites and drugs produced by endophytic fungi of Spermatophyta. *Agriculture* **2015**, *5*, 918–970. [CrossRef]

70. Camargo Dos Santos, P.J.; Savi, D.C.; Rodrigues Gomes, R.; Goulin, E.H.; Da Costa Senkiv, C.; Ossamu Tanaka, F.A.; Rodrigues Almeida, A.M.; Galli-Terasawa, L.; Kava, V.; Glienke, C. *Diaporthe endophytica* and *D. terebinthifolii* from medicinal plants for biological control of *Phyllosticta citricarpa*. *Microbiol. Res.* **2016**, *186*, 153–160. [CrossRef]

71. Hartley, S.E.; Gange, A.C. Impacts of plant symbiotic fungi on insect herbivores: Mutualism in a multitrophic context. *Ann. Rev. Entomol.* **2009**, *54*, 323–342. [CrossRef]

72. Eberl, F.; Uhe, C.; Unsicker, S.B. Friend or foe? The role of leaf-inhabiting fungal pathogens and endophytes in tree-insect interactions. *Fungal Ecol.* **2019**, *38*, 104–112. [CrossRef]

73. Martinuz, A.; Schouten, A.; Menjivar, R.D.; Sikora, R.A. Effectiveness of systemic resistance toward *Aphis gossypii* (Hom., Aphididae) as induced by combined applications of the endophytes *Fusarium oxysporum* Fo162 and *Rhizobium etli* G12. *Biol. Control* **2012**, *62*, 206–212. [CrossRef]

74. Bogner, C.W.; Kamdem, R.S.; Sichtermann, G.; Matthäus, C.; Hölscher, D.; Popp, J.; Proksch, P.; Grundler, F.M.; Schouten, A. Bioactive secondary metabolites with multiple activities from a fungal endophyte. *Microb. Biotechnol.* **2017**, *10*, 175–188. [CrossRef] [PubMed]

75. Brakhage, A.A.; Schroeckh, V. Fungal secondary metabolites–strategies to activate silent gene clusters. *Fungal Genet. Biol.* **2011**, *48*, 15–22. [CrossRef] [PubMed]

76. Deepika, V.B.; Murali, T.S.; Satyamoorthy, K. Modulation of genetic clusters for synthesis of bioactive molecules in fungal endophytes: A review. *Microbiol. Res.* **2016**, *182*, 125–140. [CrossRef] [PubMed]

77. Kumaran, R.S.; Muthumary, J.; Hur, B.K. Taxol from *Phyllosticta citricarpa*, a leaf spot fungus of the angiosperm *Citrus medica*. *J. Biosci. Bioeng.* **2008**, *106*, 103–106. [CrossRef] [PubMed]

78. Nicoletti, R.; Fiorentino, A. Antitumor metabolites of fungi. *Curr. Bioact. Comp.* **2014**, *10*, 207–244. [CrossRef]

79. Savi, D.C.; Shaaban, K.A.; Mitra, P.; Ponomareva, L.V.; Thorson, J.S.; Glienke, C.; Rohr, J. Secondary metabolites produced by the citrus phytopathogen *Phyllosticta citricarpa*. *J. Antibiot.* **2019**, *72*, 306–310. [CrossRef]

80. Buckel, I.; Andernach, L.; Schüffler, A.; Piepenbring, M.; Opatz, T.; Thines, E. Phytotoxic dioxolanones are potential virulence factors in the infection process of *Guignardia bidwellii*. *Sci. Rep.* **2017**, *7*, 8926. [CrossRef]

81. Yuan, W.H.; Liu, M.; Jiang, N.; Guo, Z.K.; Ma, J.; Zhang, J.; Song, Y.C.; Tan, R.X. Guignardones A–C: Three meroterpenes from *Guignardia mangiferae*. *Eur. J. Org. Chem.* **2010**, *33*, 6348–6353. [CrossRef]

82. Guimarães, D.O.; Lopes, N.P.; Pupo, M.T. Meroterpenes isolated from the endophytic fungus *Guignardia mangiferae*. *Phytochem. Lett.* **2012**, *5*, 519–523. [CrossRef]

83. Han, W.B.; Dou, H.; Yuan, W.H.; Gong, W.; Hou, Y.Y.; Ng, S.W.; Tan, R.X. Meroterpenes with toll-like receptor 3 regulating activity from the endophytic fungus *Guignardia mangiferae*. *Planta Med.* **2015**, *81*, 145–151. [CrossRef]

84. Sun, Z.H.; Liang, F.L.; Wu, W.; Chen, Y.C.; Pan, Q.L.; Li, H.H.; Ye, W.; Liu, H.X.; Li, S.N.; Tan, G.H.; et al. Guignardones P–S, new meroterpenoids from the endophytic fungus *Guignardia mangiferae* A348 derived from the medicinal plant smilax glabra. *Molecules* **2015**, *20*, 22900–22907. [CrossRef]

85. Chen, K.; Zhang, X.; Sun, W.; Liu, J.; Yang, J.; Chen, C.; Liu, X.; Gao, L.; Wang, J.; Li, H.; et al. Manginoids A–G: Seven monoterpene–shikimate-conjugated meroterpenoids with a spiro ring system from *Guignardia mangiferae*. *Org. Lett.* **2017**, *19*, 5956–5959. [CrossRef]

86. Li, T.X.; Yang, M.H.; Wang, X.B.; Wang, Y.; Kong, L.Y. Synergistic antifungal meroterpenes and dioxolanone derivatives from the endophytic fungus *Guignardia* sp. *J. Nat. Prod.* **2015**, *78*, 2511–2520. [CrossRef]

87. Eze, P.M.; Ojimba, N.K.; Abonyi, D.O.; Chukwunwejim, C.R.; Abba, C.C.; Okoye, F.B.C.; Esimone, C.O. Antimicrobial activity of metabolites of an endophytic fungus isolated from the leaves of *Citrus jambhiri* (Rutaceae). *Trop. J. Nat. Prod. Res.* **2018**, *2*, 145–149. [CrossRef]

88. Doty, S.L. Growth-promoting endophytic fungi of forest trees. In *Endophytes of Forest Trees*; Pirttilä, A., Frank, A., Eds.; Springer: Berlin, Germany, 2011; pp. 151–156.

89. Waqas, M.; Khan, A.L.; Kamran, M.; Hamayun, M.; Kang, S.M.; Kim, Y.H.; Lee, I.J. Endophytic fungi produce gibberellins and indoleacetic acid and promotes host-plant growth during stress. *Molecules* **2012**, *17*, 10754–10773. [CrossRef]

90. Suwannarach, N.; Kumla, J.; Bussaban, B.; Nuangmek, W.; Matsui, K.; Lumyong, S. Biofumigation with the endophytic fungus *Nodulisporium* spp. CMU-UPE34 to control postharvest decay of citrus fruit. *Crop Prot.* **2013**, *45*, 63–70. [CrossRef]

91. Gomes, A.A.M.; Queiroz, M.V.; Pereira, O.L. Mycofumigation for the biological control of post-harvest diseases in fruits and vegetables: A review. *Austin J. Biotechnol. Bioeng.* **2015**, *2*, 1051.

92. Kaddes, A.; Fauconnier, M.L.; Sassi, K.; Nasraoui, B.; Jijakli, M.H. Endophytic fungal volatile compounds as solution for sustainable agriculture. *Molecules* **2019**, *24*, 1065. [CrossRef]

93. Huang, Z.; Yang, J.; Cai, X.; She, Z.; Lin, Y. A new furanocoumarin from the mangrove endophytic fungus *Penicillium* sp. (ZH16). *Nat. Prod. Res.* **2012**, *26*, 1291–1295. [CrossRef]

94. Zaher, A.M.; Moharram, A.M.; Davis, R.; Panizzi, P.; Makboul, M.A.; Calderón, A.I. Characterisation of the metabolites of an antibacterial endophyte *Botryodiplodia theobromae* Pat. of *Dracaena draco* L. by LC–MS/MS. *Nat. Prod. Res.* **2015**, *29*, 2275–2281. [CrossRef]

95. Balakumaran, M.D.; Ramachandran, R.; Kalaichelvan, P.T. Exploitation of endophytic fungus, *Guignardia mangiferae* for extracellular synthesis of silver nanoparticles and their in vitro biological activities. *Microbiol. Res.* **2015**, *178*, 9–17. [CrossRef]

96. Busby, P.E.; Ridout, M.; Newcombe, G. Fungal endophytes: Modifiers of plant disease. *Plant Mol. Biol.* **2016**, *90*, 645–655. [CrossRef]

97. Schlaeppi, K.; Bulgarelli, D. The plant microbiome at work. *Mol. Plant Microbe Interact.* **2015**, *28*, 212–217. [CrossRef]

98. Ahlholm, J.U.; Helander, M.; Henriksson, J.; Metzler, M.; Saikkonen, K. Environmental conditions and host genotype direct genetic diversity of *Venturia ditricha*, a fungal endophyte of birch trees. *Evolution* **2002**, *56*, 1566–1573. [CrossRef]

99. Balint, M.; Tiffin, P.; Hallstrom, B.; O'Hara, R.B.; Olson, M.S.; Fankhauser, J.D.; Piepenbring, M.; Schmitt, I. Host genotype shapes the foliar fungal microbiome of balsam poplar (*Populus balsamifera*). *PLoS ONE* **2013**, *8*, e53987. [CrossRef]

 *agriculture*

Review

Endophytism of *Lecanicillium* and *Akanthomyces*

Rosario Nicoletti [1,2,*] **and Andrea Becchimanzi** [2]

[1] Council for Agricultural Research and Economics, Research Centre for Olive, Fruit and Citrus Crops, 81100 Caserta, Italy

[2] Department of Agricultural Sciences, University of Naples Federico II, 80055 Portici, Italy; andrea.becchimanzi@unina.it

[*] Correspondence: rosario.nicoletti@crea.gov.it

Received: 11 May 2020; Accepted: 5 June 2020; Published: 6 June 2020

Abstract: The rise of the holobiont concept confers a prominent importance to the endophytic associates of plants, particularly to species known to be able to exert a mutualistic role as defensive or growth-promoting agents. The finding that many entomopathogenic fungi are harbored within plant tissues and possess bioactive properties going beyond a merely anti-insectan effect has recently prompted a widespread investigational activity concerning their occurrence and functions in crops, in the aim of an applicative exploitation conforming to the paradigm of sustainable agriculture. The related aspects particularly referring to species of *Lecanicillium* and *Akanthomyces* (Sordariomycetes, Cordycipitaceae) are revised in this paper, also in light of recent and ongoing taxonomic reassessments.

Keywords: entomopathogens; endophytic fungi; crop protection; plant growth promotion; integrated pest management; bioactive compounds; Cordycipitaceae

1. Introduction

The great microbial diversity harbored in plants has just started being explored in light of a consolidated awareness that what we manage in the agricultural practice is actually the outcome of the combined expression of plant and microbial genes [1,2]. The symbiotic relationships between endophytic fungi and their host plants exteriorize in many ways, ranging from opportunistic saprophytism in senescent tissues, to latent pathogenicity disclosing after the impact of various stress factors, to genuine mutualistic interactions deriving from nutritional support and/or increased protection against pests and pathogens. The latter are particularly relevant for the holistic approach making its way in integrated pest management (IPM), considering the crop production system as a whole in the aim to contain rather than eradicate pests.

Within this conceptual rearrangement, the improvement of our knowledge on occurrence and functions of endophytic associates of plants is fundamental in view of their possible exploitation in sustainable agriculture. Endophytic entomopathogens are an important category of the plant microbiome, which is increasingly considered for applicative purposes. So far, the majority of investigations and reports concerning these organisms deal with *Beauveria bassiana* and *Metarhizium anisopliae*, with several fine reviews available in the literature [3,4]. This paper offers an overview on the current knowledge concerning endophytism in species of *Lecanicillium* and *Akanthomyces* (Sordariomycetes, Hypocreales, Cordycipitaceae).

2. Taxonomic Background

Until the early 2000s, these fungi were classified in the section *Prostrata* of the genus *Verticillium*, basically with reference to their imperfect stage producing verticillate conidiophores [5]. A few species best known for their parasitic behavior against arthropods, nematodes and/or fungi were ascribed to this section, such as *V. chlamydosporium*, *V. lecanii* and *V. psalliotae*. Afterwards, the application of biomolecular techniques enabled to shed light on the phylogenetic relationships within this heterogeneous genus. Particularly, species within the section *Prostrata* were separated in a few unrelated genera, such as *Pochonia*, *Haptocillium*, *Simplicillium* and *Lecanicillium*, and their teleomorphs identified within the genera *Cordyceps* and *Torrubiella* [6]. The species *V. fungicola*, previously ascribed to the section *Albo-erecta* in the genus *Verticillium*, was later aggregated to *Lecanicillium* [7]. As a result of this fundamental revision, about fifteen *Lecanicillium* species were recognized, a few of which (*L. attenuatum*, *L. longisporum*, *L. muscarium*, *L. nodulosum* and *L. lecanii* s.str.) enucleated from the previously collective *V. lecanii*.

However, as it often happens in fungal taxonomy, such a sound rearrangement was not destined to persist. In fact the genus *Lecanicillium* was shown to be paraphyletic [8], and some species were moved to *Akanthomyces*, a pre-existing but overlooked genus including entomogenous species [9] (Table 1). At the same time, investigations in more or less peculiar ecological contexts brought to the description of novel taxa of both *Akanthomyces* and *Lecanicillium* [10,11], while some species ascribed to the latter genus, such as *L. uredinophilum* and *L. pissodis*, were shown to actually fit in the *A. lecanii* clade [12]. Following the dismissal of the dual nomenclature system for pleomorphic fungi, a more comprehensive revision of the whole family of the Cordycipitaceae is in progress. Particularly, rejection has been proposed for the genus name *Lecanicillium*, while some *Akanthomyces* species have in turn been moved to another genus (*Hevansia*) [13]. Hence, further adjustments concerning species still classified in *Lecanicillium* are to be expected.

Table 1. Nomenclatural correspondence of accepted *Lecanicillium*/*Akanthomyces* species with sequences of internal transcribed spacers of ribosomal DNA (rDNA-ITS) available in GenBank.

Species Names *			ITS Sequence Used in
Lecanicillium	*Akanthomyces*	*Cordyceps/Torrubiella*	Phylogenetic Analysis
L. acerosum			NR11268
L. antillanum			AJ292392
L. aphanocladii			LT220701
L. aranearum	*A. aranearum*	*T. alba*	AJ292464
L. araneicola			AB378506
L. araneogenum	*A. neoaraneogenus*		NR161115
L. attenuatum	*A. attenuatus*		AJ292434
L. cauligalbarum			MH730663
L. coprophilum			MH177615
L. dimorphum			AJ292429
L. flavidum			EF641877
L. fungicola var.*aleophilum*			NR111064
L. fungicola var.*fungicola*			NR119653
L. fusisporum			AJ292428
L. kalimantanense			AB360356
L. lecanii	*A. lecanii*	*C. confragosa*	AJ292383
L. longisporum	*A. dipterigenus*		AJ292385
L. muscarium	*A. muscarius*		NR111096
L. nodulosum	*Akanthomyces* sp.		EF513012
L. primulinum			NR119418
L. psalliotae			AJ292389
L. restrictum			LT548279
L. sabanense	*A. sabanensis*		KC633232
L. subprimulinum			MG585314
L. tenuipes			AJ292391
L. testudineum			LT548278
L. uredinophilum	*Akanthomyces* sp.		MG948305
L. wallacei		*T. wallacei*	NR111267
Lecanicillium sp.		*C. militaris*	AF153264

Table 1. *Cont.*

Species Names *			ITS Sequence Used in
Lecanicillium	*Akanthomyces*	*Cordyceps/Torrubiella*	Phylogenetic Analysis
	A. aculeatus		KC519371
	A. coccidioperitheciatus	*C. coccidioperitheciata*	JN049865
	A. kanyawimiae		MF140751
	A. sphingum	*C. sphingum*	AY245641
	A. sulphureus	*Torrubiella* sp.	MF140756
	A. thailandicus	*Torrubiella* sp.	MF140755
	A. tuberculatus	*C. tuberculata*	JN049830
	A. waltergamsii		MF140747

* The currently used species names as inferred from the Mycobank database [14] are reported in bold.

3. Occurrence

The number of reports concerning endophytic isolates of *Lecanicillium* and *Akanthomyces* has increased in recent years. This is due not only to the several taxonomic reassessments introducing new species, but also to the easier access to techniques and databases for DNA sequencing, which in most instances enable one to overcome the intrinsic difficulties of morphological identification. However, more prompts have probably resulted by the awareness of the basic role that endophytic fungi play on plant fitness, introducing applicative perspectives for investigations in the field. For the above genera, literature shows a prevalence of findings concerning natural phytocoenoses (Table 2) over those inherent crops (Table 3); even more so considering that the latter series includes a few cases of endophytic colonization resulting after artificial inoculation in experimental work. Basically connected with the issue of ecosystem simplification characterizing the agricultural contexts, such a difference emphasizes the opportunity to recover the functional role of this component of the plant holobiont in view of improving crop performances.

Table 2. Endophytic occurrence of *Lecanicillium/Akanthomyces* in wild contexts.

Species	Host Plant	Country	ITS Sequence √	Reference
A. attenuatus	*Astrocaryum sciophilum*	French Guyana	MK279520	[15]
	Conifer plant	China	MN908945	GenBank
	Symplocarpus foetidus	Canada	KC916681	[16]
A. lecanii	*Ammophila arenaria*	Spain	-	[17]
	Dactylis glomerata	Spain	AM262369	[18]
	Deschampsia flexuosa	Finland	KJ529005	[19]
	Elymus farctus	Spain	AM924163	[17]
	Laretia acaulis	Chile	-	[20]
	Pinus sylvestris	Italy	KJ093501	[21]
	Pinus sylvestris	Poland	-	[22]
	Shorea thumbuggaia	India	KJ542654	GenBank
	Taxus baccata	Iran	KF573987	[23]
A. muscarius	*Acer campestre*	Italy	MT230457	This paper
	Laurus nobilis	Italy	-	[24]
	Myrtus communis	Italy	MT230435	This paper
	Nypa fruticans	Thailand	MH497223	[25]
	Quercus robur	Italy	MT230463	This paper
Akanthomyces sp. *	*Arctostaphylos uva-ursi*	Switzerland	-	[26]
	Carpinus caroliniana	USA	-	[27]
L. aphanocladii	*Ageratina adenophora*	China	MK304090 MK304173 MK304418	[28]
	Hemidesmus indicus	India	MH594215	[29]
	Huperzia serrata	China	KP689216 KP689173	[30]
	Picea mariana	Canada	-	[31]
L. fungicola	*Phragmites australis*	Korea	KP017880	[32]
L. kalimantanense	*Zingiber officinale*	Indonesia	-	[33]

Table 2. *Cont.*

Species	Host Plant	Country	ITS Sequence √	Reference
L. psalliotae	*Cerastium fischerianum*	Korea	JX238776	[34]
	Coix lachryma-jobi	China	KJ572167	GenBank
	Magnolia officinalis	China		GenBank
	Phoradendron perrottettii	Brazil	-	[35]
	Pinus radiata	New Zealand	-	[36]
	Sedum oryzifolium	Korea	KU556134	[37]
	Tapirira guianensis	Brazil	-	[35]
	Triticum dicoccoides	Israel	-	[38]
Lecanicillium sp.	*Artocarpus lacucha*	India	MH700423 MH700428	GenBank
	Bupleurum chinense	China	MG561939	GenBank
	Huperzia serrata	China	KM513600	[30]
	Liparis japonica	China	KT719186 KT719187 KT719188 KT719189 KT719192	GenBank
	Micrandra spruceana	Peru	MH267985	[39]
	Microthlaspi perfoliatum	Greece	KT269776	[40]
	Quassia indica	India	MH910098	GenBank
	Sandwithia guyanensis	French Guyana	MN514023	[41]
	Theobroma gileri	Ecuador	-	[42]

√ Missing ITS accession number implies identification based on morphological characters only, or without depositing the ITS sequence. * These strains were originally identified as *Verticillium lecanii*.

Table 3. Endophytic occurrence of *Lecanicillium/Akanthomyces* in crops.

Species	Host Plant	Country	ITS Sequence √	Reference
A. attenuatus	*Brachiaria* sp.	Kenya	KU574698	[43]
	Salvia miltiorrhiza	China	JX406555	GenBank
A. lecanii	*Cucurbita maxima*	Australia	-	[44]
	Gossypium hirsutum	Australia	-	[45]
	Gossypium hirsutum	Brazil	-	[46]
	Gossypium hirsutum	Texas, USA	KP407570	[47]
	Solanum lycopersicum	Australia	-	[44]
	Phaseolus vulgaris	Australia	-	[44]
	Phaseolus vulgaris	China	-	[48]
	Pistacia vera	Iran	MF000354	[49]
	Triticum aestivum	Australia	-	[44]
	Vitis vinifera	Spain	-	[50]
	Zea mays	Australia	-	[44]
A. muscarius	*Brassica oleracea*	New Zealand	-	[51]
	Cucumis sativus	Canada	-	[52]
	Cucumis sativus	Japan	-	[53]
	Prunus cerasus	Iran	KY472303	[54]
L. aphanocladii	*Zea mays*	Slovenia	-	[55]
L. dimorphum	*Phoenix dactylifera*	Spain	-	[56]
L. psalliotae	*Phoenix dactylifera*	Spain	-	[56]
Lecanicillium sp.	*Citrus limon*	Iran	MN448344	GenBank
	Vitis vinifera	China	MT123107	GenBank
	Zea mays	India	-	[57]

√ Missing ITS accession number implies identification based on morphological characters only, or without depositing the ITS sequence.

Overall, Tables 2 and 3 include 65 citations of endophytic strains belonging to these two genera as a result of a search considering literature in the field and the GenBank database. A widespread capacity to colonize plants from heterogeneous ecological contexts is evident considering that these citations refer to 54 species belonging to 35 botanical families. With 10 species Poaceae is the most represented family, followed by Arecaceae and Pinaceae with three species each, and Anacardiaceae, Apiaceae, Brassicaceae, Cucurbitaceae, Euphorbiaceae and Malvaceae with two species. The rest of the families

(Apocynaceae, Araceae, Asteraceae, Betulaceae, Caryophyllaceae, Crassulaceae, Dipterocarpaceae, Ericaceae, Fabaceae, Fagaceae, Lamiaceae, Lauraceae, Lycopodiaceae, Magnoliaceae, Moraceae, Myrtaceae, Orchidaceae, Rosaceae, Rutaceae, Santalaceae, Sapindaceae, Simaroubaceae, Solanaceae, Taxaceae, Vitaceae and Zingiberaceae) are represented by a single species.

Such a variety of hosts seems to contrast any hypothesis of host specialization, and is rather indicative of a possible tendency to spread horizontally within the phytocoenoses. In this respect, the recovery of *A. muscarius* from four woody species (*Acer campestre, Laurus nobilis, Quercus robur* and *Myrtus communis* in two separate stands) at the Astroni Nature Reserve near Napoli, Italy ([24] and in this paper), appears to support this ability, which may as well imply a permanent functional role in natural ecosystems. On the other hand, indications of a constant association with crop species could be favorable for possible applications in IPM. The limited available data only support preliminary clues in the case of cotton (*Gossypium hirsutum*) where, considering the economic impact of insect pests, the endophytic occurrence of strains of *A. lecanii* reported from distant countries such as Australia, Brazil and the United States might deserve further attention.

Phylogenetic Relationships of Endophytic Strains

In the evolving taxonomic scheme outlined above, the endophytic isolates provisionally classified as *Lecanicillium* sp. are to be further considered for a more definite taxonomic assignment. In this perspective, we propose a phylogenetic analysis (Figure 1) considering strains whose sequences of internal transcribed spacers of ribosomal DNA (rDNA-ITS) are deposited in GenBank (Tables 2 and 3), along with official reference strains for the currently accepted species of *Lecanicillium* and *Akanthomyces* (Table 1).

Although more DNA sequences, such as the translation elongation factor 1 alpha (*TEF*) and RNA polymerase II largest subunits 1 (*RPB1*) and 2 (*RPB2*), are considered in taxonomic assessments concerning genera in the Cordycipitaceae [12,13,25,59], provisional identification of isolates recovered in the course of biodiversity studies is routinely done on account of ITS. Therefore only these kinds of sequences are usually deposited in GenBank for such strains, representing the only possible marker available for phylogenetic reconstructions.

In the absence of opportunities for a direct examination of these isolates, the phylogenetic tree proposed in Figure 1 provides an indication for their provisional assimilation to any of the accepted taxa in the genera *Lecanicillium* and *Akanthomyces*. A major cluster in the upper part of the tree includes the type strains of the species of *Cordyceps, Akanthomyces* (except *A. aranearum*), and of *L. nodulosum* and *L. uredinophilum*, which are also credited for ascription to *Akanthomyces*, along with all the endophytic strains ascribed to the species *A. lecanii, A. muscarius* and *A. attenuatus* (clades A, B and C, respectively). However, just two out of seven endophytic isolates ascribed to *A. lecanii* are next to the type strain of this species, while five more isolates rather group with *A. muscarius*. Confirming evidence from previous phylogenetic analyses [25,59], *A. attenuatus* is very close to *A. muscarius*, but an isolate from the palm *Astrocaryum sciophilum* is displaced in clade B. Another isolate from *Brachiaria* sp. reported as *A. attenuatus* is more distant, having *L. uredinophilum* as the closest relative. While these remarks cannot be taken as an evidence of a more common endophytic occurrence of *A. muscarius*, they represent an indication that at least some isolates of this species might have been misidentified as *A. lecanii*. This is not surprising, considering that a previous study pointed out the difficulty of resolving species ascription of strains previously ascribed to *V. lecanii* by using ITS sequences only [60].

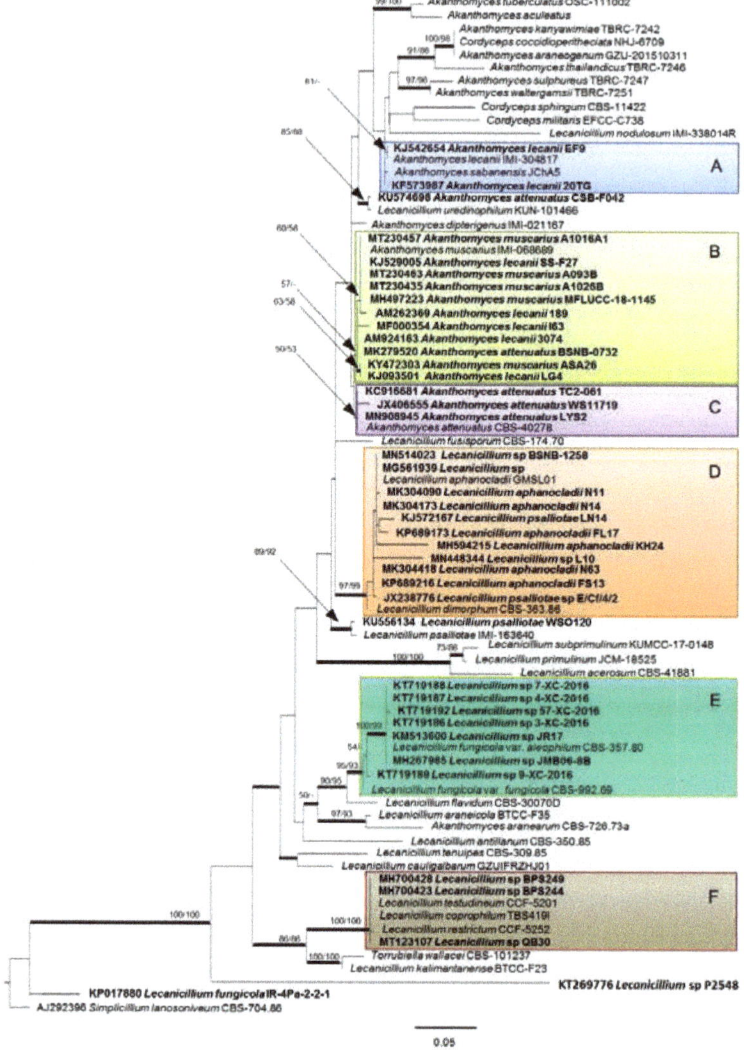

Figure 1. Phylogenetic tree based on maximum likelihood (ML) analysis of the rDNA-ITS sequences deposited in GenBank for the known species (Table 1) and the endophytic strains of *Lecanicillium* and *Akanthomyces* (in bold, Tables 2 and 3). Multiple sequence alignment comprised 592 nucleotide positions, including gaps. The analysis was carried out using RAxML software (version 8.2.12; https://cme.h-its.org/exelixis/web/software/raxml) for ML, PAUP (version 4.0a166; https://paup.phylosolutions.com) for maximum parsimony (MP), and MrBayes (version 3.2.7a; https://nbisweden.github.io/MrBayes/download.html) for Bayesian analysis. Phylogenetic tree was drawn using FigTree software (version 1.4.4; http://tree.bio.ed.ac.uk/software/figtree). Details and complete references are specified in a recent paper [58]. Bootstrap support values ≥60% for ML and MP are presented above branches as follows: ML/MP, bootstrap support values <50% are marked with '-'. Branches in bold are supported by Bayesian analysis (posterior probability ≥95%). *Simplicillium lanosoniveum* CBS 704.86 (GenBank: AJ292396) was used as outgroup reference. Main clades are indicated by colored boxes A, B, C, D, E and F.

Interestingly, no endophytic isolates provisionally identified as *Lecanicillium* sp. belong to the above major *Akanthomyces* cluster. Three of them are part of clade D, corresponding to the species *L. aphanocladii*, which also includes two strains identified as *L. psalliotae*. This is acceptable since these species and *L. dimorphum* have been reported in a close phylogenetic relationship in previous analyses [6,59]. However, *L. psalliotae* seems somehow problematic with reference to the resolution power of ITS, considering that it was reported as the closest relative (99.65% sequence identity at 100% query cover) of another isolate from *Microthlaspi perfoliatum* [40], which is in a quite distant position in our phylogenetic tree.

As many as seven unidentified strains cluster with *L. fungicola*, prevalently with the type strain of var. *aleophilum* (clade E), indicating a relevant endophytic occurrence of this species, which was not recognized so far. Another isolate reported as *L. fungicola* [32], deserves a more careful consideration with reference to its basal placement. In fact, BLAST search in GenBank indicated a 100% identity with ten strains of this species and several strains of the unrelated *Simplicillium aogashimaense*. The latter was characterized in 2013 with the support of a phylogenetic analysis based on ITS only, which anyway showed a consistent distance from *L. fungicola* [61]. Quite meaningfully, in our analysis the isolate in question was placed in proximity to the outgroup (*Simplicillium lanosoniveum*) on which our tree was rooted. Considering that sequences of six out of this group of ten *L. fungicola* strains were deposited in GenBank before 2013, it is quite possible that original misidentification of those that might rather have been *Simplicillium* strains could have determined the incorrect assignment of the more recent isolates.

Finally, three isolates (two Indian from *Artocarpus lacucha* and one Chinese from *Vitis vinifera*) are grouped in clade F together with the type strains of the recently described *L. coprophilum* [11], *L. restrictum* and *L. testudineum* [62]. A BLAST search in the GenBank database shows the first species as the closest relative, with 100% and 99.81% ITS sequence identity for the Chinese and the Indian isolates, respectively.

4. Implications in Crop Protection

As introduced above, so far there are few observations concerning the effects of endophytic strains of *Lecanicillium* and *Akanthomyces* in crops. Within the limited data available so far, cotton stands out for remarks on the endophytic occurrence of *A. lecanii* from independent cropping areas. In Australia an endophytic isolate was shown to be able to colonize cotton plants ensuring protection against the cotton aphid (*Aphis gossypii*) after artificial inoculation. Besides evidence from direct microscopic examination, the ability to colonize plant tissues was confirmed by re-isolation from leaves of the treated plants, which was successful up to 35 days after inoculation. This persistence can be taken as an indication of an endophytic life strategy, considering that endophytic colonization enables the fungus to become resident in a stable and nutritious insect-attracting environment. High humidity enhanced colonization of both plants and aphids; this expected effect is relevant for the management of the cotton aphid, which is most commonly found in the lower canopy, where humidity is high and the fungus is more protected against the adverse effects of UV radiation from sun [63]. Moreover, contact with conidia of *A. lecanii* significantly reduced the rate and period of reproduction of *A. gossypii*. The culture filtrate of the fungus significantly increased mortality and reduced reproduction, while feeding-choice experiments indicated that the aphids might be able to detect the fungal metabolites. The ethyl acetate and methanolic fractions of culture filtrate and mycelia also caused significant mortality and reduced fecundity [64]. Besides cotton, the same strain displayed the ability to colonize plants of wheat, corn, tomato, bean and pumpkin after artificial inoculation of leaves, while soil inoculation was ineffective [44].

Additional reports from cotton come from Texas [47] and Brazil, where the endophytic occurrence of *A. lecanii* was detected in leaves and roots of both normal and *Bt*-transgenic plants [46]. Although no aspects concerning interactions with pests were evaluated in these cases, it is meaningful that several strains of *A. lecanii* were recovered in each of these three contexts, indicating a possible common association of this species with cotton, which deserves to be more thoroughly verified.

The adaptation of *A. lecanii* to exert entomopathogenicity in association with plants is well attested by the finding that the fungus responds to volatile compounds produced by the plant during insect feeding. Particularly, in a model based on thale cress (*Arabidopsis thaliana*) and the mustard aphid (*Lipaphis erysimi*), compounds such as methyl salicylate and menthol were found to promote spore germination and pathogenicity of the fungus [65,66].

Besides aphids, protective effects after systemic colonization have been demonstrated against the red spider mite (*Tetranychus urticae*) in bean plants. In this case a strain of *A. lecanii* was reported to spread within the plant tissues after artificial inoculation of seeds, promoting growth and impairing survival and fecundity of the mites. These effects were even carried over the following generation of mites fed on fresh plants [48].

Pathogenicity of *A. lecanii* against a wide array of noxious arthropods is integrated by antagonism towards plant pathogenic fungi. In addition to a general antifungal activity demonstrated in vitro against polyphagous species such as *Sclerotinia sclerotiorum*, *Rhizoctonia solani* and *Aspergillus flavus* [49], possible exploitation of this double functionality has been conceived on several crops, such as coffee where *A. lecanii* behaves as both a parasite of the leaf rust (*Hemileia vastatrix*) and a pathogen of the green scale (*Coccus viridis*) [67]. The same role can be considered in crops where powdery mildews can represent a major phytosanitary problem, such as cucurbits [68,69].

Moreover, antifungal effects could derive from stimulation of the plant defense response, as reported for an endophytic strain able to promote such reaction against *Pythium ultimum* in transformed cucumber plants [52]. Additional experimental evidence in this regard is provided by observations carried out on the date palm (*Phoenix dactylifera*) where the inoculation of endophytic strains of *L. dimorphum* and *L.* cf. *psalliotae*, previously reported for entomopathogenicity against the red palm scale (*Phoenicococcus marlatti*) [56], induced proteins involved in plant defense or stress response. Proteins related with photosynthesis and energy metabolism were also upregulated, along with accumulation of a heavy chain myosin-like protein [70].

The concurrent role against plant pests and pathogens is known to operate for other *Lecanicillium* and *Akanthomyces* species, and for non-endophytic strains of various origin, as more in detail discussed in dedicated papers [71,72]. The need to combat multiple adversities has also prompted the evaluation of a possible combined use of these fungi with chemical pesticides. In this respect, it has been observed that the spread of *A. lecanii* in plant tissues is not affected by treatments with insecticides belonging to several classes [73]. Moreover, substantial safety of insecticides has been reported in in vitro assays carried out on *A. muscarius*, while several herbicides and fungicides were responsible for negative effects or even suppression of mycelial growth [74]. For the latter species, in vivo observations on the sweet potato whitefly (*Bemisia tabaci*) demonstrated the positive effects of association with chemical insecticides in view of reducing their use, particularly in the greenhouse [75]. Again with reference to application of *A. muscarius* for the control of *B. tabaci*, it is worth mentioning the synergistic effects resulting in combined treatments with matrine, a plant-derived quinolizidine alkaloid [76].

In addition to the indirect side effects deriving from protection against biotic and abiotic adversities, many endophytes have been reported to promote plant growth through essentially two mechanisms; that is the release of plant hormones, or the improvement of nutritional conditions. Of course, strains possessing both properties are likely to contribute in an additive manner, as observed for an isolate of *L. psalliotae* from cardamom (*Elettaria cardamomum*). Besides producing indole-3-acetic acid, this strain enhanced chlorophyll content of leaves as a likely result of release of siderophores, and increased availability of zinc and inorganic phosphate by promoting their solubilization [77]. Release of siderophore has also been reported for an endophytic isolate of *A. lecanii* from *Pistacia vera* [49].

5. Biochemical Factors Involved in the Tritrophic Interaction with Plants and Pests

It has been previously introduced that, at least in part, the antagonistic/pathogenic ability by *Lecanicillium* and *Akanthomyces* strains is mediated by biochemical factors, such as enzymes and secondary metabolites. Endophytic fungi are regarded as a goldmine of undescribed chemodiversity, and even diffusely reported as capable to synthesize bioactive products originally characterized from their host plants [78]. Although it is quite reasonable that they exploit this biosynthetic potential in the natural environment, more rigid opinions occasionally question a real role by these compounds until their production is demonstrated in plants. Pending a solution of this diatribe through the development of methods for ascertaining their effective release and bioactivity in plant tissues, so far research in the field has disclosed interesting properties by species of *Lecanicillium* and *Akanthomyces*, too.

The first metabolomic studies concerning these fungi were carried out with strains of *V. lecanii* before the taxonomic revision. Two isolates were found to produce 2,6-dimethoxy-*p*-benzoquinone, phenylalanine anhydride, aphidicolin and dipicolinic acid, with the latter showing insecticidal effects in bioassays on the blowfly *Calliphora erythrocephala* [79]. Afterwards, two more triterpenoid carboxylic acids with alleged insecticidal properties were reported from the same source [80]. Incompletely identified toxic products, possibly phospholipids, were extracted from another strain showing activity against *B. tabaci*, the western flower thrips (*Frankliniella occidentalis*) and a few aphid species [81]. Anti-insectan effects against the corn earworm (*Helicoverpa zea*) were later reported for vertilecanin A, the most abundant component in a group of five new phenopicolinic acid analogues [82]. Moreover, two structurally unidentified products were extracted from two Chinese strains, displaying toxic, ovicidal and antifeedant properties against *B. tabaci* [83]. Finally, the novel indolosesquiterpenes lecanindoles A-D, with quite peculiar structures and bioactivities, were characterized from another aphidiculous strain [84].

Later on more strains were found to produce novel compounds without a direct connection with their entomopathogenicity. Two inactive aromadendrane sesquiterpenes, inonotins M and N, were extracted from a strain of *L. psalliotae* [85]. An unidentified *Lecanicillium* sp. was reported to produce lecanicillolide [86], and lecanicillones A-C, three unusual dimeric spiciferones with an acyclobutane ring displaying moderate cytotoxic effects [87]. More interesting inhibitory effects on tube formation by endothelial cells, implying antiangiogenic properties, were reported for the decalin polyketide 11-norbetaenone, from a strain of *L. antillanum* [88].

Besides novel compounds, investigations on these fungi have also disclosed the production of well-known bioactive metabolites. A strain of *L. psalliotae* was found to produce oosporein, a common product of *Beauveria* spp., which displayed strong inhibitory activity against the potato late blight fungus (*Phytophthora infestans*) [89]. Likewise, several cyclic depsipeptides have been reported from miscellaneous isolates. The list includes eight destruxin analogues, well-known secondary metabolites of *M. anisopliae*, by strain KV71 of *L. longisporum* (the active principle of the mycoinsecticide Vertalec) [90]; bassianolide, previously reported from *B. bassiana*, from *A. lecanii* [91], and the antifungal verlamelins A-B, previously known from *Simplicillium lamellicola*, from an unidentified *Lecanicillium* strain [92]. Finally, stephensiolides C, D, F, G and I, originally characterized from a gram-negative bacterium (*Serratia* sp.) symbiotic with a mosquito (*Anopheles stephensi*), have been recently detected in the culture extract of an endophytic *Lecanicillium* sp. as the bioactive principles responsible for antibacterial activity against methicillin-resistant *S. aureus* [41]. Inhibitory properties against the same bacteria, along with cytotoxicity on human lung fibroblast cells, were ascribed to cyclic depsipeptides contained in the culture extracts of a strain of *A. attenuatus* [15].

Antibiotic effects against *S. aureus* were also reported for akanthomycin, extracted from cultures of *Akanthomyces gracilis* together with the closely related pyridine alkaloids 8-methylpyridoxatin and cordypyridone C [93]. Additional findings from *Akanthomyces novoguineensis* concerning the akanthopyrones [94], akanthol, akanthozine, butanamide and oxodiazanone derivatives [95] are not to be further considered in this review by reason that this species is now classified in the genus *Hevansia* [14].

This concise analysis of the pertinent literature, mostly made of independent or occasional findings, highlights the importance of carrying out more systematic work on the metabolomics of members of *Lecanicillium* and *Akanthomyces*. In fact, a thorough revision could ascertain whether some compounds eventually represent biochemical markers for selected species, and which products are effectively associated with the expression of pathogenicity towards insects, nematodes and spiders, as well as with antagonism/mycoparasitism against plant pathogens. In this respect, an interesting hypothesis has been advanced concerning the above-mentioned dipicolinic acid, which is known to act as a prophenoloxidase inhibitor and an immunosuppressive agent in insects. After its concomitant detection as a product of several entomopathogenic species belonging to the Hypocreales, including *A. muscarius*, it has been advanced that the acquired ability to synthesize this compound might have shaped evolution of these fungi from mere plant associates to the more specialized lifestyle as arthropod pathogens [96].

Literature on enzyme production by endophytic strains of *Lecanicillium* and *Akanthomyces* is more limited. Chitinolytic enzymes are not only necessary to these fungi to penetrate cuticle of insects, nematodes or spiders, but they are also involved in the activation of the disease response by the plant and induction of systemic resistance [97–99]. The same function may also be played by other enzyme complexes, such as proteases and β-glucanases, which are known to integrate the enzymatic profile of many endophytes [100–102]. Besides directly affecting survival and fecundity of the green peach aphid (*Myzus persicae*) in a concentration-dependent manner, a protein characterized from a strain of *A. lecanii* was found to concomitantly induce upregulation in tomato plants of genes associated to the salycilate and jasmonate pathways, which are involved in the systemic response to biotic stress [103].

6. Future Perspectives

As a likely heritage of old investigational schemes, there is a cultural propensity in research projects and reports to refer to plant-associated microorganisms within the boundaries of functional categories. However, it is increasingly evident that many endophytic fungi are eclectic and possess a multifaceted connotation enabling them to perform several more or less interconnected beneficial roles in the symbiotic relationship with their host plants. Such a revised concept particularly applies to species of *Lecanicillium* and *Akanthomyces*, which should not be merely regarded as entomopathogenic fungi anymore.

Strains of the species *A. lecanii*, *A. muscarius*, *A. attenuatus* and *A. longisporus* are already used as the active ingredients of several mycoinsecticides [72]. Although their inclusion in IPM appears to be an obvious approach, a more efficient employment should be pursued in light of the body of evidence disclosed by recent experimental work that, besides killing pests as a result of inundative treatments, the endophytic establishment of these fungi may have further relevance on plant fitness. That is a clear antagonistic role against plant pathogens, the capacity to stimulate plant defense reactions and various plant growth promoting effects.

These valuable properties, shared with other species of *Lecanicillium* and *Akanthomyces*, make it advisable to carry out extensive investigations in crops to verify the natural endophytic occurrence, and to increase our knowledge on ecology of these fungi. Particularly relevant is gathering additional information on the production in plants of the biochemical factors, which possibly play a role in regulating the tritrophic relationship with the host and its pests/pathogens.

At the same time it is fundamental to assess whether their endophytic establishment is possible following artificial introduction. In this respect, inoculation methods (foliar spraying, soil drenching, seed soaking, and injections) are crucial for an enduring survival within plant tissues, and their compliance should be more accurately evaluated [104]. Particularly in crops where these fungi can exert a positive impact, additional observations are appropriate to verify whether their distribution pattern is localized or systemic. Actually, a great challenge for considering endophytic fungi as a strategy in plant protection is to manage their reproducible introduction into crops, and to predict the outcome. As well, the effectiveness of this attractive phytosanitary tool needs to be proven in the field to stimulate growers to adopt it in view of gaining clear economic benefits.

Author Contributions: Conceptualization, R.N.; phylogenetic analysis, A.B.; data curation, A.B.; writing—original draft preparation, R.N.; writing—review and editing, R.N. and A.B. All authors have read and agreed to the published version of the manuscript.

Funding: This research received no external funding.

Conflicts of Interest: The authors declare no conflict of interest.

References

1. Busby, P.E.; Ridout, M.; Newcombe, G. Fungal endophytes: Modifiers of plant disease. *Plant. Mol. Biol.* **2016**, *90*, 645–655. [CrossRef]
2. Hassani, M.A.; Durán, P.; Hacquard, S. Microbial interactions within the plant holobiont. *Microbiome* **2018**, *6*, 58. [CrossRef]
3. McKinnon, A.C.; Saari, S.; Moran-Diez, M.E.; Meyling, N.V.; Raad, M.; Glare, T.R. *Beauveria bassiana* as an endophyte: A critical review on associated methodology and biocontrol potential. *BioControl* **2017**, *62*, 1–17. [CrossRef]
4. Vega, F.E. The use of fungal entomopathogens as endophytes in biological control: A review. *Mycologia* **2018**, *110*, 4–30. [CrossRef]
5. Gams, W.; Van Zaayen, A. Contribution to the taxonomy and pathogenicity of fungicolous *Verticillium* species. I. Taxonomy. *Neth. J. Plant. Pathol.* **1982**, *88*, 57–78. [CrossRef]
6. Zare, R.; Gams, W. A revision of *Verticillium* section *Prostrata*. IV. The genera *Lecanicillium* and *Simplicillium* gen. nov. *Nova Hedwig.* **2001**, *73*, 1–50.
7. Zare, R.; Gams, W. A revision of the *Verticillium* fungicola species complex and its affinity with the genus *Lecanicillium*. *Mycol. Res.* **2008**, *112*, 811–824. [CrossRef] [PubMed]
8. Sung, G.H.; Hywel-Jones, N.J.; Sung, J.M.; Luangsa-ard, J.J.; Shrestha, B.; Spatafora, J.W. Phylogenetic classification of *Cordyceps* and the clavicipitaceous fungi. *Stud. Mycol.* **2007**, *57*, 5–59. [CrossRef]
9. Mains, E.B. Entomogenous species of *Akanthomyces*, *Hymenostilbe* and *Insecticola* in North America. *Mycologia* **1950**, *42*, 566–589. [CrossRef]
10. Mongkolsamrit, S.; Noisripoom, W.; Thanakitpipattana, D.; Wutikhun, T.; Spatafora, J.W.; Luangsa-ard, J. Disentangling cryptic species with *Isaria*-like morphs in Cordycipitaceae. *Mycologia* **2018**, *110*, 230–257. [CrossRef]
11. Su, L.; Zhu, H.; Guo, Y.; Du, X.; Guo, J.; Zhang, L.; Qin, C. *Lecanicillium coprophilum* (Cordycipitaceae, Hypocreales), a new species of fungus from the feces of *Marmota monax* in China. *Phytotaxa* **2019**, *387*, 55–62. [CrossRef]
12. Wei, D.P.; Wanasinghe, D.N.; Chaiwat, T.A.; Hyde, K.D. *Lecanicillium uredinophilium* known from rusts, also occurs on animal hosts with chitinous bodies. *Asian J. Mycol.* **2018**, *1*, 63–73. [CrossRef] [PubMed]
13. Kepler, R.M.; Luangsa-ard, J.J.; Hywel-Jones, N.L.; Quandt, C.A.; Sung, G.H.; Rehner, S.A.; Aime, M.C.; Henkel, T.W.; Sanjuan, T.; Zare, R.; et al. A phylogenetically-based nomenclature for Cordycipitaceae (Hypocreales). *IMA Fungus* **2017**, *8*, 335. [CrossRef] [PubMed]
14. Mycobank Database. Available online: http://www.mycobank.org/ (accessed on 9 April 2020).
15. Barthélemy, M.; Elie, N.; Pellissier, L.; Wolfender, J.L.; Stien, D.; Touboul, D.; Eparvier, V. Structural identification of antibacterial lipids from Amazonian palm tree endophytes through the molecular network approach. *Int. J. Mol. Sci.* **2019**, *20*, 2006. [CrossRef]
16. Ellsworth, K.T.; Clark, T.N.; Gray, C.A.; Johnson, J.A. Isolation and bioassay screening of medicinal plant endophytes from eastern Canada. *Can. J. Microbiol.* **2013**, *59*, 761–765. [CrossRef]
17. Sánchez-Márquez, S.; Bills, G.F.; Zabalgogeazcoa, I. Diversity and structure of the fungal endophytic assemblages from two sympatric coastal grasses. *Fungal Divers.* **2008**, *33*, 87–100.
18. Sánchez Márquez, M.; Bills, G.F.; Zabalgogeazcoa, I. The endophytic mycobiota of the grass *Dactylis Glomerata*. *Fungal Divers.* **2007**, *27*, 171–195.
19. Poosakkannu, A.; Nissinen, R.; Kytöviita, M.M. Culturable endophytic microbial communities in the circumpolar grass, *Deschampsia flexuosa* in a sub-Arctic inland primary succession are habitat and growth stage specific. *Environ. Microbiol. Rep.* **2015**, *7*, 111–122. [CrossRef]
20. Molina-Montenegro, M.A.; Oses, R.; Torres-Díaz, C.; Atala, C.; Núñez, M.A.; Armas, C. Fungal endophytes associated with roots of nurse cushion species have positive effects on native and invasive beneficiary plants in an alpine ecosystem. *Perspect. Plant. Ecol. Evol. Syst.* **2015**, *17*, 218–226. [CrossRef]
21. Giordano, L.; Gonthier, P.; Varese, G.C.; Miserere, L.; Nicolotti, G. Mycobiota inhabiting sapwood of healthy and declining Scots pine (*Pinus sylvestris* L.) trees in the Alps. *Fungal Divers.* **2009**, *38*, e83.

22. Behnke-Borowczyk, J.; Kwaśna, H.; Kulawinek, B. Fungi associated with *Cyclaneusma* needle cast in Scots pine in the west of Poland. *For. Pathol.* **2019**, *49*, e12487. [CrossRef]

23. Ashkezari, S.J.; Fotouhifar, K.B. Diversity of endophytic fungi of common yew (*Taxus baccata* L.) in Iran. *Mycol. Progr.* **2017**, *16*, 247–256. [CrossRef]

24. Nicoletti, R.; De Filippis, A.; Buommino, E. Antagonistic aptitude and antiproliferative properties on tumor cells of fungal endophytes from the Astroni Nature Reserve, Italy. *Afr. J. Microbiol. Res.* **2013**, *7*, 4073–4083.

25. Vinit, K.; Doilom, M.; Wanasinghe, D.N.; Bhat, D.J.; Brahmanage, R.S.; Jeewon, R.; Xiao, Y.; Hyde, K.D. Phylogenetic placement of *Akanthomyces muscarius*, a new endophyte record from *Nypa fruticans* in Thailand. *Curr. Res. Environ. Appl. Mycol.* **2018**, *8*, 404–417. [CrossRef]

26. Widler, B.; Müller, E. Untersuchungen über endophytische pilze von *Arctostaphylos uva-ursi* (L.) Sprengel (Ericaceae). *Bot. Helv.* **1984**, *94*, 307–337.

27. Bills, G.F.; Polishook, J.D. Microfungi from *Carpinus caroliniana*. *Can. J. Bot.* **1991**, *69*, 1477–1482. [CrossRef]

28. Fang, K.; Miao, Y.F.; Chen, L.; Zhou, J.; Yang, Z.P.; Dong, X.F.; Zhang, H.B. Tissue-specific and geographical variation in endophytic fungi of *Ageratina adenophora* and fungal associations with the environment. *Front. Microbiol.* **2019**, *10*, 2919. [CrossRef]

29. Shobha, M.; Bharathi, T.R.; Sampath Kumara, K.K.; Prakash, H.S. Diversity and biological activities of fungal root endophytes of *Hemidesmus indicus* (L.) R.Br. *J. Pharmacogn. Phytochem.* **2019**, *8*, 273–280.

30. Wang, Y.; Lai, Z.; Li, X.X.; Yan, R.M.; Zhang, Z.B.; Yang, H.L.; Zhu, D. Isolation, diversity and acetylcholinesterase inhibitory activity of the culturable endophytic fungi harboured in *Huperzia serrata* from Jinggang Mountain, China. *World J. Microbiol. Biotechnol.* **2016**, *32*, 20. [CrossRef]

31. Summerbell, R.C. Root endophyte and mycorrhizosphere fungi of black spruce, *Picea mariana*, in a boreal forest habitat: Influence of site factors on fungal distributions. *Stud. Mycol.* **2005**, *53*, 121–145. [CrossRef]

32. Khalmuratova, I.; Kim, H.; Nam, Y.J.; Oh, Y.; Jeong, M.J.; Choi, H.R.; You, Y.H.; Choo, Y.S.; Lee, I.J.; Shin, J.H.; et al. Diversity and plant growth promoting capacity of endophytic fungi associated with halophytic plants from the west coast of Korea. *Mycobiology* **2015**, *43*, 373–383. [CrossRef] [PubMed]

33. Ginting, R.C.B.; Sukarno, N.; Widyastuti, U.; Darusman, L.K.; Kanaya, S. Diversity of endophytic fungi from red ginger (*Zingiber officinale* Rosc.) plant and their inhibitory effect to *Fusarium oxysporum* plant pathogenic fungi. *HAYATI J. Biosci.* **2013**, *20*, 127–137. [CrossRef]

34. Kim, H.; You, Y.H.; Yoon, H.; Seo, Y.; Kim, Y.E.; Choo, Y.S.; Lee, I.J.; Shin, J.H.; Kim, J.G. Culturable fungal endophytes isolated from the roots of coastal plants inhabiting Korean east coast. *Mycobiology* **2014**, *42*, 100–108. [CrossRef] [PubMed]

35. de Abreu, L.M.; Almeida, A.R.; Salgado, M.; Pfenning, L.H. Fungal endophytes associated with the mistletoe *Phoradendron perrottettii* and its host tree *Tapirira guianensis*. *Mycol. Progr.* **2010**, *9*, 559–566. [CrossRef]

36. Brownbridge, M.; Reay, S.D.; Nelson, T.L.; Glare, T.R. Persistence of *Beauveria bassiana* (Ascomycota: Hypocreales) as an endophyte following inoculation of radiata pine seed and seedlings. *Biol. Control.* **2012**, *61*, 194–200. [CrossRef]

37. You, Y.H.; Park, J.M.; Seo, Y.G.; Lee, W.; Kang, M.S.; Kim, J.G. Distribution, characterization, and diversity of the endophytic fungal communities on Korean seacoasts showing contrasting geographic conditions. *Mycobiology* **2017**, *45*, 150–159. [CrossRef]

38. Ofek-Lalzar, M.; Gur, Y.; Ben-Moshe, S.; Sharon, O.; Kosman, E.; Mochli, E.; Sharon, A. Diversity of fungal endophytes in recent and ancient wheat ancestors *Triticum dicoccoides* and *Aegilops sharonensis*. *FEMS Microbiol. Ecol.* **2016**, *92*, fiw152. [CrossRef]

39. Skaltsas, D.N.; Badotti, F.; Vaz, A.B.M.; da Silva, F.F.; Gazis, R.; Wurdack, K.; Castlebury, L.; Góes-Neto, A.; Chaverri, P. Exploration of stem endophytic communities revealed developmental stage as one of the drivers of fungal endophytic community assemblages in two Amazonian hardwood genera. *Sci. Rep.* **2019**, *9*, 12685. [CrossRef]

40. Glynou, K.; Ali, T.; Buch, A.K.; Haghi Kia, S.; Ploch, S.; Xia, X.; Çelik, A.; Thines, M.; Maciá-Vicente, J.G. The local environment determines the assembly of root endophytic fungi at a continental scale. *Environ. Microbiol.* **2016**, *18*, 2418–2434. [CrossRef]

41. Mai, P.Y.; Levasseur, M.; Buisson, D.; Touboul, D.; Eparvier, V. Identification of antimicrobial compounds from *Sandwithia guyanensis*-associated endophyte using molecular network approach. *Plants* **2020**, *9*, 47. [CrossRef]

42. Evans, H.C.; Holmes, K.A.; Thomas, S.E. Endophytes and mycoparasites associated with an indigenous forest tree, *Theobroma gileri*, in Ecuador and a preliminary assessment of their potential as biocontrol agents of cocoa diseases. *Mycol. Progr.* **2003**, *2*, 149–160. [CrossRef]

43. Kago, L.; Njuguna, J.; Njarui, D.M.G.; Ghimire, S.R. Fungal endophyte communities of Brachiaria grass (Brachiaria spp.) in Kenya. In *Climate Smart Brachiaria Grasses for Improving Livestock Production in East Africa—Kenya Experience: Proceedings of the Workshop, Naivasha, Kenya, 14–15 September 2016*; Njarui, D.M.G., Gichangi, E.M., Ghimire, S.R., Muinga, R.W., Eds.; Kenya Agricultural and Livestock Research Organization: Nairobi, Kenya, 2016; pp. 150–162.

44. Gurulingappa, P.; Sword, G.A.; Murdoch, G.; McGee, P.A. Colonization of crop plants by fungal entomopathogens and their effects on two insect pests when in planta. *Biol. Control.* **2010**, *55*, 34–41. [CrossRef]

45. McGee, P.A. Reduced growth and deterrence from feeding of the insect pest *Helicoverpa armigera* associated with fungal endophytes from cotton. *Aus. J. Experim. Agric.* **2002**, *42*, 995–999. [CrossRef]

46. de Souza Vieira, P.D.; de Souza Motta, C.M.; Lima, D.; Torres, J.B.; Quecine, M.C.; Azevedo, J.L.; de Oliveira, N.T. Endophytic fungi associated with transgenic and non-transgenic cotton. *Mycology* **2011**, *2*, 91–97. [CrossRef]

47. Castillo-Lopez, D. Ecological roles of two entomopathogenic endophytes: *Beauveria bassiana* and *Purpureocillium lilacinum* in cultivated cotton. Ph.D. Thesis, Texas A&M University, College Station, TX, USA, 11 May 2015.

48. Dash, C.K.; Bamisile, B.S.; Keppanan, R.; Qasim, M.; Lin, Y.; Islam, S.U.; Hussain, M.; Wang, L. Endophytic entomopathogenic fungi enhance the growth of *Phaseolus vulgaris* L. (Fabaceae) and negatively affect the development and reproduction of *Tetranychus urticae* Koch (Acari: Tetranychidae). *Microb. Pathog.* **2018**, *125*, 385–392. [CrossRef]

49. Dolatabad, H.K.; Javan-Nikkhah, M.; Shier, W.T. Evaluation of antifungal, phosphate solubilisation, and siderophore and chitinase release activities of endophytic fungi from *Pistacia vera*. *Mycol. Progr.* **2017**, *16*, 777–790. [CrossRef]

50. González, V.; Tello, M.L. The endophytic mycota associated with *Vitis vinifera* in central Spain. *Fungal Divers.* **2011**, *47*, 29–42. [CrossRef]

51. Kuchár, M.; Glare, T.R.; Hampton, J.G.; Dickie, I.A.; Christey, M.C. Virulence of the plant-associated endophytic fungus *Lecanicillium muscarium* to diamondback moth larvae. *New Zeal. Plant. Prot.* **2019**, *72*, 253–259. [CrossRef]

52. Benhamou, N.; Brodeur, J. Pre-inoculation of Ri T-DNA transformed cucumber roots with the mycoparasite, *Verticillium lecanii*, induces host defense reactions against *Pythium ultimum* infection. *Physiol. Mol. Plant. Pathol.* **2001**, *58*, 133–146. [CrossRef]

53. Hirano, E.; Koike, M.; Aiuchi, D.; Tani, M. Pre-inoculation of cucumber roots with *Verticillium lecanii* (*Lecanicillium muscarium*) induces resistance to powdery mildew. *Res. Bull. Obihiro Univ.* **2008**, *29*, 82–94.

54. Aghdam, S.A.; Fotouhifar, K.B. Introduction of some endophytic fungi of sour cherry trees (*Prunus cerasus*) in Iran. *Rostaniha* **2017**, *18*, 77–94.

55. Rijavec, T.; Lapanje, A.; Dermastia, M.; Rupnik, M. Isolation of bacterial endophytes from germinated maize kernels. *Can. J. Microbiol.* **2007**, *53*, 802–808. [CrossRef] [PubMed]

56. Gómez-Vidal, S.; Lopez-Llorca, L.V.; Jansson, H.B.; Salinas, J. Endophytic colonization of date palm (*Phoenix dactylifera* L.) leaves by entomopathogenic fungi. *Micron* **2006**, *37*, 624–632. [CrossRef] [PubMed]

57. Bhagyasree, S.; Ghosh, S.; Thippaiah, M.; Rajgopal, N. Survey on natural occurrence of endophytes in maize (*Zea mays* L.) ecosystem. *Int. J. Curr. Microbiol. Appl. Sci.* **2018**, *7*, 2526–2533. [CrossRef]

58. Zimowska, B.; Okoń, S.; Becchimanzi, A.; Krol, E.D.; Nicoletti, R. Phylogenetic characterization of *Botryosphaeria* strains associated with *Asphondylia* galls on species of Lamiaceae. *Diversity* **2020**, *12*, 41. [CrossRef]

59. Mitina, G.; Kazartsev, I.; Vasileva, A.; Yli-Mattila, T. Multilocus genotyping based species identification of entomopathogenic fungi of the genus *Lecanicillium* (=*Verticillium lecanii* sl). *J. Basic Microbiol.* **2017**, *57*, 950–961. [CrossRef]

60. Kouvelis, V.N.; Sialakouma, A.; Typas, M.A. Mitochondrial gene sequences alone or combined with ITS region sequences provide firm molecular criteria for the classification of *Lecanicillium* species. *Mycol. Res.* **2008**, *112*, 829–844. [CrossRef]

61. Nonaka, K.; Kaifuchi, S.; Ōmura, S.; Masuma, R. Five new *Simplicillium* species (Cordycipitaceae) from soils in Tokyo, Japan. *Mycoscience* **2013**, *54*, 42–53. [CrossRef]

62. Crous, P.W.; Wingfield, M.J.; Burgess, T.I.; Hardy, G.S.J.; Gené, J.; Guarro, J.; Baseia, I.G.; García, D.; Gusmão, L.F.P.; Souza-Motta, C.M.; et al. Fungal planet description sheets: 716–784. *Persoonia* **2018**, *40*, 240–393. [CrossRef]

63. Anderson, C.M.; McGee, P.A.; Nehl, D.B.; Mensah, R.K. The fungus *Lecanicillium lecanii* colonises the plant *Gossypium hirsutum* and the aphid *Aphis gossypii*. *Australas. Mycol.* **2007**, *26*, 65–70.

64. Gurulingappa, P.; McGee, P.A.; Sword, G. Endophytic *Lecanicillium lecanii* and *Beauveria bassiana* reduce the survival and fecundity of *Aphis gossypii* following contact with conidia and secondary metabolites. *Crop. Prot.* **2011**, *30*, 349–353. [CrossRef]

65. Lin, Y.; Hussain, M.; Avery, P.B.; Qasim, M.; Fang, D.; Wang, L. Volatiles from plants induced by multiple aphid attacks promote conidial performance of *Lecanicillium lecanii*. *PLoS ONE* **2016**, *11*, e0151844. [CrossRef] [PubMed]

66. Lin, Y.; Qasim, M.; Hussain, M.; Akutse, K.S.; Avery, P.B.; Dash, C.K.; Wang, L. The herbivore-induced plant volatiles methyl salicylate and menthol positively affect growth and pathogenicity of entomopathogenic fungi. *Sci. Rep.* **2017**, *7*, 40494. [CrossRef] [PubMed]

67. Jackson, D.; Skillman, J.; Vandermeer, J. Indirect biological control of the coffee leaf rust, *Hemileia vastatrix*, by the entomogenous fungus *Lecanicillium lecanii* in a complex coffee agroecosystem. *Biol. Control.* **2012**, *61*, 89–97. [CrossRef]

68. Romero, D.; De Vicente, A.; Zeriouh, H.; Cazorla, F.M.; Fernández-Ortuño, D.; Torés, J.A.; Pérez-García, A. Evaluation of biological control agents for managing cucurbit powdery mildew on greenhouse-grown melon. *Plant. Pathol.* **2007**, *56*, 976–986. [CrossRef]

69. Ownley, B.H.; Gwinn, K.D.; Vega, F.E. Endophytic fungal entomopathogens with activity against plant pathogens: Ecology and evolution. *BioControl* **2010**, *55*, 113–128. [CrossRef]

70. Gómez-Vidal, S.; Salinas, J.; Tena, M.; Lopez-Llorca, L.V. Proteomic analysis of date palm (*Phoenix dactylifera* L.) responses to endophytic colonization by entomopathogenic fungi. *Electrophoresis* **2009**, *30*, 2996–3005. [CrossRef]

71. Kim, J.J.; Goettel, M.S.; Gillespie, D.R. Potential of *Lecanicillium* species for dual microbial control of aphids and the cucumber powdery mildew fungus, *Sphaerotheca Fuliginea*. *Biol. Control* **2007**, *40*, 327–332. [CrossRef]

72. Goettel, M.S.; Koike, M.; Kim, J.J.; Aiuchi, D.; Shinya, R.; Brodeur, J. Potential of *Lecanicillium* spp. for management of insects, nematodes and plant diseases. *J. Invertebr. Pathol.* **2008**, *98*, 256–261. [CrossRef]

73. Gurulingappa, P.; Mc Gee, P.; Sword, G.A. In vitro and in planta compatibility of insecticides and the endophytic entomopathogen, *Lecanicillium lecanii*. *Mycopathologia* **2011**, *172*, 161–168. [CrossRef]

74. Ondráčková, E.; Seidenglanz, M.; Šafář, J. Effect of seventeen pesticides on mycelial growth of *Akanthomyces, Beauveria, Cordyceps* and *Purpureocillium* strains. *Czech. Mycol.* **2019**, *7*, 123–135. [CrossRef]

75. Cuthbertson, A.G.S.; Blackburn, L.F.; Northing, P.; Luo, W.; Cannon, R.J.C.; Walters, K.F.A. Chemical compatibility testing of the entomopathogenic fungus *Lecanicillium muscarium* to control *Bemisia tabaci* in glasshouse environment. *Int. J. Environ. Sci. Technol.* **2010**, *7*, 405–409. [CrossRef]

76. Ali, S.; Zhang, C.; Wang, Z.; Wang, X.M.; Wu, J.H.; Cuthbertson, A.G.; Shao, Z.; Qiu, B.L. Toxicological and biochemical basis of synergism between the entomopathogenic fungus *Lecanicillium muscarium* and the insecticide matrine against *Bemisia tabaci* (Gennadius). *Sci. Rep.* **2017**, *7*, 46558. [CrossRef] [PubMed]

77. Kumar, C.S.; Jacob, T.K.; Devasahayam, S.; Thomas, S.; Geethu, C. Multifarious plant growth promotion by an entomopathogenic fungus *Lecanicillium psalliotae*. *Microbiol. Res.* **2018**, *207*, 153–160. [CrossRef] [PubMed]

78. Nicoletti, R.; Fiorentino, A. Plant bioactive metabolites and drugs produced by endophytic fungi of Spermatophyta. *Agriculture* **2015**, *5*, 918–970. [CrossRef]

79. Claydon, N.; Grove, J.F. Insecticidal secondary metabolic products from the entomogenous fungus *Verticillium lecanii*. *J. Invertebr. Pathol.* **1982**, *40*, 413–418. [CrossRef]

80. Claydon, N.; Grove, J.F.; Pople, M.; Begley, M.J. New metabolic products of *Verticillium lecanii*. Part 1. 3β-Hydroxy-4,4,14α-trimethyl-5α-pregna-7,9(11)-diene-20S-carboxylic acid and the isolation and characterisation of some minor metabolites. *J. Chem. Soc. Perkin Trans. 1* **1984**, 497–502. [CrossRef]

81. Gindin, G.; Barash, I.; Harari, N.; Raccah, B. Effect of endotoxic compounds isolated from *Verticillium lecani* on the sweet potato whitefly, *Bemisia tabaci*. *Phytoparasitica* **1994**, *22*, 189–196. [CrossRef]

82. Soman, A.G.; Gloer, J.B.; Angawi, R.F.; Wicklow, D.T.; Dowd, P.F. Vertilecanins: New phenopicolinic acid analogues from *Verticillium lecanii*. *J. Nat. Prod.* **2001**, *64*, 189–192. [CrossRef]

83. Wang, L.; Huang, J.; You, M.; Guan, X.; Liu, B. Toxicity and feeding deterrence of crude toxin extracts of *Lecanicillium (Verticillium) lecanii* (Hyphomycetes) against sweet potato whitefly, *Bemisia tabaci* (Homoptera: Aleyrodidae). *Pest. Manag. Sci.* **2007**, *63*, 381–387. [CrossRef]

84. Roll, D.M.; Barbieri, L.R.; Bigelis, R.; McDonald, L.A.; Arias, D.A.; Chang, L.P.; Singh, M.P.; Luckman, S.W.; Berrodin, T.J.; Yudt, M.R. The lecanindoles, nonsteroidal progestins from the terrestrial fungus *Verticillium lecanii* 6144. *J. Nat. Prod.* **2009**, *72*, 1944–1948. [CrossRef] [PubMed]

85. Qiu, A.; Dang, L.Z.; Zhao, P.J.; Zhang, K.Q.; Li, G.H. Two new aromadendrane sesquiterpenes from *Verticillium psalliotae*. *Nat. Prod. Res.* **2019**, *33*, 1257–1261. [CrossRef] [PubMed]

86. Wang, Z.; Chen, S.; Sang, X.; Pan, H.; Li, Z.; Hua, H.; Han, A.; Bai, J. Lecanicillolide, an α-pyrone substituted spiciferone from the fungus *Lecanicillium* sp. PR-M-3. *Tetrahedron Lett.* **2017**, *58*, 740–743. [CrossRef]

87. Wang, Z.Y.; Sang, X.N.; Sun, K.; Huang, S.D.; Chen, S.S.; Xue, C.M.; Ban, L.F.; Li, Z.L.; Hua, H.M.; Pei, Y.H.; et al. Lecanicillones A–C, three dimeric isomers of spiciferone A with a cyclobutane ring from an entomopathogenic fungus *Lecanicillium* sp. PR-M-3. *RSC Advan.* **2016**, *6*, 82348–82351. [CrossRef]

88. Li, C.Y.; Lo, I.W.; Wang, S.W.; Hwang, T.L.; Chung, Y.M.; Cheng, Y.B.; Tseng, S.P.; Liu, Y.H.; Hsu, Y.M.; Chen, S.R.; et al. Novel 11-norbetaenone isolated from an entomopathogenic fungus *Lecanicillium antillanum*. *Bioorg. Med. Chem. Lett.* **2017**, *27*, 1978–1982. [CrossRef]

89. Nagaoka, T.; Nakata, K.; Kouno, K. Antifungal activity of oosporein from an antagonistic fungus against *Phytophthora infestans*. *Z. Naturforsch. C. Biosci.* **2004**, *59*, 302–304. [CrossRef]

90. Butt, T.M.; Hadj, N.B.E.; Skrobek, A.; Ravensberg, W.J.; Wang, C.; Lange, C.M.; Vey, A.; Umi-Kulsoom, S.; Dudley, E. Mass spectrometry as a tool for the selective profiling of destruxins; their first identification in *Lecanicillium longisporum*. *Rapid Commun. Mass Spectrom.* **2009**, *23*, 1426–1434. [CrossRef]

91. Ravindran, K.; Sivaramakrishnan, S.; Hussain, M.; Dash, C.K.; Bamisile, B.S.; Qasim, M.; Wang, L. Investigation and molecular docking studies of bassianolide from *Lecanicillium lecanii* against *Plutella xylostella* (Lepidoptera: Plutellidae). *Comp. Biochem. Physiol. Part. C Toxicol. Pharmacol.* **2018**, *206–207*, 65–72. [CrossRef]

92. Ishidoh, K.I.; Kinoshita, H.; Igarashi, Y.; Ihara, F.; Nihira, T. Cyclic lipodepsipeptides verlamelin A and B, isolated from entomopathogenic fungus *Lecanicillium* sp. *J. Antibiot.* **2014**, *67*, 459–463. [CrossRef]

93. Wagenaar, M.M.; Gibson, D.M.; Clardy, J. Akanthomycin, a new antibiotic pyridone from the entomopathogenic fungus *Akanthomyces gracilis*. *Org. Lett.* **2002**, *4*, 671–673. [CrossRef]

94. Kuephadungphan, W.; Helaly, S.E.; Daengrot, C.; Phongpaichit, S.; Luangsa-ard, J.J.; Rukachaisirikul, V.; Stadler, M. Akanthopyrones A–D, α-pyrones bearing a 4-O-methyl-β-D-glucopyranose moiety from the spider-associated ascomycete *Akanthomyces novoguineensis*. *Molecules* **2017**, *22*, 1202. [CrossRef] [PubMed]

95. Helaly, S.E.; Kuephadungphan, W.; Phongpaichit, S.; Luangsa-ard, J.J.; Rukachaisirikul, V.; Stadler, M. Five unprecedented secondary metabolites from the spider parasitic fungus *Akanthomyces novoguineensis*. *Molecules* **2017**, *22*, 991. [CrossRef] [PubMed]

96. Watanabe, N.; Hattori, M.; Yokoyama, E.; Isomura, S.; Ujita, M.; Hara, A. Entomogenous fungi that produce 2,6-pyridine dicarboxylicacid (dipicolinic acid). *J. Biosci. Bioeng.* **2006**, *102*, 365–368. [CrossRef]

97. Gan, Z.; Yang, J.; Tao, N.; Liang, L.; Mi, Q.; Li, J.; Zhang, K.Q. Cloning of the gene *Lecanicillium psalliotae* chitinase Lpchi1 and identification of its potential role in the biocontrol of root-knot nematode *Meloidogyne incognita*. *Appl. Microbiol. Biotechnol.* **2007**, *76*, 1309–1317. [CrossRef] [PubMed]

98. Rocha-Pino, Z.; Vigueras, G.; Shirai, K. Production and activities of chitinases and hydrophobins from *Lecanicillium lecanii*. *Bioproc. Biosyst. Eng.* **2011**, *34*, 681–686. [CrossRef] [PubMed]

99. Nguyen, H.Q.; Quyen, D.T.; Nguyen, S.L.T.; Vu, V.H. An extracellular antifungal chitinase from *Lecanicillium lecanii*: Purification, properties, and application in biocontrol against plant pathogenic fungi. *Turk. J. Biol.* **2015**, *39*, 6–14. [CrossRef]

100. Saksirirat, W.; Hoppe, H.H. Degradation of uredospores of the soybean rust fungus (*Phakopsora pachyrhizi* Syd.) by cell-free culture filtrates of the mycoparasite *Verticillium psalliotae* Treschow. *J. Phytopathol.* **1991**, *132*, 33–45. [CrossRef]

101. Bye, N.J.; Charnley, A.K. Regulation of cuticle-degrading subtilisin proteases from the entomopathogenic fungi, *Lecanicillium* spp: Implications for host specificity. *Arch. Microbiol.* **2008**, *189*, 81–92. [CrossRef]

102. Balasubramanian, V.; Vashisht, D.; Cletus, J.; Sakthivel, N. Plant β-1,3-glucanases: Their biological functions and transgenic expression against phytopathogenic fungi. *Biotechnol. Lett.* **2012**, *34*, 1983–1990. [CrossRef]

103. Hanan, A.; Basit, A.; Nazir, T.; Majeed, M.Z.; Qiu, D. Anti-insect activity of a partially purified protein derived from the entomopathogenic fungus *Lecanicillium lecanii* (Zimmermann) and its putative role in a tomato defense mechanism against green peach aphid. *J. Invertebr. Pathol.* **2020**, *170*, 107282. [CrossRef]
104. Bamisile, B.S.; Dash, C.K.; Akutse, K.S.; Keppanan, R.; Afolabi, O.G.; Hussain, M.; Qasim, M.; Wang, L. Prospects of endophytic fungal entomopathogens as biocontrol and plant growth promoting agents: An insight on how artificial inoculation methods affect endophytic colonization of host plants. *Microbiol. Res.* **2018**, *217*, 34–50. [CrossRef] [PubMed]

Article

Pathogenic and Non-Pathogenic Fungal Communities in Wheat Grain as Influenced by Recycled Phosphorus Fertilizers: A Case Study

Magdalena Jastrzębska [1,*], Urszula Wachowska [2] and Marta K. Kostrzewska [1]

[1] Department of Agroecosystems, Faculty of Environmental Management and Agriculture, University of Warmia and Mazury in Olsztyn, Plac Łódzki 3, 10-718 Olsztyn, Poland; marta.kostrzewska@uwm.edu.pl
[2] Department of Entomology, Phytopathology and Molecular Diagnostics, University of Warmia and Mazury in Olsztyn, 10-721 Olsztyn, Poland; urszula.wachowska@uwm.edu.pl
* Correspondence: jama@uwm.edu.pl

Received: 23 April 2020; Accepted: 17 June 2020; Published: 19 June 2020

Abstract: Waste-based fertilizers provide an alternative to fertilizers made from non-renewable phosphate rock. Fungal communities colonizing the grain of spring wheat fertilized with preparation from sewage sludge ash and dried animal blood (Rec) and the same fertilizer activated by *Bacillus megaterium* (Bio) were evaluated against those resulting from superphosphate (SP) and no phosphorus (control, C0) treatments. The Illumina MiSeq sequencing system helped to group fungal communities into three clades. Clade 1 (communities from C0, Bio 60 and 80, Rec 80 and SP 40 kg P_2O_5 ha^{-1} treatments) was characterized by a high prevalence of *Alternaria infectoria, Monographella nivalis* and *Gibberella tricincta* pathogens. Clade 2 (Bio 40 kg, Rec 40 and 60 kg, and SP 60 kg P_2O_5 ha^{-1}) was characterized by the lowest amount of the identified pathogens. Commercial SP applied at 80 kg P_2O_5 ha^{-1} (clade 3) induced the most pronounced changes in the fungal taxa colonizing wheat grain relative to non-fertilized plants. The above was attributed mainly to the lower amount of *A. infectoria* and higher counts of species of the family *Nectriaceae*, mostly epiphytic pathogens *Fusarium culmorum* and *Fusarium poae*.

Keywords: *Alternaria; Fusarium;* Illumina MiSeq; secondary raw materials

---

1. Introduction

Spring wheat (*Triticum aestivum* L.) is infected by several dozens of pathogenic fungi. Species such as *Mycosphaerella graminicola, Pyrenophora tritici-repentis, Tilletia caries,* and *Ustilago tritici*, as well as numerous species of the genus *Fusarium* are the most dangerous pathogens of wheat that are transmitted with grain [1]. Fungal species of the genera *Alternaria, Cladosporium,* and *Epiccocum* are regarded as weak pathogens or saprotrophs [2]. Research has demonstrated close interactions between plants and microbes [2]. Wheat grain is infected by fungal pathogens, but it is also colonized by non-pathogenic fungi which inhibit the proliferation of pathogens and promote the growth and development of wheat plants [3,4]. The interactions between these fungal groups determine grain health and improve the consumer value of grain by reducing its mycotoxin content [3,5,6]. Growing conditions and nutrient availability can exert both positive and negative effects on the occurrence of pathogenic and non-pathogenic fungi [5–7].

Phosphorus is essential for root growth, healthy development of stems and ears, a desirable growth rate, high yield and quality, and resistance to abiotic and biotic stress factors [8]. Among the latter, fungal pathogens deserve special attention. A high supply of plant-available phosphorus has been linked with increased levels of fungistatic components, such as phenolic compounds and flavonoids, in different plant parts [9]. The indirect effect of phosphorus on increased plant growth seems to outweigh the direct effect of fungi by increasing the synthesis of phenolic compounds which contribute to resistance against fungal pathogens [10]. Since the natural amount of available phosphorus in arable soils does not fully cater to the nutritional needs of plants [11], crops have to be fertilized [12]. Phosphate rock is the raw material for the production of phosphorus fertilizers [13] which are indispensable in modern agriculture [14].

Rational phosphorus management poses a contemporary global challenge [15–17]. Primary sources of phosphorus are being massively wasted in the production process and it is estimated that only 20–25% of mined phosphorus reaches the produced food [17]. The above raises significant concerns about the availability of phosphorus for agriculture in the future [16]. Global phosphorus resources have not yet reached critical levels [18], but they are undeniably limited and non-renewable. Phosphate rock is distributed unevenly around the world [18] and many countries are dependent on phosphorus imports [19]. This problem applies to the European Union, which has recently added phosphate rock to the list of 20 critical raw materials [20].

Recycled phosphorus provides an alternative to non-renewable phosphate rock deposits [14]. The most abundant secondary sources of phosphorus include sewage and sludge from municipal and industrial wastewater treatment plants [21,22] and waste products from the meat processing industry [23].

Unprocessed phosphorus compounds from both primary and secondary sources are characterized by low solubility [24]. Fertilizer efficiency can be improved through the use of phosphorus solubilizing microbes (PSMs) which transform insoluble phosphorus compounds (PO_4^{3-}) into highly bioavailable forms (HPO_4^{2-} and $H_2PO_4^-$) [4,25,26]. PSMs are a natural component of the soil edaphon [11]. *Bacillus megaterium* is one of the most effective PSMs [27]. These bacteria solubilize phosphorus by producing weak organic acids (gluconic, lactic, acetic, and succinic) [28]. Through solubilization and other biological mechanisms, PSMs can also act as plant growth-promoting microorganisms (PGPMs) [25,29]. It could be expected that by solubilizing phosphorus from soil and fertilizers, PSMs could contribute to a reduction in the fertilizer rate. The production of phosphorus biofertilizers from cheap renewables resources by PSMs promotes sustainable phosphorus management [16] and contributes to a circular economy [30].

Research into the production of phosphorus biofertilizers has been conducted by a Polish scientific consortium established by the Wrocław University of Science and Technology, the New Chemical Syntheses Institute in Puławy, and the University of Warmia and Mazury in Olsztyn [31]. Innovative biofertilizers are expected to deliver similar yield-forming effects to chemical fertilizers and to guarantee the safety of the produced crops. One of the most recent research concepts postulates the use of sewage sludge ash, dried animal blood, and *B. megaterium* in the production of biofertilizers.

This research aimed to determine the effect of the fertilizers produced from sewage sludge ash and dried animal blood on the species composition and structure of fungal communities colonizing wheat grain. The recycled fertilizer (Rec) and biofertilizer (Bio), i.e., Rec activated by *B. megaterium*, were assessed against commercial superphosphate. Mycological analyses were conducted using culture-dependent methods based on fungal sporulation as well as next-generation sequencing in the Illumina MiSeq system.

2. Materials and Methods

2.1. Field Experiment

A field experiment was carried out in 2016 in Bałcyny (Poland, 53°60′ N, 19°85′ E). The experimental plant was spring wheat (*Triticum aestivum* ssp. *vulgare*) cv. Monsun sown on 21 April at 450 plants m^{-2}, at a depth of 3–4 cm, at a row spacing of 15 cm.

The experimental factor was phosphorus fertilization (Table 1). Granular recycled phosphorus fertilizer (Rec) and biofertilizer (Bio) were compared with commercial superphosphate (SP; Gdańskie Zakłady Nawozów Fosforowych Fosfory Sp. z o.o., Gdańsk, Poland). Phosphorus fertilizers were applied before sowing at 40, 60, and 80 kg P_2O_5 per ha. The fertilizers from recyclables (Rec and Bio) were produced by the New Chemical Syntheses Institute in Puławy based on the formula developed by the Department of Advanced Material Technologies of the Wrocław University of Science and Technology. Sewage sludge ash was obtained from the Łyna Municipal Wastewater Treatment Plant in Olsztyn, and dried animal blood was obtained from the meat industry. The bacterial strain of *B. megaterium* was obtained from the Polish Collection of Microorganisms of the Institute of Immunology and Experimental Therapy of the Polish Academy of Sciences in Wrocław (Poland). The procedure of obtaining fertilizer formulations was described by Rolewicz et al. [32].

Table 1. Elemental composition of phosphorus fertilizers.

P-Fertilizer	P_2O_5 Rate, kg ha^{-1}	Treatment Symbol	Fertilizer Characteristics (Elemental Composition of Fertilizers)
Control	0	C0	No P fertilization
Superphosphate	40	SP40	Fosdar^{TM}40 commercial superphosphate fertilizer (P_2O_5 40%; CaO 10%; SO_3 5%; trace presence: Fe, Zn, Cu, B, Co, Mn, Mo) [1]
	60	SP60	
	80	SP80	
Recycled fertilizer	40	Rec40	Granular fertilizer made from ash from the incineration of biological sewage sludge (third level of treatment), and dried animal blood (P_2O_5 19.9%; N 2.89%; K_2O 1.31%; CaO 18.71%; MgO 2.56%; SO_3 1.40%; C 13.92%, Fe 27 g kg^{-1}; Al 23.8 g kg^{-1}; Zn 3.14 g kg^{-1}; As 31.39 mg kg^{-1}; Cd < LD; Cu 777.7 mg kg^{-1}; Ni 54.78 mg kg^{-1}, Pb 19.91 mg kg^{-1}; B 71.27 mg kg^{-1}; Ba 349.6 mg kg^{-1}; Co 14.02 mg kg^{-1}; Mn 561.7 mg kg^{-1}; Mo 35.31 mg kg^{-1}) [2]
	60	Rec60	
	80	Rec80	
Recycled biofertilizer	40	Bio40	Granular biofertilizer made from sewage sludge ash (as above), dried animal blood, and cultured *Baccilus megaterium* (P_2O_5 21.9%C; N 2.87%; K_2O 1.40%; CaO 20.51%; MgO 2.82%; SO_3 1.40%; C 13.92%; Fe 29.0 g kg^{-1}; Al 25.5 g kg^{-1}; Zn 3.29 g kg^{-1}; As 19.99 mg kg^{-1}; Cd 0.345 mg kg^{-1}; Cu 850.1 mg kg^{-1}; Ni 62.65 mg kg^{-1}. Pb 21.76 mg kg^{-1}; B 74.12 mg kg^{-1}; Ba 381.5 mg kg^{-1}. Co 16.19 mg kg^{-1}; Mn 609.4 mg kg^{-1}; Mo 23.75 mg kg^{-1})[2]
	60	Bio60	
	80	Bio80	

[1] according to the information provided on the label, [2] according to the Department of Advanced Material Technologies of the Wrocław University of Science and Technology, LD—level of detection.

The field experiment had a randomized block design with four replications. The experimental plots had an area of 20 m^2 each. Winter oilseed rape was the preceding crop. In addition to phosphorus fertilization, wheat in all plots was fertilized with nitrogen at 130 kg N ha^{-1} (34% ammonium nitrate, Grupa Azoty Puławy, Poland) and potassium at 100 kg K_2O ha^{-1} (60% potash salt, Luvena, Luboń, Poland). Potassium was applied at a single rate before sowing, and nitrogen was split into three applications: 60 kg before sowing, 50 kg in the stem elongation stage (BBCH 30) [33], and 20 kg in the heading stage (BBCH 55).

Wheat was protected against diseases, weeds, and pests (Table 2) and was harvested with a plot harvester on August 12.

Table 2. Plant protection treatments applied in the field experiment.

Pesticide Type	Trade Name (Manufacturer)	Active Ingredient (g dm^{-3})	Rate (dm^3 ha^{-3})	Application Time
Herbicides	Mustang 309 SE (Dow AgroSciences [1])	Florasulam (6.25) + 2,4-D (300)	0.5	Flag leaf stage (BBCH 39; 29 May)
Fungicides	Yamato 303 SE (Sumi Agro [1])	Thiophanate-methyl (233) + Tetraconazole (70)	1.5	Early boot stage (BBCH 41; 9 June)
	Amistar 250 SC (Syngenta [1])	Azoxystrobin (250)	0.8	End of flowering (BBCH 69; 8 July)
Insecticides	Karate Zeon 050 CS (Syngenta [1])	Lambda-cyhalothrin (50)	0.1	Early boot stage (BBCH 41; 6 June)

[1] Warsaw, Poland.

2.2. Soil and Meteorological Conditions

Wheat was grown on luvisol [34] formed from sandy clay loam. The arable layer was slightly acidic (average pH of 6.28 in 1 M KCl). At the beginning of the experiment in 2016, soil contained 8.53 g kg^{-1} C, 1.42 g kg^{-1} N, 2975 mg kg^{-1} K, and 607 mg kg^{-1} P (total content). Soil phosphorus content after spring wheat harvest is presented in Table 3.

Table 3. Total P content of soil after spring wheat harvest (mean ± standard error).

P-Treatment	Total P, mg kg^{-1}
C0	540.3 ± 5.9
SP40	590.7 ± 18.1
SP60	603.1 ± 9.7
SP80	612.9 ± 23.9
Rec40	604.3 ± 4.7
Rec60	613.2 ± 11.9
Rec80	626.3 ± 36.6
Bio40	597.4 ± 17.7
Bio60	611.2 ± 16.4
Bio80	621.5 ± 13.7

Abbreviations are explained in Table 1.

Mean annual precipitation was 62.5 mm, with 66.3 mm in June, 138.6 mm in July, 71.9 mm in August, and 17.1 mm in September. Mean annual temperature was 8.8 °C, and the mean monthly temperature ranged from −3.8 °C in January to 18.5 °C in July.

2.3. Isolation of Fungi from Grain

Grain was harvested in the over-ripe stage (BBCH 92) with a plot harvester on 12 August 2016. Fungal colonization of grain was analyzed, and fungal DNA was isolated immediately after harvest. Grain samples of 10 g each were placed in 250 cm^3 flasks containing 90 cm^3 of sterile water and 0.01 cm^3 of Tween ®40 (Merck, Darmstadt, Germany). The flasks were shaken for 60 min on an Elpin Plus 358 S table shaker (180 rpm, Elpin Plus, Lubawa, Poland) to remove microorganisms from grain. Using a pipette, 0.1 cm^3 of the propagule suspension was transferred to Petri plates with a diameter of 9 cm and flooded with selective Martin medium [35] cooled to 42 °C. The experiment was conducted in four replications. Yeasts and filamentous fungi cultured on the Martin medium were incubated at 24 °C in darkness for 7 days (En 120 Incubator, Nuve, Ancara, Turkey). Yeast and fungal colonies were counted on plates, and different colonies of filamentous fungi were transferred to Petri plates filled with potato dextrose agar (PDA, Merck, Warsaw, Poland) for species identification under a microscope. The number of colony forming units (CFUs) was log-transformed (CFU+1). One hundred disinfected and non-disinfected kernels from each treatment were placed on PDA. Kernels were disinfected by immersion in 1% sodium hypochlorite (NaOCl, ABO, Gdańsk, Poland) solution for 5 min and they were then rinsed three times in sterile water and dried on blotting paper. Colonies of filamentous fungi were identified at the species level based on the sporulation characteristics described in the literature [36,37].

2.4. Isolation of Fungal DNA and PCR Amplification

Fungal DNA was isolated directly from grain with the Bead-Beat Micro AX Gravity Kit (A&A Biotechnology, Gdynia, Poland) according to the manufacturer's protocol. The quantity and quality of the isolated DNA were tested by measuring absorbance at 260 and 280 nm (NanoDrop 2000, Thermo Scientific, Wilmington, DE, USA). A metagenomic analysis of the fungal community was carried out in the ITS2 hypervariable region. The selected region was amplified and the library was prepared with the use of three specific primer sequences: fITS7 (GTGARTCATCGAATCTTTG), ITS4 (TCCTCCGCTTATTGATATGC) and an additional adapter sequence at the 5′ end. PCR was conducted with the Q5 Hot Start High-Fidelity 2X Master Mix under the conditions recommended by the manufacturer. The Nextera Index Kit was used to add specific index adapter sequences to both ends of the analyzed DNA fragment.

2.5. Illumina MiSeq Sequencing

The samples were sequenced in the Illumina MiSeq system (Poland) in paired-end (PE) mode, 2×250 nt, with the Illumina v2 kit (Genomed S.A., Warsaw, Poland). A preliminary analysis of the results was performed automatically in the MiSeq system with MiSeq Reporter (MSR) v2.6 software (Illumina, USA). The analysis was conducted in two steps: (1) automatic demultiplexing of samples, and (2) generation of fastq files with raw read data. A bioinformatics analysis with operational taxonomic unit (OTU) picking was conducted in the QIIME (Quantitative Insights Into Microbial Ecology) program based on the reference sequences in UNITE v7 [38]. The bioinformatics analysis was conducted in the following steps: (1) analysis of read quality and removal of low-quality sequences (quality < 20, minimal length—30)—cutadapt, (2) joining pair-ended sequences—fastq-join, (3) clustering based on a selected database of reference sequences—uclust, (4) removal of chimeric sequences with the usearch61 algorithm [39], and (5) taxonomic identification based on the UNITE-BLAST [40].

2.6. Statistical Analysis

The analysis of variance (ANOVA) was performed in the Statistica 13 program [41]. The significance of differences between mean values was determined by the Newman–Keuls test or Tukey's test ($p < 0.01$). The taxonomic status of fungi obtained by sequencing in the Illumina MiSeq system was presented in heat maps for each product [42]. Hierarchical cluster analysis was carried out on ln-transformed DNA data for OTU 1–10. The Ward clustering method [43] was used based on a dissimilarity matrix representing Euclidean distances between OTUs relative to their prevalence in seed samples of different origin. To examine the correlations between OTUs more closely, the DNA data for OTU 1–10 were subjected to principal component analysis (PCA), and the results were visualized in a biplot.

3. Results

3.1. Fungal Colony Counts on Wheat Grain

Five pathogenic species of the genus *Fusarium* (*F. culmorum, F. poae, F. graminearum, F. avenaceum* and *F. solani*), species of the genus *Alternaria* (*Alternaria* sect. *alternata* and *Alternaria* sect. *infectoriae*) and, sporadically, *Pyrenophora- tritici-repentis* and *Rhizoctonia cerealis* were isolated from wheat grain (Table 4). The CFUs of epiphytic *Alternaria* spp. were significantly higher in nearly all grain samples (excluding grain from treatments fertilized with SP60, Rec60, and Rec80) relative to control grain (C0) where the above pathogen was not detected. The colony counts of Alternaria spp. were highest in wheat kernels from treatments supplied with the biofertilizer (Bio40). *Fusarium culmorum* and

F. graminearum were detected in eight out of the 10 analyzed grain samples. *Fusarium culmorum* was the predominant species in treatments with the highest rate of the commercial fertilizer. The colony counts of *F. graminearum* were significantly higher in treatments supplied with the biofertilizer (Bio40, 60, and 80), Rec80, and SP60 than in the control treatment. The pathogenic species *P. tritici-repentis* and *R. cerealis* were identified only in the Bio80 treatment.

Table 4. Pathogens contaminating wheat grain.

P-Treatment	*Alternaria* spp.	*Fusarium culmorum*	*Fusarium poae*	*Fusarium graminearum*	*Fusarium avenaceum*	*Fusarium solani* Species Complex	Other [1]
	Log (CFU + 1) per 1 g of grain						
C0	0 [d]	0 [c]	1.28 [a]	0.35 [bc]	0 [c]	0	0 [b]
SP40	1.23 [abc]	0.84 [abc]	0 [b]	0.94 [ab]	0.35 [bc]	0	0 [b]
SP60	0.44 [cd]	0 [c]	1.42 [a]	1.19 [a]	0 [c]	0	0 [b]
SP80	1.38 [abc]	1.57 [a]	0 [b]	0.35 [bc]	0 [c]	0	0 [b]
Rec40	1.23 [abc]	1.04 [ab]	0 [b]	0 [c]	0 [c]	0	0 [b]
Rec60	0.35 [d]	0.69 [abc]	0 [b]	0 [c]	0 [c]	0	0 [b]
Rec80	0.44 [cd]	0.35 [bc]	0 [b]	1.43 [a]	0 [c]	0.44	0 [b]
Bio40	1.64 [a]	0.88 [abc]	0 [b]	1.49 [a]	0.69 [ab]	0	0 [b]
Bio60	1.03 [abc]	1.19 [ab]	0 [b]	1.33 [a]	0 [c]	0	0 [b]
Bio80	1.48 [a]	0.35 [bc]	0 [b]	1.40 [a]	1.04 [a]	0	1.14 [a]

[1] *Pyrenophora tritici-repentis, Rhizoctonia cerealis.* Values in columns that did not differ significantly in the Newman–Keuls test ($p < 0.01$) are marked with identical letters; values not marked with letters do not differ significantly (abbreviations are explained in Table 1).

The most prevalent non-pathogenic fungi were yeasts (2.34–2.87 Log(CFU + 1)) and *Mycosphaerella tassiana* (2.05–2.71 Log(CFU + 1)) (Table 5). Yeast counts were significantly higher on grain harvested from treatments fertilized with Bio40 and Bio80 in comparison with the SP80 treatment. Species of the genus *Acremonium* were also relatively abundant in all analyzed grain samples. The colony counts of *Penicillium* spp. were significantly higher in treatment SP80 than in the control treatment (C0). The method of isolation from non-disinfected grains allowed to detect huge yeast communities and six species of pathogenic fungi.

Table 5. Non-pathogenic fungi colonizing wheat grain.

P-Treatment	Yeasts	*Mycosphaerella tassiana*	*Acremonium* spp.	*Mucor* spp.	*Aspergillus* spp.	*Penicillium* spp.
	Log (CFU + 1) per 1 g of grain					
C0	2.58 [ab]	2.41 [abc]	1.55 [abc]	0	0.44 [ab]	0 [b]
SP40	2.55 [ab]	2.56 [abc]	1.76 [abc]	0.34	1.04 [a]	0 [b]
SP60	2.64 [ab]	2.33 [c]	1.84 [abc]	0	0.35 [b]	0 [b]
SP80	2.34 [b]	2.05 [c]	0.94 [c]	0	0 [b]	2.21 [a]
Rec40	2.75 [ab]	2.20 [c]	1.97 [ab]	0	0 [b]	0 [b]
Rec60	2.62 [ab]	2.71 [a]	1.38 [abc]	0	0 [b]	0 [b]
Rec80	2.79 [ab]	2.35 [c]	2.33 [a]	0	0 [b]	0 [b]
Bio40	2.84 [a]	2.56 [abc]	1.18 [bc]	0	0 [b]	0 [b]
Bio60	2.79 [ab]	2.64 [ab]	2.34 [a]	0	0 [b]	0 [b]
Bio80	2.87 [a]	2.54 [abc]	1.92 [ab]	0	0.35 [b]	0 [b]

Values in columns that did not differ significantly in the Newman–Keuls test ($p < 0.01$) are marked with identical letters; values not marked with letters do not differ significantly (abbreviations are explained in Table 1).

3.2. Percentage of Pathogenic and Saprotrophic Fungi Colonizing Grain on PDA

Dark fungal colonies of the genus *Alternaria* were prevalent on non-disinfected kernels cultured on PDA, and they were identified in 14.81% of grain samples from treatments SP80 and Rec40 to 27.78% of grain samples from treatment Rec80 (Table 6). *Fusarium* fungi were encountered most frequently on kernels from plots fertilized with superphosphate (SP) and control plots (C0). Four *Fusarium* species—*F. avenaceum*, *F. graminearum*, *F. poae*, and *F. sporotrichioides*—were identified on 14.82% of control kernels. Three *Fusarium* species were also abundant on grain samples from treatments supplied with the commercial phosphorus fertilizer (14.82% in treatment SP40, 12.96% in treatments SP60 and SP80). The second method of isolation from disinfected grain appeared to yield more *Fusarias*.

Table 6. Percentage of non-disinfected wheat grain colonized by epiphytic fungi.

P-Treatment	*Alternaria* spp.	*Fusarium avenaceum*	*Fusarium graminearum*	*Fusarium poae*	*Fusarium sporotrichioides*	*Epicoccum nigrum*	*Botrytis cinerea*
C0	20.37	5.56	1.85	5.56	1.85	1.85	0
SP40	16.67	5.56	0	1.85	7.41	0	1.85
SP60	24.07	1.85	0	3.70	7.41	0	0
SP80	14.81	3.70	3.70	5.56	0	1.85	1.85
Rec40	14.81	1.85	0	5.57	1.85	3.70	0
Rec60	25.93	3.70	0	0	3.70	0	0
Rec80	27.78	0	1.85	3.70	1.85	0	0
Bio40	25.93	0	0	1.85	3.70	0	0
Bio60	22.22	0	0	1.85	0	5.56	0
Bio80	20.37	0	0	3.70	0	0	0

No significant differences between treatments (abbreviations are explained in Table 1).

The percentage of disinfected kernels contaminated with fungi of the genus *Alternaria* ranged from 18.52% (SP60, Bio40) to 31.48% (Rec40) (Table 7). *Fusarium* fungi colonized less than 4% of disinfected kernels. The only exception was disinfected grain from treatment Bio40 which was colonized by *F. sporotrichioides* at 5.56%.

Table 7. Percentage of disinfected wheat kernels colonized by endophytic fungi.

P-Treatment	*Alternaria* spp.	*Fusarium avenaceum*	*Fusarium graminearum*	*Fusarium oxysporum*	*Fusarium poae*	*Fusarium solani* Species Complex	*Fusarium sporotrichioides*	*Epicoccum nigrum*	*Botrytis cinerea*
C0	27.78	0	3.70	0	1.85	0	0 [b]	1.85	0
SP40	24.07	1.85	1.85	0	0	0	1.85 [ab]	1.85	1.85
SP60	18.52	0	0	1.85	0	1.85	0 [b]	0	1.85
SP80	22.22	3.70	0	1.85	0	0	0 [b]	0	0
Rec40	31.48	3.70	0	0	0	1.85	0 [b]	0	0
Rec60	29.63	0	0	0	3.70	0	0 [b]	0	0
Rec80	29.63	1.85	0	0	0	0	1.85 [ab]	0	0
Bio40	18.52	0	0	0	0	0	5.56 [a]	1.85	0
Bio60	25.93	0	0	0	1.85	0	0 [b]	0	0
Bio80	24.07	1.85	1.85	0	1.85	0	1.85 [ab]	0	0

Values in columns that did not differ significantly in Tukey's test ($p < 0.01$) are marked with identical letters; values not marked with letters do not differ significantly (abbreviations are explained in Table 1).

3.3. Structure and Composition of Fungal Communities

The biodiversity of fungal communities was analyzed by next-generation sequencing in the Illumina MiSeq system. The sequence of the ITS region was compared with the sequences from the UNITE-BLAST database to reveal that fungi of the phylum Ascomycota predominated in all grain samples and accounted for 91.99% (Bio40) to 98.92% of OTUs (Rec40). Fungi of the phylum Basidiomycota accounted for 0.38% (Rec60) to 1.7% (SP60) of sequence reads. Species of the genus *Alternaria*, family Pleosporaceae, order Pleosporares, class Dothideomycetes accounted for 58.06% (SP80) to 95.35% (Rec60) of reading frames in the ITS2 region. A very high percentage of *Alternaria* fungi were classified as *A. infectoria* (43.41–92.79%), whereas only 2.43–16.3% were identified as *A. betae-kenyensis* (Figure 1).

Fungi of the genus *Gibberella*, family Nectriaceae, order Hypocreales were identified in all grain samples (Table 8, Figure 1). They were represented mainly by the pathogenic species *Gibberella tricincta* which was most abundant in grain samples from treatments SP80 (4.5% OTUs), Bio60 (4.5%), and Bio80 (5.48%). Grain samples from treatments C0, SP60, and Bio40 were also colonized by unidentified *Gibberella* species. Unidentified pathogenic species of the genus *Fusarium* (family Nectriaceae) were identified in grain samples from treatments SP40 and Bio40. The pathogenic species *Monographella nivalis* of the order Xylariales, class Sordariomycetes was detected in seven grain samples, excluding samples from treatments SP40, Rec40, Rec60. *Monographella nivalis* accounted for 11% reading frames in control grain (C0). The pathogenic species *P. tritici-repentis* of the family Pleosporaceae, order Pleosporales, class Dothideomycetes was detected in grain from treatments Rec40 (1.73% OTUs), Rec80 (3.33% OTUs), Bio40 (2.41% OTUs), Bio60 (5.89% OTUs), and Bio80 (2.46% OTUs). A metagenomic analysis also demonstrated the presence of biotrophic species of the genus *Ustilago*, family Ustilaginaceae, order Ustilaginales, class Ustilaginomycetes, phylum Basidiomycota (Table 8, Figure 1). These fungi were identified only on grain from treatment Rec80 (0.41% OTUs). Fungi of the genus *Ustilago* cannot be isolated on synthetic media in a laboratory. The saprotrophic species *M. tassiana* of the family Mycosphaerellaceae, order Capnodiales, class Dothideomycetes colonized seven out of the 10 analyzed grain samples, and it accounted for 0.6% (Rec60) to 8% (Bio80) of reading frames. Unidentified species of the genus *Mycosphaerella* represented 1.6% (SP80), 0.6% (Bio40), and 1% (Bio60) of reading frames.

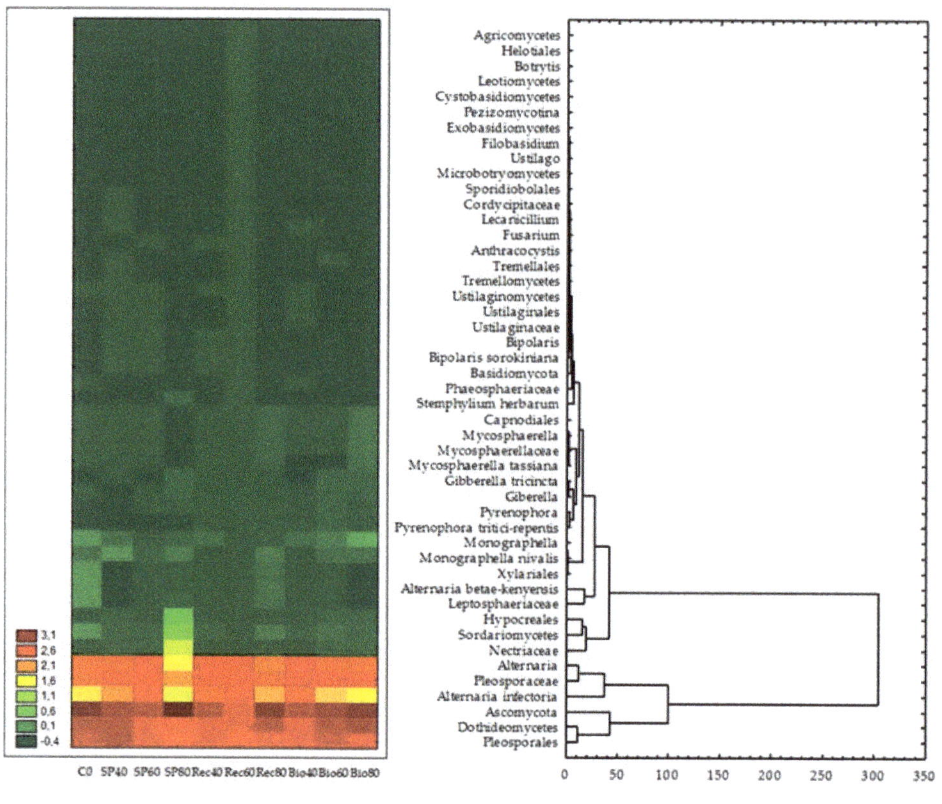

(**a**)

Figure 1. *Cont.*

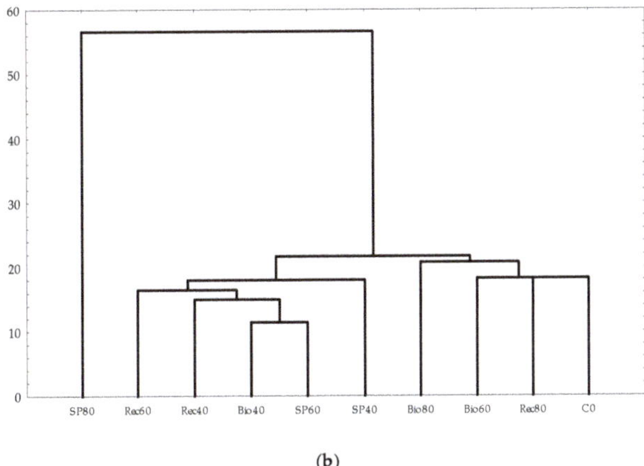

(b)

Figure 1. Heat map of operational taxonomic units (OTUs) in each experimental unit, classified at the class, order, family, genus, and species level (abbreviations are explained in Table 1). Red corresponds to high amount and green to low amount. Scale: −0.4—>5.6% OTUs, 0.1—5.7–12.6% OTUs, 0.6—12.7–27.6% OTUs, 1.1—27.7–49.8% OTUs, 1.6—49.9–58.0% OTUs, 2.1—58.1–68.2% OTUs, 2.6—68.3–93.6% OTUs, and 3.1—<93.7% OTUs. Dendrogram from hierarchical cluster analysis (Ward method using a dissimilarity matrix of Euclidean distances) on ln-transformed DNA-data of OTU 1 to OTU 10 combined for (**a**) fungi and (**b**) type of phosphorus fertilizers.

Table 8. Structure of fungal genera in wheat grain (percentage of OTUs).

Phylum	Class	Order	Family	Genus	CO *	SP40	SP60	SP80	Rec40	Rec60	Rec80	Bio40	Bio60	Bio80
Ascomycota	Dothideomycetes	Pleosporales	Pleosporaceae	*Alternaria*	67.58	78.49	83.75	49.99	89.68	95.35	66.21	81.5	70.61	71.05
				Pyrenophora	0.33	0	0.54	2.59	1.73	0	3.33	2.41	5.89	2.46
				Bipolaris	0	1,7	0	0	0	0	0	0	0	0
				Stemphylium	0	0	0	5.12	0	0	0	0	0	0
		Capnodiales	Mycosphaerellaceae	*Mycosphaerella*	2.15	1.22	1.49	1.64	1.11	0.61	5.58	0.66	1.01	8.23
	Sordariomycetes	Xylariales	Amphisphaeriaceae	*Monographella*	11.97	0	1.82	4.17	0.88	0.27	7.14	2.57	2.85	0
		Hypocreales	Nectriaceae	*Gibberella*	1.6	1.22	0.94	4.5	3.06	0.62	2.9	0.62	4.45	5.49
				Fusarium	0	0.56	0	0	0	0	0	0.76	0	0
			Cordycipitaceae	*Lecanicillium*	0	0.5	0	0	0	0	0	0	0	0.23
	Leotiomycetes	Helotiales	Sclerotiniaceae	*Botrytis*	0	0	0	0	0	0.08	0	0	0	0
Basidiomycota	Tremellomycetes	Filobasidiales	Filobasidiaceae	*Filobasidium*	0	0	0	0	0	0	0	0	0	0.19
	Ustilaginomycetes	Ustilaginales	Ustilaginaceae	*Ustilago*	0	0	0	0	0	0	0.41	0	0	0
				Anthracocystis	0.76	0	0.72	0	0.34	0.11	0	0	0	0

*—abbreviations are explained in Table 1.

The fungal community colonizing wheat grain from the treatment fertilized with superphosphate (clade 3, SP80) differed from the fungal communities identified in the remaining treatments (Figure 1). This difference was attributed to the lower amount of *A. infectoria*, sporadic appearance of *Stemphylium herbarum*, and higher amount of species of the family Nectriaceae. Fungal communities from the remaining treatments were grouped in two clades. Clade 1 was composed of fungal communities from treatments C0, Rec80, SP40, Bio60, and Bio 80, and clade 2 comprised fungal communities from treatments SP60, Rec40, Rec60, and Bio40. Clade 1 was characterized by a high frequency of *A. infectoria* and *M. nivalis* (C0, Rec80), and *G. tricincta* (Bio60, Bio80). The identified pathogens were less abundant in the communities forming clade 2. The method of next-generation sequencing in the Illumina MiSeq system allowed to identify of rare species and biotrophic fungi unable to grow on agar media.

The applied phosphorus fertilization modified the amount of fungal genera, as demonstrated by the PCA biplot (Figure 2). Treatments Rec40, Bio40, and Bio60 were grouped closest to the Tukey median (in the bagplot), and treatments SP60, SP80, Rec60, and Rec80 were located further away (in the bagplot cover region). An analysis of the PCA biplot revealed that the control treatment was separated by a significant distance from the Tukey median, and it was located in the opposite direction from treatments SP40 and Bio80.

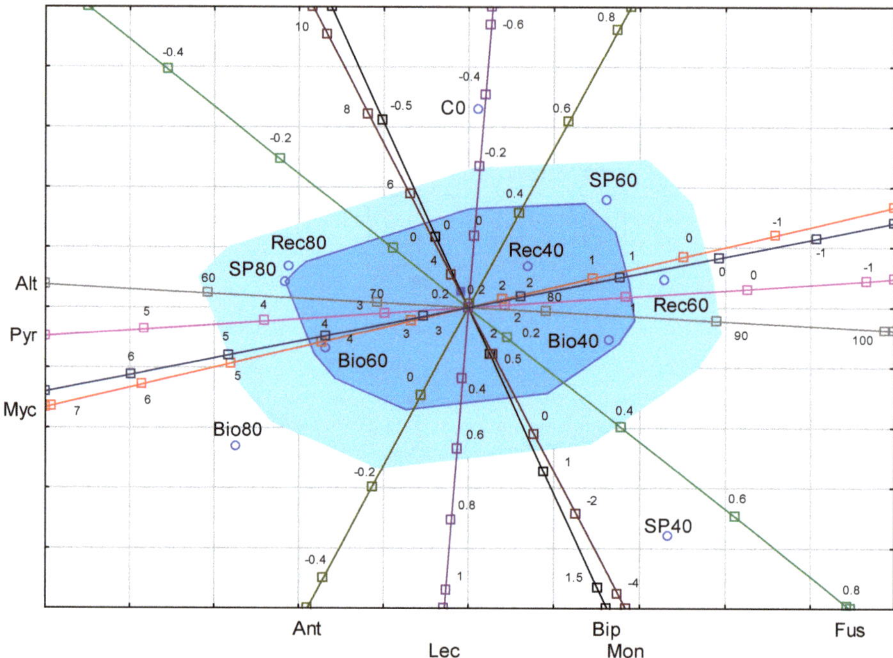

Figure 2. Principal component analysis (PCA) biplot of the microbiome in wheat grain based on fungal genera. The dark blue square denotes the Tukey median, the blue square is the bagplot, the light blue square is the bagplot cover. Alt—*Alternaria* spp., Pyr—*Pyrenophora* spp., Myc—*Mycosphaerella* spp., Ant—*Anthracocystis* spp., Lec—*Lecanicillium* spp., Bip—*Bipolaris* spp., Mon—*Monographella* spp., Fus—*Fusarium* spp.; C0, SP40, SP60, SP80, Rec40, Rec60, Rec80, Bio40, Bio60, Bio80—abbreviations are explained in Table 1.

4. Discussion

Although only a small percentage (0.1–10%) of microorganisms can be grown on synthetic media in a laboratory, they can be predominant in the analyzed microbial communities [44,45]. The results of the culture-dependent method, as well as the modern high-throughput sequencing approach, indicate that wheat grain is an ecological niche which is colonized by relatively few fungal species with low amount [1]. The genera of filamentous fungi, *Alternaria*, *Cladosporium*, *Epicoccum*, *Botrytis*, and *Fusarium*, as well as yeast genera *Cryptococcus* and *Sporobolomyces* are characteristic of this environment [1,46,47]. In the present study, the co-existence patterns could be condensed into three distinct clusters of OTUs. Clade 1 was composed of fungal communities colonizing grain from non-fertilized plants and grain from plants supplied with the recycled biofertilizer with the addition of *B. megaterium* (Bio) bacteria at 60 and 80 P_2O_5 ha^{-1}, recycled biofertilizer at 80 kg ha^{-1} (Rec), and superphosphate (SP) at 40 kg ha^{-1}, and was characterized by higher counts of pathogenic species *Monographella nivalis* and *G. tricincta*, as well as species of the genera *Pyrenophora* and *Mycosphaerella*. Clade 3 comprised a fungal community colonizing grain from plants fertilized with the highest superphosphate rate (80 kg ha^{-1}), characterized by above-average proportions of pathogenic species of the genus *Fusarium*, unidentified species of the class Sordariomycetes, with the possible presence of the pathogenic genus *Claviceps*, and the saprotrophic species *Stemphylium herbarum*. Clade 2 grouped fungal communities colonizing grain from treatments with low and moderate fertilizer rates (Bio 40 kg, Rec 40 and 60 kg, and SP 60 kg P_2O_5 ha^{-1}). The fungal communities in clade 2 were characterized by a very high prevalence of *A. infectoria*, while the proportions of the remaining pathogens were low. In a study by Supronienė et al. [48], fungi of the genus *Fusarium* were more prevalent in wheat grain grown in non-fertilized treatments and treatments fertilized with a moderate rate of NPK than in grain from treatments fertilized with a high rate of NPK. According to the literature, nitrogen fertilization exerts a negative effect on the health status of wheat plants and contributes to grain colonization by pathogens. The above can be attributed mainly to changes in stand structure: fertilized stands are dense, and they retain more moisture, which promotes the growth and sporulation of pathogenic fungi [49]. Higher rates of nitrogen fertilizers also prolong flowering and plant maturation, and wheat is most susceptible to infections during flowering [49].

The influence of phosphorus fertilizers on plant health is significantly more complex. In a study by Karimzadeh et al. [50], wheat plants fertilized with phosphorus were characterized by higher root and above-ground biomass, higher chlorophyll and proline concentrations in tissues, as well as higher yields than plants not fertilized with this nutrient. Proline is an amino acid with a secondary amine that functions as an osmolyte during stress and plays a significant role in protecting plants against stress related to the infection process [51]. Phosphorus uptake by plants from soil is also modified by bacteria and soil moisture content [51]. In the work of Arif et al. [52], phosphorus uptake was significantly higher in soybean plants inoculated with *Bacillus cereus* GS6 than in control plants. In the present experiment, recycled phosphorus fertilizers were as effective sources of plant-available phosphorus in soil as superphosphate.

Phosphorus fertilizers probably enhanced plant growth and increased stand density, but they also promoted the production of compounds which increased wheat resistance against pathogens. However, the influence of the tested types of phosphorus fertilizers, including those containing *B. megaterium* that can act as PGPM [26], on the prevalence of pathogens in the field was sometimes ambiguous and modified by other factors. Similar results have never been reported in the literature, and further research is needed to explore these ambiguities.

In this study, wheat grain was mainly colonized by fungi of the genus *Alternaria*. High-throughput sequencing in the Illumina MiSeq system revealed that *Alternaria* fungi accounted for 45–95% of OTUs (subject to treatment). The colony counts of *Alternaria* grown on PDA ranged from 0.35 to 1.48 Log(CFU + 1) per 1 g of grain. *Alternaria* fungi were also isolated from 14.81–31.48% of wheat kernels plated on PDA. Dark colonies growing on PDA and the Martin medium were identified as *A. alternata*, and similar observations were made by other authors [46,47]. *Alternaria alternata* is a ubiquitous saprotroph which infects cereal spikes and causes black scab and black point disease in cereals [53]. The species produces more than 10 allergizing proteins (www.allergen.org). The most frequently described protein Alt a 1 (AAM90320.1. NCBI. Protein Database [54] has been linked with asthma. Alt a 1 is a glycoprotein with a molecular mass of 29 kDa. *Alternaria alternata* also produces around 70 secondary metabolites, including mycotoxins that are potentially dangerous for humans and animals [55].

In traditional analyses of the plant microbiome, microorganisms are isolated and cultured on various media with the use of different methods. However, microbial communities isolated from wheat by culture-dependent methods are characterized by lower diversity than those detected with the use of culture-independent molecular techniques [53]. In the present study, a higher number of pathogenic fungi, in particular pathogens of the genus *Ustilago*, were obtained by next-generation sequencing in the Illumina MiSeq system. *Ustilago tritici* causes loose smut which is widely distributed with grain and can decrease wheat yields by up to 40%. The disease is particularly dangerous for seed farms and undressed grain [56].

In the current study, several pathogenic species that are sporadically carried by wheat grain or are less frequently isolated from grain were obtained with the use of culture-dependent methods. *Rhizoctonia cerealis*, a fungus which causes sharp eyespot, was first identified in Poland in the late

1990s [57]. *Pyrenophora tritici-repentis*, the causal agent of tan spot, was isolated from 21.31% of kernels by Bankina et al. [46]. Next-generation sequencing also supported the identification of the slow-growing pathogen *M. nivalis* which is not detected with the use of culture-dependent methods. Kernels infected with *M. nivalis* and *Fusarium* species are characterized by lower plumpness and pink discoloration. *Fusarium* fungi can cause head blight and stalk rot when distributed with infected grain. *Fusarium* fungi obtained by the culture-dependent method in this study are characteristic of the cooler regions of north-eastern Europe and Canada, and *F. culmorum* was the predominant species [3]. *Fusarium graminearum* is most prevalent in warmer, humid areas of the world such as North America, Europe, and South America [58], and it was also relatively frequently isolated in this study. The growing season of 2016 was characterized by favorable weather conditions for the growth of spring wheat, but high precipitation during grain setting and filling (total precipitation in July was 71% higher than the long-term average) delayed ripening. The above contributed to the spread of infections caused by *Fusarium* fungi.

Fungi colonizing crops can exert both positive and negative effects on the growth of host plants. The former include secreting plant growth hormones and producing compounds that inhibit the development of pathogens and increase plant resistance to infections [59,60]. In the current study, the cultured yeast communities were not significantly influenced by the tested fertilizers. The authors' previous research demonstrated that yeasts inhibit the development of *Fusarium* pathogens [3].

High-throughput sequencing in the Illumina MiSeq system supports more detailed analyses of the structure and diversity of microbial communities than conventional isolation techniques. Fungi respond more rapidly to environmental changes than other living organisms [61,62], and changes in the structure and diversity of microbial communities influence plant health. In this study, the structure and diversity of fungal communities colonizing spring wheat grain were influenced by changes in soil P content caused by the tested fertilizers. However, the observed changes were determined mainly by the P-rate rather than fertilizer type. The highest rate of commercial fertilizer induced the most adverse changes in the balance between pathogenic and non-pathogenic fungi. In a study by Eschen et al. [61], the composition of endophytic fungal communities colonizing the leaves and stems of *Cirsium arvense* varied subject to soil P content. The above authors attributed these changes to differences in fungal species' demand for leaf nutrients which can be affected by the availability of soil nutrients. Pellissier et al. [62] analyzed the composition of fungal communities in grain dust and aerosols released during wheat harvest and did not report significant correlations between total soil P and the taxonomic and phylogenetic beta diversity of fungi.

5. Conclusions

Recycled phosphorus fertilizers can at least partly replace commercial fertilizers in wheat production. They are less abundant in phosphorus than commercial mineral fertilizers, but they contain numerous macronutrients and micronutrients. Lower rates of recycled phosphorus fertilizers are adequate sources of plant-available phosphorus in soil, and they exert a beneficial impact on the structure of fungal communities colonizing the grain. Wheat grain from the treatments supplied with recycled fertilizer at 40 and 60 kg P_2O_5 ha^{-1} and the *B. megaterium* biofertilizer at 40 kg P_2O_5 ha^{-1}, was colonized by fungal communities with the most desirable composition and the lowest proportion of plant pathogens. However, the influence of recycled fertilizers on the physiology of field-grown plants and possible interactions with other environmental factors have not been fully elucidated and require further research.

Author Contributions: Conceptualization, M.J., U.W. and M.K.K.; methodology, M.J. and U.W.; formal analysis, U.W.; investigation, M.J., U.W. and M.K.K.; resources, M.J., U.W. and M.K.K.; writing—original draft preparation, U.W. and M.J.; writing—review and editing, M.J. and M.K.K.; visualization, M.J. and U.W.; funding acquisition, M.J. and M.K.K. All authors have read and agree to the published version of the manuscript.

Funding: This research was funded by the National Center for Research and Development, Poland, grant number PBS 2/A1/11/2013.

Acknowledgments: The Institute of New Chemical Synthesis in Puławy is highly acknowledged for providing Rec and Bio-fertilizers for the field experiment. The authors kindly acknowledge the technical support of Kinga Treder and Przemysław Makowski from the University of Warmia and Mazury in Olsztyn.

Conflicts of Interest: The authors declare no conflict of interest.

References

1. Nicolaisen, M.; Justesen, A.F.; Knorr, K.; Wang, J.; Pinnschmidt, H.O. Fungal communities in wheat grain show significant co-existence patterns among species. *Fungal Ecol.* **2014**, *11*, 145–153. [CrossRef]
2. Larran, S.; Perelló, A.; Simón, M.R.; Moreno, V. The endophytic fungi from wheat (*Triticum aestivum* L.). *World J. Microbiol. Biotechnol.* **2007**, *23*, 565–572. [CrossRef]
3. Wachowska, U.; Waśkiewicz, A.; Jedryczka, M. Using a protective treatment to reduce *Fusarium* pathogens and mycotoxins contaminating winter wheat grain. *Pol. J. Environ. Stud.* **2017**, *26*, 2277–2286. [CrossRef]
4. Wang, X.; Wang, C.; Sui, J.; Liu, Z.; Li, Q.; Ji, C.; Song, X.; Hu, Y.; Wang, C.; Sa, R. Isolation and characterization of phosphofungi, and screening of their plant growth-promoting activities. *AMB Express* **2018**, *8*, 63. [CrossRef] [PubMed]
5. Hassegawa, R.H.; Fonseca, H.; Fancelli, A.L.; da Silva, V.N.; Schammass, E.A.; Reis, T.A.; Corrêa, B. Influence of macro- and micronutrient fertilization on fungal contamination and fumonisin production in corn grains. *Food Control* **2008**, *19*, 36–43. [CrossRef]
6. Wachowska, U.; Stasiulewicz-Paluch, A.D.; Głowacka, K.; Mikołajczyk, W.; Kucharska, K. Response of epiphytes and endophytes isolated from winter wheat grain to biotechnological and fungicidal treatments. *Pol. J. Environ. Stud.* **2013**, *22*, 267–273.
7. Taghinasab, M.; Imani, J.; Steffens, D.; Glaeser, S.P.; Kogel, K.H. The root endophytes *Trametes versicolor* and *Piriformospora indica* increase grain yield and P content in wheat. *Plant Soil* **2018**, *426*, 339–348. [CrossRef]
8. Grzebisz, W.; Potarzycki, J.; Biber, M.; Szczepaniak, W. Reakcja roślin uprawnych na nawożenie fosforem. *J. Elem.* **2003**, *8*, 83–93.
9. Haneklaus, S.H.; Schnug, E. Assessing the plant phosphorus status. In *Phosphorus in Agriculture: 100 % Zero*; Springer: Dordrecht, The Netherlands, 2016; pp. 95–125.
10. Prabhu, A.S.; Fageria, N.K.; Berni, R.F.; Rodrigues, F.A. Phosphorus and plant disease. In *Mineral Nutrition and Plant Disease*; American Phytopathological Society: St. Paul, MN, USA, 2007; pp. 45–55.
11. Mohammadi, K. Phosphorus solubilizing bacteria: Occurrence, mechanisms and their role in crop production. *Resour. Environ.* **2012**, *2*, 80–85.
12. Weigand, H.; Bertau, M.; Hübner, W.; Bohndick, F.; Bruckert, A. RecoPhos: Full-scale fertilizer production from sewage sludge ash. *Waste Manag.* **2013**, *33*, 540–544. [CrossRef]
13. Van Kauwenbergh, S.J.; Stewart, M.; Mikkelsen, R. World reserves of phosphate rock … a dynamic and unfolding story. *Better Crops* **2013**, *97*, 18–20.
14. Tenkorang, F.; Lowenberg-Deboer, J. Forecasting long-term global fertilizer demand. *Nutr. Cycl. Agroecosyst.* **2009**, *83*, 233–247. [CrossRef]
15. Cordell, D.; White, S. Peak phosphorus: Clarifying the key issues of a vigorous debate about long-term phosphorus security. *Sustainability* **2011**, *3*, 2027–2049. [CrossRef]
16. Schroder, J.J.; Cordell, D.; Smit, A.L.; Rosemarin, A. *Sustainable use of Phosphorus: EU Tender ENV. B1/ETU/2009/0025*; Plant Research International: Wageningen, The Netherlands, 2010.
17. Rosemarin, A.; Jensen, L.S. The phosphorus challenge. In Proceedings of the European Sustainable Phosphorus Conference, Brussels, Belgium, 6–7 March 2013.
18. USGS. *Mineral Commodity Summaries 2019. Phosphate Rock*; U.S. Geological Survey: Reston, VA, USA, 2019; pp. 122–123.
19. Geissler, B.; Mew, M.C.; Steiner, G. Phosphate supply security for importing countries: Developments and the current situation. *Sci. Total Environ.* **2019**, *677*, 511–523. [CrossRef]
20. EC. Report on critical raw materials for the EU. European Commission. In *Report of the Ad-Hoc Working Group on Defining Critical Raw Materials*; EC: Brussels, Belgium, 2014; p. 41.
21. Jaskulski, D.; Jaskulska, I. Possibility of using waste from the polyvinyl chloride production process for plant fertilization. *Pol. J. Environ. Stud.* **2011**, *20*, 351–354.

22. Smol, M.; Kulczycka, J.; Kowalski, Z. Sewage sludge ash (SSA) from large and small incineration plants as a potential source of phosphorus—Polish case study. *J. Environ. Manag.* **2016**, *184*, 617–628. [CrossRef]

23. Staroń, A.; Kowalski, Z.; Banach, M.; Wzorek, Z. Sposoby termicznej utylizacji odpadów z przemysłu mięsnego. *Czas. Tech. Chem.* **2010**, *107*, 323–332.

24. Saeid, A.; Labuda, M.; Chojnacka, K.; Górecki, H. Use of microorganisms in the production of phosphorus fertilizers. *Przem. Chem.* **2012**, *91*, 956–958.

25. Karpagam, T.; Nagalakshmi, P.K. Isolation and characterization of phosphate solubilizing microbes from agricultural soil. *Int. J. Curr. Microbiol. Appl. Sci.* **2014**, *3*, 601–614.

26. Awasthi, R.; Tewari, R.; Nayyar, H. Synergy between plants and P-solubilizing microbes in soils: Effects on growth and physiology of crops. *Int. Res. J. Microbiol.* **2011**, *2*, 484–503.

27. El-Komy, H.M.A. Coimmobilization of *Azospirillum lipoferum* and *Bacillus megaterium* for successful phosphorus and nitrogen nutrition of wheat plants. *Food Technol. Biotechnol.* **2005**, *43*, 19–27.

28. Saeid, A.; Prochownik, E.; Dobrowolska-Iwanek, J. Phosphorus solubilization by *Bacillus* species. *Molecules* **2018**, *23*. [CrossRef] [PubMed]

29. Etesami, H.; Beattie, G.A. Mining halophytes for plant growth-promoting halotolerant bacteria to enhance the salinity tolerance of non-halophytic crops. *Front. Microbiol.* **2018**, *9*. [CrossRef] [PubMed]

30. Nesme, T.; Withers, P.J.A. Sustainable strategies towards a phosphorus circular economy. *Nutr. Cycl. Agroecosys.* **2016**, *104*, 259–264. [CrossRef]

31. Saeid, A.; Wyciszkiewicz, M.; Jastrzebska, M.; Chojnacka, K.; Gorecki, H. A concept of production of new generation of phosphorus-containing biofertilizers. BioFertP project. *Przem. Chem.* **2015**, *94*, 361–365. [CrossRef]

32. Rolewicz, M.; Rusek, P.; Borowik, K. Obtaining of granular fertilizers based on ashes from combustion of waste residues and ground bones using phosphorous solubilization by bacteria *Bacillus megaterium*. *J. Environ. Manag.* **2018**, *216*, 128–132. [CrossRef]

33. Meier, U. *Growth Stages of Mono- and Dicotyledonous Plants: BBCH-Monograph*; Blackwell Wissenschafts-Verlag: Berlin, Germany; Boston, MA, USA, 1997; p. 622.

34. World Reference Base (WRB); IUSS Working Group. World reference base for soil resources 2014. In *International Soil Classification System for Naming Soils and Creating Legends for Soil Maps*; Food and Agriculture Organization of the United Nations: Rome, Italy, 2014; update 2015.

35. Martin, J.P. Use of acid, rose bengal, and streptomycin in the plate method for estimating soil fungi. *Soil Sci.* **1950**, *69*, 215–232. [CrossRef]

36. Ellis, M.B. *Dematiaceous Hyphomycetes*; CMI: Kew, UK, 1971; p. 608.

37. Leslie, J.F.; Summerell, B.A. *The Fusarium Laboratory Manual*; Blackwell Publishing: Hoboken, NJ, USA, 2007; pp. 1–388.

38. Caporaso, J.G.; Kuczynski, J.; Stombaugh, J.; Bittinger, K.; Bushman, F.D.; Costello, E.K.; Fierer, N.; Pêa, A.G.; Goodrich, J.K.; Gordon, J.I.; et al. QIIME allows analysis of high-throughput community sequencing data. *Nat. Methods* **2010**, *7*, 335–336. [CrossRef]

39. Edgar, R.C. Search and clustering orders of magnitude faster than BLAST. *Bioinformatics* **2010**, *26*, 2460–2461. [CrossRef]

40. Nilsson, R.H.; Larsson, K.H.; Taylor, A.F.S.; Bengtsson-Palme, J.; Jeppesen, T.S.; Schigel, D.; Kennedy, P.; Picard, K.; Glöckner, F.O.; Tedersoo, L.; et al. The UNITE database for molecular identification of fungi: Handling dark taxa and parallel taxonomic classifications. *Nucleic Acids Res.* **2019**, *47*, D259–D264. [CrossRef]

41. StatSoft, I. *Statistica (Data Analysis Software System), Version 13*; Statsoft Inc.: Tulsa, OK, USA, 2016.

42. Wilkinson, L.; Friendly, M. History corner the history of the cluster heat map. *Am. Stat.* **2009**, *63*, 179–184. [CrossRef]

43. Ward, J.H., Jr. Hierarchical grouping to optimize an objective function. *J. Am. Stat. Assoc.* **1963**, *58*, 236–244. [CrossRef]

44. Ellis, R.J.; Morgan, P.; Weightman, A.J.; Fry, J.C. Cultivation-dependent and -independent approaches for determining bacterial diversity in heavy-metal-contaminated soil. *Appl. Environ. Microbiol.* **2003**, *69*, 3223–3230. [CrossRef] [PubMed]

45. Littlefield-Wyer, J.G.; Brooks, P.; Katouli, M. Application of biochemical fingerprinting and fatty acid methyl ester profiling to assess the effect of the pesticide Atradex on aquatic microbial communities. *Environ. Pollut.* **2008**, *153*, 393–400. [CrossRef] [PubMed]

46. Bankina, B.; Bimšteine, G.; Neusa-Luca, I.; Roga, A.; Fridmanis, D. What influences the composition of fungi in wheat grains? *Acta Agrobot.* **2017**, *70*. [CrossRef]

47. Xu, K.G.; Jiang, Y.M.; Li, Y.K.; Xu, Q.Q.; Niu, J.S.; Zhu, X.X.; Li, Q.Y. Identification and pathogenicity of fungal pathogens causing black point in wheat on the North China Plain. *Indian J. Microbiol.* **2018**, *58*, 159–164. [CrossRef] [PubMed]

48. Suproniene, S.; Mankeviciene, A.; Kadžiene, G.; Kacergius, A.; Feiza, V.; Feiziene, D.; Semaškiene, R.; Dabkevicius, Z.; Tamošiunas, K. The impact of tillage and fertilization on *Fusarium* infection and mycotoxin production in wheat grains skaidre suproniene. *Zemdirbyste* **2012**, *99*, 265–272.

49. Lemmens, M.; Haim, K.; Lew, H.; Ruckenbauer, P. The effect of nitrogen fertilization on *Fusarium* head blight development and deoxynivalenol contamination in wheat. *J. Phytopathol.* **2004**, *152*, 1–8. [CrossRef]

50. Karimzadeh, J.; Alikhani, H.A.; Etesami, H.; Pourbabaei, A.A. Improved phosphorus uptake by wheat plant (*Triticum aestivum* L.) with rhizosphere fluorescent Pseudomonads strains under water-deficit stress. *J. Plant Growth Regul.* **2020**. [CrossRef]

51. Zhang, X.; Ervin, E.H.; Evanylo, G.K.; Haering, K.C. Impact of biosolids on hormone metabolism in drought-stressed tall fescue. *Crop Sci.* **2009**, *49*, 1893–1901. [CrossRef]

52. Arif, M.S.; Riaz, M.; Shahzad, S.M.; Yasmeen, T.; Ali, S.; Akhtar, M.J. Phosphorus-mobilizing rhizobacterial strain *Bacillus cereus* GS6 improves symbiotic efficiency of soybean on an aridisol amended with phosphorus-enriched compost. *Pedosphere* **2017**, *27*, 1049–1061. [CrossRef]

53. Wang, J.; Qin, J.; Li, Y.; Cai, Z.; Li, S.; Zhu, J.; Zhang, F.; Liang, S.; Zhang, W.; Guan, Y.; et al. A metagenome-wide association study of gut microbiota in type 2 diabetes. *Nature* **2012**, *490*, 55–60. [CrossRef]

54. Protein Database. National Center for Biotechnology Information. Proteins. U.S. National Library of Medicine. Available online: https://www.ncbi.nlm.nih.gov/protein (accessed on 7 July 2018).

55. Jarolim, K.; Del Favero, G.; Pahlke, G.; Dostal, V.; Zimmermann, K.; Heiss, E.; Ellmer, D.; Stark, T.D.; Hofmann, T.; Marko, D. Activation of the Nrf2-ARE pathway by the *Alternaria alternata* mycotoxins altertoxin I and II. *Arch. Toxicol.* **2017**, *91*, 203–216. [CrossRef] [PubMed]

56. Menzies, J.G. Virulence of isolates of *Ustilago tritici* collected in Manitoba and Saskatchewan, Canada, from 1999 to 2007. *Can. J. Plant Pathol.* **2016**, *38*, 470–475. [CrossRef]

57. Mikolajska, J.; Wachowska, U. Characterization of binucleate isolates of *Rhizoctonia cerealis* with respect to cereals. In Proceedings of the Symposium on New Direction in Plant Pathology, Kraków, Poland, 11–13 September 1996.

58. Weidenbörner, M. Mycotoxin contamination of cereals and cereal products. In *Mycotoxins in Plants and Plant Products*; Springer: Berlin/Heidelberg, Germany, 2017; pp. 1–715.

59. Lakshmanan, V.; Selvaraj, G.; Bais, H.P. Functional soil microbiome: Belowground solutions to an aboveground problem. *Plant Physiol.* **2014**, *166*, 689–700. [CrossRef]

60. Berendsen, R.L.; Pieterse, C.M.J.; Bakker, P.A.H.M. The rhizosphere microbiome and plant health. *Trends Plant Sci.* **2012**, *17*, 478–486. [CrossRef]

61. Eschen, R.; Hunt, S.; Mykura, C.; Gange, A.C.; Sutton, B.C. The foliar endophytic fungal community composition in *Cirsium arvense* is affected by mycorrhizal colonization and soil nutrient content. *Fungal Biol.* **2010**, *114*, 991–998. [CrossRef]

62. Pellissier, L.; Oppliger, A.; Hirzel, A.H.; Savova-Bianchi, D.; Mbayo, G.; Mascher, F.; Kellenberger, S.; Niculita-Hirzel, H. Airborne and grain dust fungal community compositions are shaped regionally by plant genotypes and farming practices. *Appl. Environ. Microbiol.* **2016**, *82*, 2121–2131. [CrossRef]

 *agriculture*

Review

Linking Endophytic Fungi to Medicinal Plants Therapeutic Activity. A Case Study on Asteraceae

Gianluca Caruso [1], **Magdi T. Abdelhamid** [2], **Andrzej Kalisz** [3] and **Agnieszka Sekara** [3,*]

[1]   Department of Agricultural Sciences, University of Naples Federico II, 80055 Portici (Naples), Italy; gcaruso@unina.it

[2]   Botany Department, National Research Centre, Cairo 12622, Egypt; mt.abdelhamid@nrc.sci.eg

[3]   Department of Horticulture, Faculty of Biotechnology and Horticulture, University of Agriculture, 31-120 Krakow, Poland; andrzej.kalisz@urk.edu.pl

*   Correspondence: agnieszka.sekara@urk.edu.pl

Received: 12 June 2020; Accepted: 9 July 2020; Published: 10 July 2020

Abstract: Endophytes are isolated from every plant species investigated to date, so the metabolome coevolution has been affecting the plants' (microbiota) ethnobotanic, especially therapeutic, usage. Asteraceae fulfill the rationale for plant selection to isolate endophytes since most of the species of this family have a long tradition of healing usage, confirmed by modern pharmacognosy. The present review compiles recent references on the endophyte–Asteraceae spp. interactions, targeting the secondary metabolites profile as created by both members of this biological system. Endophyte fungi associated with Asteraceae have been collected globally, however, dominant taxa that produce bioactive compounds were specific for the plant populations of different geographic origins. Endophytic fungi richness within the host plant and the biological activity were positively associated. Moreover, the pharmacological action was linked to the plant part, so differential forms of biological interactions in roots, stem, leaves, inflorescences were developed between endophytic fungi and host plants. The comparative analysis of the Asteraceae host and/or fungal endophyte therapeutic activity showed similarities that need a future explanation on the metabolome level.

Keywords: compositae; fungi; herbs; secondary metabolites; symbiosis

---

1. Introduction

Each plant coexists with microorganisms residing within tissues and producing their metabolites, which are defined as endophytes if their occurrence does not cause apparent injuries [1,2]. Wilson [3] defined "endophytes" (from Greek endon—within; and phyton—plant) as microorganisms, commonly fungi and bacteria, spending their life cycle inter- and/or intra-cell space of the tissues of host plants, which do not show any symptoms of disease. Endophytes were isolated from plants belonging to all taxa investigated to date, occurring in all the world's ecosystems. In recent years, there has been an increased interest in explaining the endophytes/host plant cross-talk because the effects of these relationships could be beneficial to humans [1,4–6]. Host plants abide endophytes due to symbiotic relationships, profitable for microbes due to the availability of habitat and nutrients in the plant, while plants acquire a wide spectrum of microbial metabolites, including vitamins, hormones, and antibiotics [7,8]. Endophyte–host relationships can be so close, that microbes can even biosynthesize the same chemical compounds as the host, as myrtucommulones from *Myrtus communis*, camptothecin from *Camptotheca acuminata*, paclitaxel from *Taxus brevifolia*, or deoxypodophyllotoxin from *Juniperus communis* for better adaptation to the microenvironment of plant tissues [7,9–13]. It is an unresolved hypothesis that the production of secondary metabolites in plants is not achieved only by endophytes but arises from concomitant plant and fungal biosynthesis [13]. Endophytes occupy a unique ecological niche, their relationship with a host plant a balance between mutualistic, parasitic, or commensal symbiosis, which is largely controlled via chemicals. That is the reason why endophytes produce highly specific metabolites [14]. Indeed, these microorganisms are being increasingly investigated as they play an important role in natural product discovery, especially when the source plant is used for medicinal purposes. In the latter respect, the healing action can be the result not only of the host plant metabolome but also the microorganism-derived active compounds and their interactions [4]. Moreover, organic extracts obtained from isolated endophytes show a wide spectrum of biological action and may be applied as antidiabetic, antimicrobial, antiviral, larvicidal, antimalarial, cytotoxic, and plant growth promoters [15,16]. The problem is that some endophyte genes responsible for secondary metabolite biosynthesis were found to be significantly expressed in planta but silent in vitro cultures. Plant and coexisting microbial signal molecules are required to induce particular pathways of endophyte metabolism leading to a balance of sexual to asexual reproduction and biochemical profile modification as well [17–20]. Moreover, the secondary metabolites are energy-consuming compounds, so endophytes can increase/decrease their production depending on specific needs, like competition with the other microorganisms or host plant communication and protection [9,21–23]. However, some fungal endophytes were shown to produce the desired compounds without a host plant association. Sustainable synthesis of tanshinone IIA and taxol by the axenic culture of endophytic fungi have been reported by Ma et al. [24] and Zhao et al. [25]. Karuppusamy [26] presented the possible origin of secondary metabolites in plant-endophyte systems, namely (i) parallel coevolution of plants and their microbiota possessing pathways to produce bioactive compounds; (ii) horizontal gene transfer between plants and microbes during their coevolution; (iii) plants or endophytic fungi synthesize and transfer metabolites to each other. Recent studies provided strong indications that endophytic fungi dispose host-independent machinery for secondary metabolite production [27–29]. Metabolites of fungal endophytes which were isolated from medicinal plants possess diverse and unique structural groups. That is the reason why they are good sources of novel secondary metabolic products contributing to the therapeutic activity [30–32]. Among medicinal plants, the members of Asteraceae family have been reported to be a source of natural remedies in all traditional medicine systems since their secondary metabolites exhibit strong antioxidant, antibacterial or anti-inflammatory activities [33].

The production of bioactive secondary metabolites by endophytic fungi colonizing medicinal plants has been largely ignored. The main idea of this review is that the Asteraceae evolutionary success is the effect of interaction between the host plant and fungal endophytic microbiota. We focused on determining the possible contribution of fungal biosynthesis to the secondary metabolome of Asteraceae, as a leading family of medicinal plants, to present the additional explanation for the distribution of bioactive compounds, including alkaloids, cardiac glycosides, and anthraquinones in the plant kingdom. We reviewed the available literature to assess therapeutic activity that had been reported previously from medicinal plants of the Asteraceae family that may likewise originate from endophytic fungi that coexist with these plants. We tried to estimate if the plants' taxonomic affinity affects the endophytic microbiome biodiversity and metabolic pathways.

2. Asteraceae Ecology and Biochemistry

The family Asteraceae (Compositae) is the largest and most cosmopolitan group of angiosperms covering 32,913 accepted species, grouped in 1911 genera and 13 subfamilies [34]. Asteraceae comprise more than 40 economically important crops, including food crops (*Lactuca sativa*, *Cichorium* spp., *Cynara scolymus*, *Smallanthus sonchifolius*, and *Helianthus tuberosus*), oil crops (*Helianthus annuus*, *Carthamus tinctorius*), medicinal and aromatic plants (*Matricaria chamomilla*, *Chamaemelum nobile*, *Calendula* spp., *Echinacea* spp., and *Artemisia* spp.), ornamentals (*Chrysanthemum* spp., *Gerbera* spp., *Dendranthema* spp., *Argyranthemum* spp., *Dahlia* spp., *Tagetes* spp., and *Zinnia* spp.), and nectar producers (*Centaurea* spp., *H. annuus*, and *Solidago* spp.) [35]. Species of this family represent a great variation regarding the habit: annual, perennial, herbs, shrubs, vines, trees, epiphytes; with the inflorescence composed of one to more than a thousand florets; and chromosome numbers range from $n = 2$ to $n = 114$ [36]. The Asteraceae store energy in the form of inulin [37], they can produce acetylenes, alcohols, alkaloids, organic acids, pentacyclic triterpenes, sesquiterpene lactones, and tannins [38–40]. They are globally distributed although most are native to temperate climatic zones, the Mediterranean zone, or higher-elevation, cooler regions of the tropics [41]. The unique success of Asteraceae in worldwide distribution has been attributed to many factors, including diversity of secondary metabolites that improve overall fitness, a highly specialized inflorescence that maximizes fertilization, and a morphology promoting outcrossing [42]. Many species of the Asteraceae family have been used as medicinal plants, although the secondary metabolites responsible for the pharmacological efficiency were not always defined. The chemical diversity of bioactive compounds and pathways of their biosynthesis is dependent on a broad spectrum of biotic and abiotic factors and their interactions. Sometimes the benefits of plant-derived pharmacological products are controversial despite standard chemical composition with the use of commonly accepted pharmacopeia's methods [43]. Numerous papers have described the pharmacological activity and chemical constituents isolated from plants of the Asteraceae, covering polyphenols, sesquiterpenes, organic and fatty acids which have been associated with the successful treatment of cardiovascular diseases, cancer, microbial and viral infections, inflammation, and other diseases [43]. Most of the Asteraceae taxa, like *Artemisia*, are well known for their resistance to herbivores, bacterial and fungal pathogens [44]. Secondary metabolites are chemicals of a very diversified structure, not fundamental in the plant metabolism, but crucial for protection against pathogens and herbivores [45]. With the use of principal component analysis, Alvarenga et al. [46] showed the relationships between chemical composition and botanical classification of Asteraceae family, based on a huge group of 4000 species and 11 main chemical classes of secondary metabolites. *Barnadesieae* tribe revealed an anomalous position owing to the poor diversity of its secondary metabolites, particularly flavonoids. *Liabeae* and *Vernonieae* tribes were localized closely because of similar lactone composition, while *Asteridae* was separated because of monoterpenes, diterpenes, sesquiterpenes content. Moreover, the correlation matrix of Asteraceae secondary metabolites showed that benzofuranes and acetophenones, as well as diterpenes and phenylpropanoids, were highly correlated with each other [46]. The role of fungal endophytes in Asteraceae's evolutionary success has been recently recognized by the scientific community, although there is still a need for complex investigations in this area. The multifarious metabolome of Asteraceae is a dynamic patchwork of chemicals synthesized solely by the plant, by the microbial inhibiting the host species, or by both elements of this ecological system.

3. Fungal Endophytes Associated with Asteraceae—Biodiversity, and Ecology

The high diversity of endophytes indicates their multiple and variable relations with the host plants and ecological functions. The widest research program to find endophytes in medicinal Asteraceae has been performed in countries which are localized in the most important biodiversity hotspots, like Brazil, China, the Mediterranean region, Iran, or Thailand [47]. In Brazil, like the other South American countries, medicinal plants have been used as a traditional, cheap, and easily available alternative to drugs. Only a few tropical herbs were investigated with respect to endophytic fungal communities with bioactivity [48–50]. Another region of Asteraceae collection as host plants for fungal endophytes is the Panxi plateau in China [51] with xerothermic climate, diversified soil, and landscape conditions contributing to the high biodiversity in the area, concerning also medicinal plants having a long history of application by local communities [52]. The global screening reflected in the present review showed minimal knowledge on Asteraceae in this respect (Figure 1).

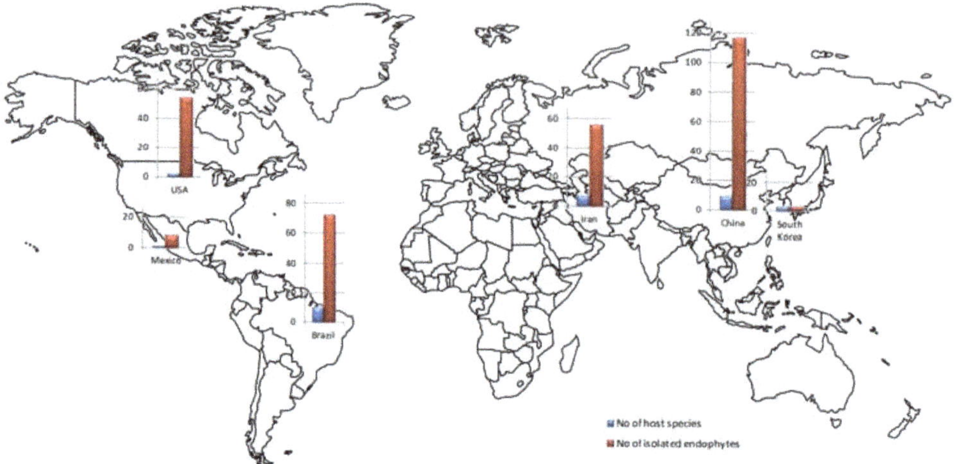

Figure 1. The Asteraceae hosts and endophyte fungi isolated from them in chosen countries (based on the references cited in this review).

Despite the high diversity and abundance of the Asteraceae worldwide, fungal endophytes associated with the plants of this family represented common or cosmopolitan species [53]. In light of the present review, about 23% of fungi taxa isolated from Asteraceae were associated with one host (Figure 2). They were mentioned in the footnote of Figure 2 as "The others". The most abundant fungi genera, *Colletotrichum, Alternaria, Penicillium*, etc., were ubiquitous and isolated from most plant species and environments [10].

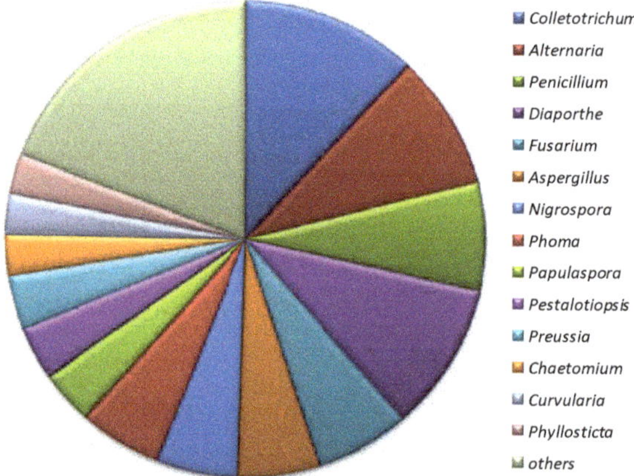

Figure 2. The frequency of isolation of endophytes (%) from Asteraceae host plants. The others: *Acremonium, Ampelomyces, Bipolaris, Botryosphaeria, Botrytis, Calonectria, Cercospora, Coniochaeta, Cylindrocarpon, Epicoccum, Exserohilum, Memnoniella, Paecilomyces, Periconia, Podospora, Pezicula, Pyrenophora, Scopulariopsis, Seiridium, Trichoderma, Xylaria* (based on references cited in this review).

To date, most of the research was focused on the overall spectrum of endophytes of the particular host plant or the particular endophyte taxon isolated from a wide range of host plants. To validate, Rodríguez–Rodríguez et al. [49] compared microorganism diversity and abundance in *Aster grisebachii* (synonym of *Neja marginata*), *Erigeron bellidiastroides, Erigeron cuneifolius, Pectis juniperina*, and *Sachsia polycephala* (Asteraceae), native to Cuba, collected in an area with a low-in-nutrients, acid, sandy soil with alternating dry, and rainy seasons. The colonization rate was higher than 50% in both the dry and rainy period for all species which is typical for changing and stressful ecosystems, with strong competition for soil resources. *Pestalotiopsis* spp. were isolated as dominant from the different medicinal plants originated to tropical and subtropical climatic zones [54]. *Preussia* spp. isolated from leaves of medicinal plants *Baccharis trimera* (Asteraceae) and *Stryphnodendron adstringens* (Fabaceae) are native to Brazilian savannah [55]. A study performed by Hatamzadeh et al. [56], on native Asteraceae medicinal plants of Iran, allowed to isolate 241 endophyte species from *Cota segetalis* (syn. *Anthemis altissima*), 163 from *Achillea millefolium*, 121 from *Anthemis triumfettii* (synonym of *Cota triumfettii* subsp. *triumfettii*), 132 from *Cichorium intybus*, 90 from *Achillea filipendulina*, and 59 from *M. chamomilla*. A few endophytic fungi such as *Acremonium sclerotigenum*, *Alternaria burnsii*, *Bjerkandera adusta*, *Colletotrichum tanaceti*, *Epicoccum nigrum*, *Fusarium acuminatum*, *Paraphoma chrysanthemicola*, *Plectosphaerella cucumerina*, and *Stemphylium amaranthi* were isolated from all host species [56], most of them colonizing the stem of the plant. Although Cheng et al. [57] concluded that the structure of the endophytic communities differed within plant tissues and habitats, similarities in the taxa of the endophytic fungi were rarely observed at the phylum order or even the host plant family level. Endophyte communities were characterized by ecological variation, different host preference, tissue specificity, spatial heterogeneity, and seasonal changes in terms of composition and quantity of fungal endophytic strains which can affect medicinal plant biochemical composition [58]. Investigations of endophytes coexisting with *Ageratina altissima* showed that the fungal microbiome was driven by host individual and geographic location. Moreover, the endophyte community of a single host collected in the urban zone was less abundant compared to the forest probably due to human disturbance and spatial isolation [59]. The expansion of the invasive species *Ageratina adenophora* was studied concerning the distribution of endophytes in tissues in surrounding environments [60–62]. The enrichment of *A. adenophora* endophytes was root tissue-specific, moreover, fungi rarely grew systemically within the plant. The roots were the habitat of *Fusarium*, the stems of *Allophoma*, the mature leaves of *Colletotrichum*, and *Diaporthe*. Additionally, some fungi might migrate tissue-to-tissue via the vascular system of the shoot, and this was the way airborne fungi infected roots, and soilborne fungi, shoots, and leaves. Leaf endophytes showed more fluctuations in the number of taxa than those in roots and stems, because of the stronger pressure of environmental factors [62]. Presented studies indicated that fungal endophyte communities varied based on host genotype or even specimen, plant tissue, growth stage, and growth conditions. The research referenced in this review were focused on the taxonomical analysis of endophytes collected in a particular area from different Asteraceae taxa, or one species, or from different tissues of that species. Another main field of investigation were secondary metabolites produced by endophytes in situ or in vitro. Table 1 summarizes the biological action of Asteraceae plant extracts and endophytes isolated from them. The evident similarities indicate that the therapeutic activity of Asteraceae plants used traditionally as herbal remedies can also be referred to associated fungal endophytes. Almost all internal symbiotic fungi showed in vitro similar activity to those of their host plant extract. However, the present review of the literature published during the last twenty years showed insufficient experimental evidence to describe the endophyte/host plant interactions on the metabolome level, so the biosynthetic pathway might be differently regulated in the fungus and the host plant.

Table 1. Endophytes isolated from Asteraceae species with a therapeutic activity referred to host and/or endophyte taxon.

Asteraceae Species and the Tissue of Endophytes Isolation	Main Therapeutic acTivities of the Host	Dominating Endophyte Genera	Main Activities of the Endophyte	Main Metabolites/Enzymes Linked to Endophyte Bioactivities
Achillea millefolium (stem, leaf, root)	Antioxidant, anti-inflammatory, antimicrobial, antitumor [63]	*Didymella, Septoria, Stemphylium, Cladosporium, Fusarium, Alternaria, Nemania*	Antitumor against lymphoblastic leukemia	L-asparaginase [56] *
Achillea filipendulina (stem, leaf)	Antioxidant, antidiabetic, anti-inflammatory, antimicrobial, antitumor, lubricant, antiparasitic [64,65]	*Plectosphaerella, Fusarium*	Antitumor against lymphoblastic leukemia	L-asparaginase [56]
Anthemis segetali (synonym of *Cota segetalis*) (stem, leaf, root, inflorescence)	Antioxidant anti-inflammatory, antitumor, antimicrobial, hepatoprotective [66]	*Alternaria, Aspergillus, Bjerkandera, Schizophyllum, Fusarium, Plenodomus, Cladosporium, Didymella, Stemphylium, Nemania, Phoma, Plectosphaerella, Sarocladium*	Antitumor against lymphoblastic leukemia	L-asparaginase [56]
Anthemis triumfettii (synonym of *Cota triumfettii* subsp. *Triumfettii* (stem, leaf)	Antioxidant anti-inflammatory, antitumor, antimicrobial, hepatoprotective [66]	*Chaetosphaeronema, Stemphylium, Alternaria*	Antitumor against lymphoblastic leukemia	L-asparaginase [56]
Artemisia annuua (stem)	Anti-inflammatory, antipyretic, antitumor, antifungal, antiparasitic, antiulcerogenic, cytotoxic [67–69]	*Colletotrichum*	Antibacterial against *Bacillus subtilis, Staphylococcus aureus, Sarcina lutea,* and *Pseudomonas* sp., and antifungal against *Candida albicans* and *Aspergillus niger*	3β,5α-dihydroxy-6β-acetoxy-ergosta-7,22-diene; 3β,5α-dihydroxy-6β-phenylacetyloxy-ergosta-7,22-diene; 3β-hydroxy-ergosta-5-ene; 3-oxo-ergosta-4,6,8(14),22-tetraene; 3β-hydroxy-5α,8α-epidioxy-ergosta-6,22-diene [70]
Artemisia vulgaris	Antimalarial, anti-inflammatory, antihypertensive, antioxidant, antitumor, immunomodulatory, hepatoprotective, antispasmodic, antiseptic [71]	*Chalara*	Antibacterial against *B. subtilis* and antifungal against *C. albicans*	Isofusidienol A, B, C, and D [72]
Artemisia mongolica (stem)	Antimicrobial, insecticidal, antioxidant [73]	*Colletotrichum*	Antibacterial against *B. subtilis, S. aureus,* and *S. lutea*; antifungal against *Bipolaris sorokiniana*	Colletotric acid [44]
Atractylodes lancea	Anti-inflammatory, hepatoprotective [74]	*Gilmaniella*	Antimicrobial	Jasmonic acid [75]
Ayapana triplinervis	Antimicrobial, anti-inflammatory [76]	*Paecilomyces, Aspergillus, Fusarium, Trichoderma, Penicillium, Curvularia*	Not investigated	Not investigated [77]

Table 1. *Cont.*

Asteraceae Species and the Tissue of Endophytes Isolation	Main Therapeutic acTivities of the Host	Dominating Endophyte Genera	Main Activities of the Endophyte	Main Metabolites/Enzymes Linked to Endophyte Bioactivities
Baccharis dracunculifolia	Immunostimulatory, anti-inflammatory, cytotoxic, antitumor, hepatoprotective [78]	*Penicillium, Aspergillus, Fusarium, Colletotrichum*	Not investigated	Not investigated [79]
Baccharis dracunculifolia (leaf)	Antioxidant, anti-inflammatory, antiviral, antimicrobial, antiparasitic [80]	*Epicoccum, Pestalotiopsis, Cochliobolus, Nigrospora*	Antimicrobial	Not investigated [53]
Bidens pilosa	Antimalarial, anti-allergic, antihypertensive, antitumor, antidiabetic, anti-inflammatory, antimicrobial, antioxidant [81]	*Botryosphaeria*	Antifungal, cytotoxic, antiproliferative against carcinoma cell lines	Botryorhodine A and B [82]
Cichorium intybus (stem, leaf, root)	Antioxidant, anti-inflammatory, cardiovascular, hypolipidemic, antitumor, antidiabetic, antimicrobial, antiparasitic [83]	*Cladosporium, Epicoccum, Septoria, Plectosphaerella, Alternaria*	Antitumor against lymphoblastic leukemia	L-asparaginase [56]
Gynura hispida	Anti-inflammatory, antiviral, hepatotoxic [84]	*Bipolaris*	Antifungal against *Cladosporium cladosporioides, C. cucumerinum, Saccharomyces cerevisiae, Aspergillus niger,* and *Rhisopus oryzae*	Bipolamide B [85]
Helianthus annuus (root)	Antibacterial, antioxidant, hepato-, nephro- and cardioprotective [86,87]	*Penicillium, Aspergillus*	Antifungal against *Sclerotium rolfsii*	Gibberellins (GA1, GA3, GA4, GA9, GA12, andGA20); organic acids (jasmonic, malic, quinic, salicylic, and succinic acid); siderophores [88,89]
Laggera alata	Anti-inflammatory, antioxidative, antibacterial, larvicidal [90]	*Podospora*	Larvicidal against *Anopheles gambiae*	Sterigmatocystin; 13-hydroxyversicolorin B [91]
Matricaria chamomilla (stem, leaf)	Anti-inflammatory, analgesic, antimicrobial, antispasmodic, sedative [92]	*Epicoccum, Didymella, Phoma*	Against lymphoblastic leukemia	L-asparaginase [56]
Mikania glomerata (leaf)	Anti-inflammatory, antispasmodic, anti-hemorrhagic, antiophidic, antiviral, antimicrobial [93]	*Diaporthe*	Antifungal against *Fusarium solani* and *Didymella bryoniae*; antimicrobial against *Staphylococcus aureus*	Not investigated [94]
Mikania laevigata (leaf)	Anti-inflammatory, antispasmodic, antihemorrhagic, antiophidic, antiviral, antimicrobial [93]	*Hypoxylon*	Not investigated	Not investigated [50]

Table 1. *Cont.*

Asteraceae Species and the Tissue of Endophytes Isolation	Main Therapeutic acTivities of the Host	Dominating Endophyte Genera	Main Activities of the Endophyte	Main Metabolites/Enzymes Linked to Endophyte Bioactivities
Notobasis syriaca	Antioxidant, antimicrobial [95]	*Phomopsis*	Antimicrobial against *Legionella pneumophila* and *Escherichia coli*	Phomosine K; 2-hydroxymethyl-4β,5α,6β-trihydroxycyclohex-2-en, (-)-phyllostine; (+)-epiepoxydon; (+)-epoxydon monoacetate [96]
Smallanthus sonchifolius (root, stem, leaf)	Antidiabetic, nutritious, fertility-enhancing, antioxidant, antimicrobial [97]	*Curvularia*	Antiparasitic against *Trypanosoma cruzi*	Stemphyperylenol [98]
		Papulaspora	Cytotoxic against melanoma, colon, glioblastoma, and promyelocytic leukemia cell lines; antimicrobial	(24R)-stigmast-4-en-3-one; (22E,24R)-ergosta-4,6,8(14),22-tetraen-3-one; (22E,24R)-8,14-epoxyergosta-4,22-diene-3,6-dione [99]
		Alternaria, Nigrospora, Phoma, Fusarium, Papulaspora	Antifungal	Stemphyperylenol, alterperylenol, altertoxin I, alternariol, alternariol monomethyl ether [100]
		Coniochaeta	Antifungal against *Colletotrichum acutatum, C. fragariae* and *C. gloeosporioides*	Fatty acids: caproic, caprylic, cis-10-pentadecenoic, heptadecanoic, lauric, linoleic, myristic, oleic, palmitic, palmitoleic, pentadecanoic, stearic [101]
Silybum marianum	Antidiabetic, hepatoprotective, hypocholesterolemic, antihypertensive, anti-inflammatory, antitumor, antioxidant [102]	*Aspergillus*	Hepatoprotective	Silybin A, silybin B, isosilybin A [103]
Tithonia diversifolia	Anti-inflammatory, antimalarial, cytotoxic, gastroprotective, antimicrobial, antihyperglycemic [104]	*Colletotrichum*	Cytotoxic against the Jurkat tumor cell line	Nectriapyrone, tyrosol [105]
		Phoma	Cytotoxic	Anthraquinones [106]
Trixis vauthieri (leaf)	Antiparasitic [107]	*Alternaria*	Trypanocidal compound with inhibitory activity of trypanothione reductase	Altenusin [108]
Urospermum picroides (flower)	Anti-inflammatory, immunomodulatory, antioxidant, antimicrobial [109]	*Ampelomyces*	Cytotoxic against L5178Y cells; antibacterial against *Staphylococcus aureus, S. epidermidis* and *Enterococcus faecalis*	3-O-methylalaternin, altersolanol A [110]

Table 1. *Cont.*

Asteraceae Species and the Tissue of Endophytes Isolation	Main Therapeutic acTivities of the Host	Dominating Endophyte Genera	Main Activities of the Endophyte	Main Metabolites/Enzymes Linked to Endophyte Bioactivities
Viguiera arenaria (synonym of *Aldama arenaria*)	Antiparasitic, analgesic, anti-inflammatory, antitumor, antimicrobial [111]	*Phomopsis*	Antiparasitic against *T. cruzi*	3,4-dimethyl-2-(4′-hydroxy-3′,5′-dimethoxyphenyl)-5-methoxy-tetrahydrofuran [112]
		Colletotrichum	Cytotoxic against leukemia tumor cells [106]	Nectriapyrone, tyrosol [105]
Viguiera robusta (synonym of *Aldama robusta*)	Anti-inflammatory, analgesic, antitumor, antiparasitic, antimicrobial [113]	*Chaetomium*	Cytotoxic against the Jurkat (leukemia) and B16F10 (melanoma) tumor cells; antibacterial against *S. aureus* and *E. coli*	Chaetoglobosin B [114]

* references in last column are linked to endophyte genera, activities, and metabolites.

4. Fungal Endophytes Associated with Asteraceae—Biochemistry

4.1. Plant Growth Promoting Secondary and Anti-Stress Metabolites

Asteraceae are leading examples of the synergistic effect of fungal endophytes in improving biotic and abiotic stress resistance and promoting plant growth because numerous species of this family possess extraordinary tolerance and competition skills. For example, Khan et al. [115] determined the growth-promoting ability of endophytic *Penicillium citrinum* in helping its plant host *Ixeris repens* in rapid colonization of the sand dunes. *P. citrinum* stimulated competition skills of the host plant through the production of secondary metabolites promoting plant growth, like gibberellins, and protective compounds, like mycotoxins, citrinin, and cellulose digesting enzymes [115]. *P. citrinum* and *Aspergillus terreus* were found to stimulate *H. annuus* growth and improve disease resistance due to the higher content of plant-defense hormones, salicylic, and jasmonic acids. The mentioned endophytes regulated oxidative stress of the host plant through activation of glutathione and polyphenol oxidases, alteration of catalase and peroxidase, as well as secretion of organic acids [88]. The individual or co-inoculation of endophytes increased amino acid content in sunflower (*H. annuus*) diseased leaves, delaying cell death, and consequently disturbing pathogen progression in plant tissues [88]. Ren et al. [75] showed that endophyte *Gilmaniella* sp. induced jasmonic acid production, which was recognized to be a signal compound promoting the accumulation of volatile oils in the Chinese medicinal plant *Atractylodes lancea*. The jasmonic acid acted as a downstream signal of nitric oxide and hydrogen peroxide-mediated production of volatile oil in the host. Various strains of *Penicillium* and *Aspergillus* species associated with Asteraceae were reported for gibberellins production [116]. *Penicillium* strains, especially MH7, produced nine gibberellins which significantly increased the growth and development of the host plant crown daisy (*Chrysanthemum coronarium*, synonym of *Glebionis coronaria*) [117]. The reactive oxygen species (ROS) production together with increased siderophore excretion by endophytes contributed towards improved growth and resistance against sunflower pathogens. Endophyte-origin ROS in plant roots are tackled by internal physiological plant apparatus resulting in an acute resistance against present and future stresses [89]. Huang et al. [58] compared the antioxidant capacity of plants used in Chinese traditional medicine, including mugworts: *Artemisia capillaris*, *A. indica*, and *A. lactiflora* (Asteraceae) and their endophytes. A fungal endophyte strain isolated from the flower of *A. capillaris* showed the strongest total antioxidant capacity. The antioxidant compounds detected in the highest amounts in both endophytic fungus and its host *A. indica* were chlorogenic and di-O-caffeoylquinic acids, and the volatile compound artemisinin. Both chlorogenic acid and artemisinin acted as antioxidant, antimutagenic, immunomodulatory, and antiviral. The production of the same bioactive natural compounds, as well as some of those found in *A. indica* and its fungal endophytes, was suggested. In general, phenolic compounds, including phenolic acids, flavonoids, tannin constituents, hydroxyanthraquinones, and phenolic terpenoids as well as volatile or aliphatic constituents were major substances in the fungal endophyte cultures and host plant extracts responsible for high antioxidant activity of all investigated Chinese medicinal plants [58]. In terms of abiotic and biotic stress, fungal endophytes conferred resistance against drought, salinity, heat stress, and enhanced resistance against pathogens and insects. The different mechanisms can stay behind the competitive success of invasive Asteraceae species like crofton weed (*A. adenophora*). The most abundant endophytic fungus isolated from this species was *Colletotrichum* sp. which has pathogenic effects on other plants. Spreading *Colletotrichum* spores could be a competitive advantage for *A. adenophora* as it was hypothesized by Fang et al. [62]. The recognition of endophyte roles in host plant expansion and competition mechanisms enables the application or modification of cultivation techniques dedicated to particular medicinal Asteraceae species, especially those with promising therapeutic and economical potential.

4.2. Antibacterial Secondary Metabolites

The best criterion for host plant selection in order to investigate the endophytes with potential antimicrobial activity is the plant traditionally used for the treatment of infections [118]. Plant-associated fungi may interact using, inter alia, antibiotic molecules, so the production of antibiotics and the parallel development of antibiotic-resistance mechanisms can spread in dynamic microbiota/plant systems by bacterial mobilization and horizontal gene transfer [119,120]. In recent years, the number of multidrug-resistant microorganisms have been a growing concern for public health worldwide. The key determinants of bacteria drug resistance are inactivation of the antibiotics, changes in bacterial targets, and restricted entry of antibiotics by less permeable drug transporters [121]. Asteraceae/fungal endophytes consortia could be a source of active compounds targeted against many drug-resistant microorganisms [122,123]. A fungus *Colletotrichum* sp. was isolated from the stems of *Artemisia annua* and characterized as a source of ergosterol derivatives (Figure 3), with inhibitory potential against both Gram-negative and -positive bacteria, such as *Pseudomonas* sp. and *Bacillus subtilis* with minimal inhibitory concentrations (MICs) ranging from 25 to 75 g mL^{-1} [70]. *Colletotrichum* sp. can also produce plant hormones such as indole-3-acetic acid (IAA), up-regulating host growth. Both mechanisms of action, namely antibiosis and growth promotion, can enhance adaptability and pathogen resistance of a host plant. At the same time, Zou et al. [44] isolated from the stem of *Artemisia mongolica* an endophytic fungus *Colletotrichum gloeosporioides*, synthesizing colletotric acid with antibacterial activity against *B. subtilis*, *Staphylococcus aureus*, *Sarcina lutea*, and *Pseudomonas* sp. with MICs of 25, 50, and 50 µg mL^{-1}, respectively, and inhibited a pathogenic fungus *Helminthosporium sativum* (current name *Bipolaris sorokiniana*) with a MIC of 50 µg mL^{-1}. This was the first report of *C. gloeosporioides* as a fungal endophyte in the Asteraceae, although it was previously mentioned as an endophyte of plants belonging to the other families. The isocoumarins and naphthalene derivatives produced by *Papulaspora immersa*, a fungal endophyte isolated from the Andean tuber crop, the yacon (*S. sonchifolius*), presented antimicrobial activities and could act synergistically [99]. Interestingly, some fungal metabolites were identified as constituents of an extract derived from a healthy Asteraceae, prickly goldenfleece (*Urospermum picroides*), indicating that the production of bactericides by the fungal endophyte *Ampelomyces* sp., proceeds also in situ within the host plant [110]. Among seven phomosine derivatives isolated from *Phomopsis* sp., an endophyte of the Syrian thistle (*Notobasis syriaca*), phomosine K had strong antibacterial activity against *Legionella pneumophila* Corby, *Escherichia coli* K12 with MIC 25 and 100 µg mL^{-1}, respectively [96]. Endophyte colonization offers protection from various stressors, such as toxins which affect plant pathogens by disrupting the cellular membrane and inducing cell death. Such ecological relationships were recorded for the mentioned Asteraceae/endophyte systems.

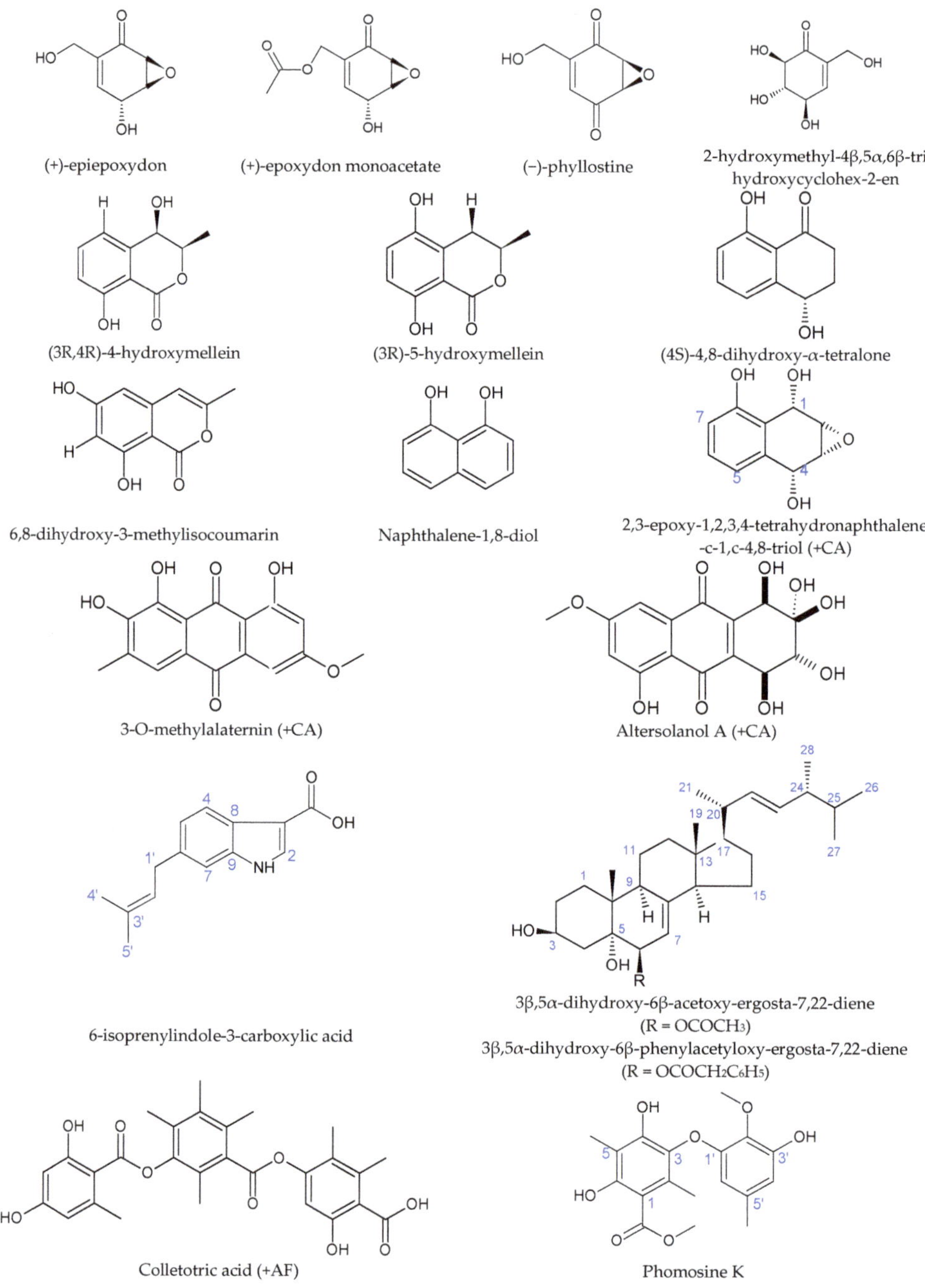

Figure 3. The molecular structure of chosen specific compounds with antibacterial activity synthesized by fungal endophytes associated with Asteraceae species [44,70,96,99,110]; +AF—antifungal activity; +CA—cytotoxic activity.

4.3. Antifungal Secondary Metabolites

Colonization of the host plant by endophytes and pathogens depends on their adaptations to the host environment but also the innate host defense mechanism and variation in virulence. A few reports on endophytic fungi, protecting against other fungal infection, found in association with Asteraceae species, especially *Viguiera* spp. (syn. *Aldama* spp.) were published, and several new compounds were described but their biological action needs future research [17,22,68,75,78,101]. *Ampelomyces* spp. were widely studied as the first fungi used as biocontrol agents of powdery mildews [122]. Chagas et al. [100] investigated the interactions between the fungal endophytes that cohabit *S. sonchifolius*. They found that *Alternaria tenuissima* synthesized some polyketides, including antifungal stemphyperylenol in the presence of endophytic *Nigrospora sphaerica* (Figure 4). *A. tenuissima* is characterized by a slower growth rate than *N. sphaerica*, so specific antifungal compounds might control the growth rate of *N. sphaerica* during host plant colonization, without any damage to the host plant tissues. The competition of fungal endophytes colonizing the same host plant stimulates the production of metabolites that could decrease the growth of particular fungi species without damaging the host plant and maintaining the symbiosis [100]. A closer metabolome relationship was found for *S. sonchifolius* and endophytic fungus *Coniochaeta ligniaria*. Both symbionts produced the same antifungal fatty acids: caproic, caprylic, and palmitic acids at high concentrations which might raise the resistance of *S. sonchifolius* to fungal pathogenic attacks and *C. ligniaria* to fungi competing within the host tissues [101]. *B. trimera* is a native medicinal plant of the Brazilian savannah. Vieira et al. [53] isolated from the leaves of this species 23 fungal taxa, inter alia, *Epicoccum* sp., *Pestalotiopsis* sp., *Cochliobolus lunatus*, and *Nigrospora* sp., which showed antifungal activity against *Paracoccidioides brasiliensis*. Additionally, the fungi isolated from different host plants displayed distinct antimicrobial activities, so the endophytic richness and the antimicrobial activity were closely correlated. The endophyte fungus *Preussia* sp. revealed strong antifungal activity, related to the synthesis of anthraquinones, auranticins, culpin, cycloartane triterpenes diphenyl ether, spirobisnaphthalenes, and thiopyranchromenones [53,124]. However, metabolome analysis of *Preussia* sp. isolated from Asteraceae herb carqueja (*B. trimera*) confirmed antioxidant but not antifungal activity of isolated compounds, namely preussidone, 1′,5-dimethoxy-3,5′-dimethyl-2,3′-oxybiphenyl-1,2′-diol, 5-methoxy-3,5′-dimethyl-2,3′-oxybiphenyl-1,1′,2′-triol, and cyperin [124]. Waqas et al. [88,89] determined the inhibitory effect of two fungal endophytes, *P. citrinum* and *A. terreus*, against *Sclerotium rolfsii*, a soilborne plant pathogen which causes root rot, stem rot, collar rot, wilt, and foot rot diseases in *H. annuus*. The antifungal activity of *Penicillium* and *Aspergillus* strains was linked with synthesis of gibberelins, organic acids, and siderophores. Two new fatty acid amides, bipolamides A and B, were isolated from endophytic fungus *Bipolaris* sp., but only bipoliamide B revealed bioactivity against *Cladosporium cladosporioides*, *C. cucumerinum*, *Saccharomyces cerevisiae*, *Aspergillus niger*, and *Rhisopus oryzae* [85]. Fungal endophytes possess multiple balanced antagonisms, namely with the other microbial inhabitants of the host plant and with the host plant itself, to support the growth conditions enabling reproduction. Most genes involved in secondary metabolite synthesis in fungi are activated while being co-cultured in plant and/or with other microbes, but they are generally silent in cultures, confirming that multiple antagonisms are involved in endophytism [22]. Three strains of endophytic fungus *Diaporthe citri* isolated from Brazilian medicinal vine, guaco (*Mikania glomerata*) presented 60% inhibition index of mycelia growth against *Fusarium solani* and 66% against *Didymella bryoniae* [94]. The mechanisms of inhibition were not tested in the cited reference, but the authors stated that endophytic microorganisms with the highest inhibition indices were considered candidates for tests involving the production of secondary metabolites with potential antimicrobial activity.

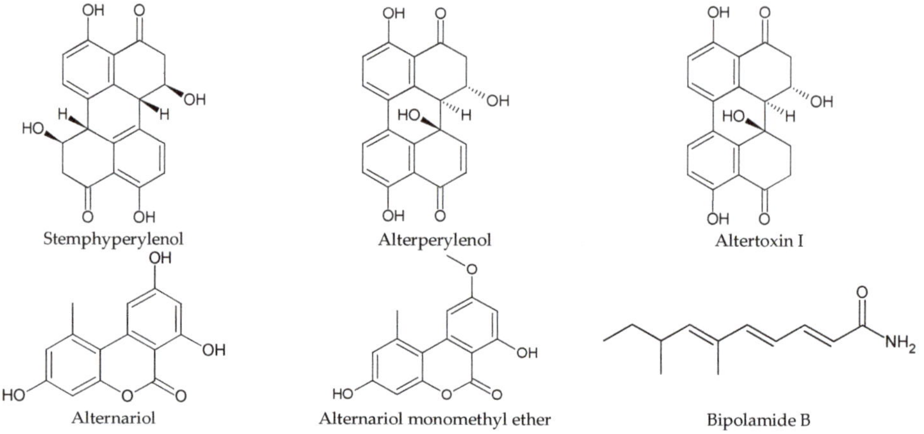

Figure 4. The molecular structure of chosen specific compounds with antifungal activity synthesized by fungal endophytes associated with Asteraceae species [85,100].

4.4. Antiparasitic Secondary Metabolites

Cota et al. [108] isolated altenusin from an *Alternaria* sp. endophytic in *Trixis vauthieri* collected in Brazil (Figure 5). This medicinal plant was reported as containing trypanocidal compounds of trypanothione reductase inhibitory activity. Meanwhile, the organic extract of the culture of *Alternaria* sp. inhibited trypanothione reductase by 99%, when tested at 20 mg mL^{-1}. The mentioned report was the first one concerning fungal metabolites with trypanothione reductase inhibitory activity, which can be used for the development of new chemotherapeutic agents to treat trypanosomiasis and leishmaniasis. *Trypanosoma cruzi* is a parasitic euglenoid causing Chagas disease in humans, and *Leishmania tarentolae* is a protozoan parasite of geckos, which might also be capable of infecting mammals [125]. Verza et al. [112] determined that endophytic fungus *Phomopsis* sp., obtained from *Viguiera arenaria* (synonym of *Aldama arenaria*), led to the formation of a new compound able to transform the tetrahydrofuran lignan, (−)-grandisin to 3,4-dimethyl-2-(4′-hydroxy-3′,5′-dimethoxy phenyl)-5-methoxy-tetrahydrofuran, which also showed trypanocidal activity against *T. cruzi*. Guimarães et al. [105] isolated 30 endophytic fungi from the leaves and four from the roots of *V. arenaria* and five endophytes were isolated from the leaves of *Tithonia diversifolia*, collected in Brazil. The ethyl acetate extract of the *Diaporthe phaseolorum* isolate's fermentation broth showed strong inhibition of glyceraldehyde 3-phosphate dehydrogenase of *T. cruzi* and adenine phosphoribosyltransferase of *L. tarentolae*. The mosquito *Culex quinquefasciatus* acts as a vector of *Wuchereria bancrofti* which causes the disease lymphatic filariasis, commonly known as elephantiasis. Belonging to the Asteraceae family, *Ageratum conyzoides*, native to Pakistan, has antilarvicidal effects against the mosquito larvae of *C. quinquefasciatus*, *Aedes aegypti*, and *Anopheles stephensi*. Endophytic actinomycetes, *Streptomyces* spp., isolated from mentioned Asteraceae species showed strong larvicidal activity at the fourth instar stage [126]. Xanthones, sterigmatocystin, and anthraquinone derivative, 13-hydroxyversicolorin B from the culture broth of the endophytic fungus *Podospora* sp., isolated from the Kenyan medicinal plant *Laggera alata*, might be used as natural mosquito larvicides [91]. The easily biodegradable endophyte metabolites could be a base for the development of modern techniques providing efficient insect control, without negative effects on the non-target population and environment.

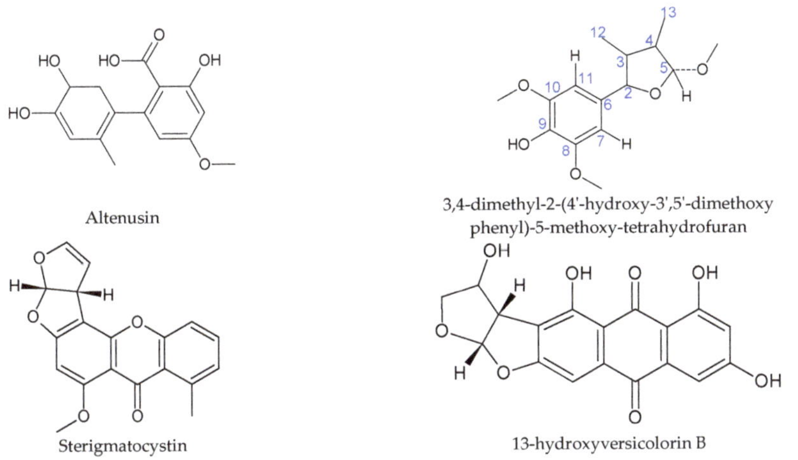

Altenusin

3,4-dimethyl-2-(4'-hydroxy-3',5'-dimethoxy
phenyl)-5-methoxy-tetrahydrofuran

Sterigmatocystin

13-hydroxyversicolorin B

Figure 5. The molecular structure of chosen specific compounds with antiparasitic activity synthesized by fungal endophytes associated with Asteraceae species [91,108,112].

4.5. Cytotoxic Secondary Metabolites

The major difficulty in the treatment of cancer is the increase in drug resistance of commonly used chemotherapeutic agents, so the crucial task is to find out the novel compounds with high efficacy and low toxicity. The research and registration of new antitumor drugs are mostly based on the compounds extracted from medicinal plants including those of endophyte origin [127]. Martinez–Klimova et al. [5] found the endophytes that produce antibiotic metabolites belonging to phylum Ascomycota, which were isolated from the Asteraceae, Fabaceae, Lamiaceae, and Araceae families. The therapeutic activity of fungal endophytes was related to the production of compounds inhibiting the drug transporters of tumor cells. Moreover, the use of secondary metabolites produced by endophytes could mediate drug resistance reversal in cancer cells. A few reports are pointing out the use of endophytes isolated from Asteraceae species as a source of antitumor compounds targeted in the most common lines of cancer cells (Figure 6). Nectriapyrone, produced by the endophytic fungus *Glomerella cingulata*, a teleomorph stage of *C. gloeosporioides*, isolated from *V. arenaria* and *T. diversifolia* showed relevant cytotoxic activity towards tumor cells [105]. In the case of *Chaetomium globosum*, a fungal endophyte associated with *Viguiera robusta*, chaetoglobosins showed inhibition of Jurkat (leukemia) and B16F10 (melanoma) tumor cells with 89.55% and 57.1% inhibition at 0.1 mg mL^{-1}, respectively [114]. Gallo et al. [99] isolated a fungus *P. immersa* from roots and leaves of *S. sonchifolius*. *P. immersa* extracts displayed strong cytotoxicity due to newly described secondary metabolites, i.e., 2,3-epoxy-1,2,3,4-tetrahydronaphthalene-c-1,c-4,8-triol, which showed highest activity against the human tumor cell lines MDA-MB435 (melanoma), HCT-8 (colon), SF295 (glioblastoma), and HL-60 (promyelocytic leukemia), with he half maximal inhibitory concentration (IC50) values of 3.3, 14.7, 5, and 1.6 mm, respectively. Moreover, sitostenone and tyrosol, other *P. immersa* secondary metabolites, showed anticancer effects when applied with isocoumarin [99]. The fungal endophytes of Asteraceae, especially the members of genera *Fusarium*, *Plectosphaerella*, *Stemphylium*, *Septoria*, *Alternaria*, *Didymella*, *Phoma*, *Chaetosphaeronema*, *Sarocladium*, *Nemania*, *Epicoccum*, and *Cladosporium* can produce the anticancer enzyme L-asparaginase used in the treatment of acute lymphoblastic leukemia. The isolates of fungi *Fusarium proliferatum* and *Plectosphaerella tracheiphilus*, obtained from an Asteraceae host *C. segetalis*, exhibited a maximum enzyme activity with 0.492 and 0.481 unit mL^{-1}, respectively [56]. The milk thistle (*Silybum marianum*) is known as a source of silymarin, a mixture of flavonolignans used in cancer chemoprevention and hepatoprotection. El-Elimat et al. [103] showed that a fungal endophyte, *Aspergillus iizukae* (current name *Fennellia flavipes*), isolated from leaves of *S. marianum* can synthesize similar compounds as a host plant, namely silybin A, silybin

B, and isosilybin, the constituent compounds of silymarin. Endophytic fungi that can produce the same compounds of their associated host plants could be a sustainable and alternative source for secondary metabolites.

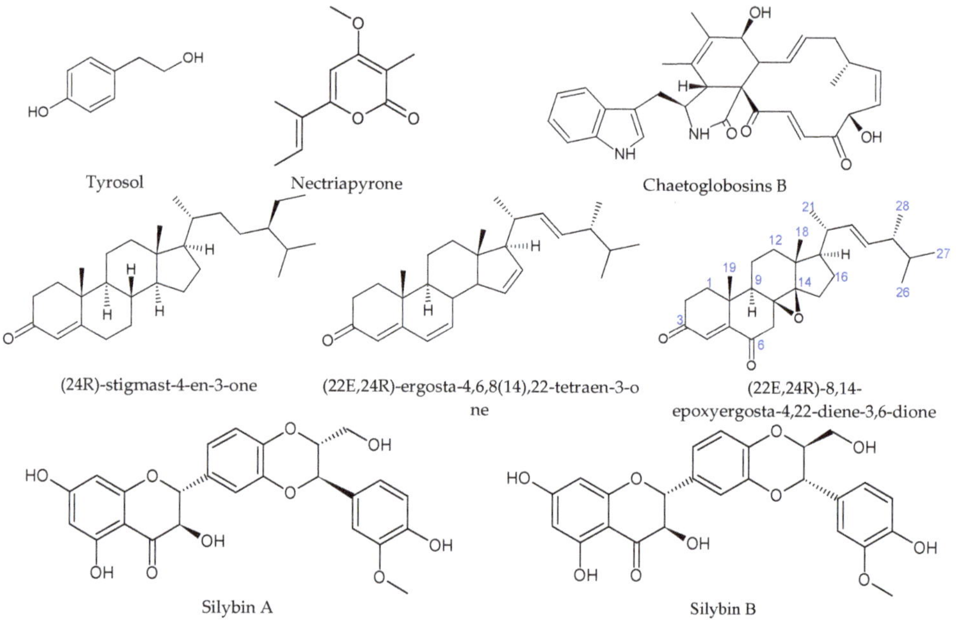

Figure 6. The molecular structure of chosen specific compounds with cytotoxic activity synthesized by fungal endophytes associated with Asteraceae species [99,103,105,110,114].

5. Review Methodology

The leading scientific databases dedicated to multidisciplinary as well as agricultural, biological, biomedical, and pharmacological sciences were screened. Relevant literature dated to the period 2000–2020 was collected, analyzed, and selected considering (i) the reports on endophyte isolation from the species of Asteaceae family, (ii) the reports on therapeutic utilization of the host plant or/and an endophyte, (iii) the reports on in vitro and in vivo bioactivity of chemical compounds produced by a host plant or/and an endophyte. Plant names were verified according to the Global Biodiversity Information Facility [128] and The Plant List [34], endophyte taxa were verified according to MycoBank database [129]. For clarity, the validated endophyte names used in the referenced literature were implemented in the text. In the tables and figures, the current taxa classification and nomenclature were used. Chemical structures were elaborated on the basis of referred publications, for new isolated compounds the number of C atoms was presented.

6. Conclusions

A growing spectrum of literature indicates that endophyte fungi colonizing different species of Asteraceae are responsible to some degree for their therapeutic potential reported in ethnobotanical and modern literature. Endophyte fungi are elements of a complex web of interactions of the plant host/endophyte/phytopathogen, and hence all elements of this system are expected to produce bioactive compounds that can improve their ability to survive in such a dynamic environment. Endophytes were involved in the superior adaptability and competitiveness reported for Asteraceae hosts and their evolutionary success. Plant/endophyte interactions regulated the energy costly process of production of secondary metabolites possessing therapeutic properties. In the case of the Asteraceae species analyzed, the host tissue's environment was more crucial than plant taxonomy for shaping the diversity and metabolite profile of fungal endophytes. Most endophyte fungi isolated from Asteraceae plants were wide-spreading. Despite that, they produced very specific secondary metabolites in planta and in vitro. The interactions between the endophyte and its host controlled by specific chemical compounds are dynamic and difficult to analyze but crucial for the composition of the medicinal plant extracts and their standardization.

Author Contributions: Conceptualization, A.S., and G.C.; writing—original draft preparation, A.S., and G.C.; writing—review and editing, M.T.A., A.K.; visualization, A.S.; supervision, A.S., and G.C. All authors have read and agreed to the published version of the manuscript.

Funding: This study was partially supported by the Ministry of Science and Higher Education of the Republic of Poland.

Conflicts of Interest: The authors declare no conflict of interest.

References

1. Aswani, R.; Vipina Vinod, T.; Ashitha, J. Benefits of plant–endophyte interaction for sustainable agriculture. In *Microbial Endophytes: Functional Biology and Applications*; Kumar, A., Radhakrishnan, E., Eds.; Elsevier INC.: Amsterdam, The Netherlands, 2020; pp. 35–55.

2. Mani, V.M.; Soundari, A.P.G.; Karthiyaini, D.; Preethi, K. Bioprospecting endophytic fungi and their metabolites from medicinal tree Aegle marmelos in Western Ghats, India. *Mycobiology* **2015**, *43*, 303–310. [CrossRef] [PubMed]

3. Wilson, D. Endophyte: The evolution of a term, and clarification of its use and definition. *Oikos* **1995**, *73*, 274–276. [CrossRef]

4. Golinska, P.; Wypij, M.; Agarkar, G.; Rathod, D.; Dahm, H.; Rai, M. Endophytic actinobacteria of medicinal plants: Diversity and bioactivity. *Antonie Van Leeuwenhoek* **2015**, *108*, 267–289. [CrossRef] [PubMed]

5. Martinez–Klimova, E.; Rodríguez–Peña, K.; Sánchez, S. Endophytes as sources of antibiotics. *Biochem. Pharm.* **2017**, *134*, 1–17. [CrossRef] [PubMed]

6. Santoyo, G.; Moreno–Hagelsieb, G.; del Carmen Orozco–Mosqueda, M.; Glick, B.R. Plant growth–promoting bacterial endophytes. *Microbiol. Res.* **2016**, *183*, 92–99. [CrossRef]

7. Alvin, A.; Miller, K.I.; Neilan, B.A. Exploring the potential of endophytes from medicinal plants as sources of antimycobacterial compounds. *Microbiol. Res.* **2014**, *169*, 483–495. [CrossRef]

8. Madhurama, G.; Nagma, S.; Preeti, S. Indole–3–acetic acid production by *Streptomyces* sp. isolated from rhizospheric soils of medicinal plants. *J. Pure Appl. Microbiol.* **2014**, *8*, 965–971.

9. Aly, A.H.; Debbab, A.; Kjer, J.; Proksch, P. Fungal endophytes from higher plants: A prolific source of phytochemicals and other bioactive natural products. *Fungal Diver.* **2010**, *41*, 1–16. [CrossRef]

10. Nicoletti, R.; Ferranti, P.; Caira, S.; Misso, G.; Castellano, M.; Di Lorenzo, G.; Caraglia, M. Myrtucommulone production by a strain of *Neofusicoccum australe* endophytic in myrtle (*Myrtus communis*). *World J. Microbiol. Biotechnol.* **2014**, *30*, 1047–1052. [CrossRef]

11. Nicoletti, R.; Fiorentino, A. Plant bioactive metabolites and drugs produced by endophytic fungi of *Spermatophyta*. *Agriculture* **2015**, *5*, 918–970. [CrossRef]

12. Zhao, J.; Shan, T.; Mou, Y.; Zhou, L. Plant–derived bioactive compounds produced by endophytic fungi. *Mini-Rev. Med. Chem.* **2011**, *11*, 159–168. [CrossRef] [PubMed]

13. Kusari, S.; Lamshöft, M.; Spiteller, M. *Aspergillus fumigatus* Fresenius, an endophytic fungus from *Juniperus communis* L. Horstmann as a novel source of the anticancer pro-drug deoxypodophyllotoxin. *J. Appl. Microbiol.* **2009**, *107*, 1019–1030. [CrossRef]

14. Guimarães, D.O.; Lopes, N.P.; Pupo, M.T. Meroterpenes isolated from the endophytic fungus *Guignardia mangiferae*. *Phytochem. Lett.* **2012**, *5*, 519–523. [CrossRef]

15. Nicoletti, R.; Ciavatta, M.L.; Buommino, E.; Tufano, M.A. Antitumor extrolites produced by *Penicillium* species. *Int. J. Biomed. Pharm. Sci.* **2008**, *2*, 1–23.

16. Stammati, A.; Nicoletti, R.; De Stefano, S.; Zampaglioni, F.; Zucco, F. Cytostatic properties of a novel compound derived from *Penicillium pinophilum*: An *in vitro* study. *ALTA Alter. Lab. Anim.* **2002**, *30*, 69–75. [CrossRef] [PubMed]

17. Rosenblueth, M.; Martínez–Romero, E. Bacterial endophytes and their interactions with hosts. *Mol. Plant–Microbe Interact.* **2006**, *19*, 827–837. [CrossRef] [PubMed]

18. Tsitsigiannis, D.I.; Keller, N.P. Oxylipins as developmental and host–fungal communication signals. *Trends Microbiol.* **2007**, *15*, 109–118. [CrossRef]

19. Yang, Z.; Rogers, L.M.; Song, Y.; Guo, W.; Kolattukudy, P. Homoserine and asparagine are host signals that trigger in planta expression of a pathogenesis gene in *Nectria haematococca*. *Proc. Natl. Acad. Sci. USA* **2005**, *102*, 4197–4202. [CrossRef]

20. Young, C.A.; Felitti, S.; Shields, K.; Spangenberg, G.; Johnson, R.D.; Bryan, G.T.; Saikia, S.; Scott, B. A complex gene cluster for indole–diterpene biosynthesis in the grass endophyte *Neotyphodium lolii*. *Fungal Gen. Biol.* **2006**, *43*, 679–693. [CrossRef]

21. Schroeckh, V.; Scherlach, K.; Nützmann, H.-W.; Shelest, E.; Schmidt-Heck, W.; Schuemann, J.; Martin, K.; Hertweck, C.; Brakhage, A.A. Intimate bacterial–fungal interaction triggers biosynthesis of archetypal polyketides in *Aspergillus nidulans*. *Proc. Natl. Acad. Sci. USA* **2009**, *106*, 14558–14563. [CrossRef]

22. Schulz, B.; Haas, S.; Junker, C.; Andrée, N.; Schobert, M. Fungal endophytes are involved in multiple balanced antagonisms. *Curr. Sci.* **2015**, *109*, 39–45.

23. Mazur, S.; Nadziakiewicz, M.; Kurzawińska, H.; Nawrocki, J. Effectiveness of mycorrhizal fungi in the protection of juniper, rose, yew and highbush blueberry against *Alternaria alternata*. *Folia Hort.* **2019**, *31*, 117–127. [CrossRef]

24. Ma, C.; Jiang, D.; Wei, X. Mutation breeding of *Emericella foeniculicola* TR21 for improved production of tanshinone IIA. *Process Biochem.* **2011**, *46*, 2059–2063. [CrossRef]

25. Zhao, K.; Sun, L.; Ma, X.; Li, X.; Wang, X.; Ping, W.; Zhou, D. Improved taxol production in *Nodulisporium sylviforme* derived from inactivated protoplast fusion. *Afr. J. Biotechnol.* **2013**, *10*, 4175–4182.

26. Karuppusamy, S. A review on trends in production of secondary metabolites from higher plants by in vitro tissue, organ and cell cultures. *J. Med. Plants Res.* **2009**, *3*, 1222–1239.

27. Kusari, S.; Singh, S.; Jayabaskaran, C. Rethinking production of Taxol (paclitaxel) using endophyte biotechnology. *Trends Biotechnol.* **2014**, *32*, 304–311. [CrossRef]

28. Yang, H.; Wang, Y.; Zhang, Z.; Yan, R.; Zhu, D. Whole-genome shotgun assembly and analysis of the genome of *Shiraia* sp. strain Slf14, a novel endophytic fungus producing huperzine A and hypocrellin A. *Genome Announc.* **2014**, *2*, e00011–e00014. [CrossRef]

29. Venugopalan, A.; Srivastava, S. Endophytes as *in vitro* production platforms of high value plant secondary metabolites. *Biotechnol. Adv.* **2015**, *33*, 873–887. [CrossRef] [PubMed]

30. Chutulo, E.C.; Chalannavar, R.K. Endophytic mycoflora and their bioactive compounds from *Azadirachta indica*: A comprehensive review. *J. Fungi* **2018**, *4*, 42. [CrossRef] [PubMed]

31. Sharma, D.; Pramanik, A.; Agrawal, P.K. Evaluation of bioactive secondary metabolites from endophytic fungus *Pestalotiopsis neglecta* BAB-5510 isolated from leaves of *Cupressus torulosa* D. Don. *3 Biotech* **2016**, *6*, 210. [CrossRef]

32. Yu, H.; Zhang, L.; Li, L.; Zheng, C.; Guo, L.; Li, W.; Sun, P.; Qin, L. Recent developments and future prospects of antimicrobial metabolites produced by endophytes. *Microbiol. Res.* **2010**, *165*, 437–449. [CrossRef]

33. Karsch-Völk, M.; Barrett, B.; Kiefer, D.; Bauer, R.; Ardjomand-Woelkart, K.; Linde, K. *Echinacea* for preventing and treating the common cold. *Cochrane Database Syst. Rev.* **2014**, *2*, CD000530. [CrossRef] [PubMed]

34. The Plan List. Available online: www.theplantlist.org (accessed on 12 June 2020).

35. Bessada, S.M.; Barreira, J.C.; Oliveira, M.B.P. Asteraceae species with most prominent bioactivity and their potential applications: A review. *Ind. Crop. Prod.* **2015**, *76*, 604–615. [CrossRef]

36. Funk, V.A.; Bayer, R.J.; Keeley, S.; Chan, R.; Watson, L.; Gemeinholzer, B.; Schilling, E.; Panero, J.L.; Baldwin, B.G.; Garcia-Jacas, N. Everywhere but Antarctica: Using a supertree to understand the diversity and distribution of the Compositae. *Biol. Skr.* **2005**, *55*, 343–374.

37. Petkova, N.T.; Ivanov, I.; Raeva, M.; Topuzova, M.; Todorova, M.; Denev, P. Fructans and antioxidants in leaves of culinary herbs from Asteraceae and Amaryllidaceae families. *Food Res.* **2019**, *3*, 407–415. [CrossRef]

38. Martucci, M.E.P.; De Vos, R.C.; Carollo, C.A.; Gobbo-Neto, L. Metabolomics as a potential chemotaxonomical tool: Application in the genus *Vernonia* Schreb. *PLoS ONE* **2014**, *9*, e93149. [CrossRef] [PubMed]

39. Lusa, M.G.; Martucci, M.E.; Loeuille, B.F.; Gobbo-Neto, L.; Appezzato-da-Glória, B.; Da Costa, F.B. Characterization and evolution of secondary metabolites in Brazilian Vernonieae (Asteraceae) assessed by LC–MS fingerprinting. *Bot. J. Linn. Soc.* **2016**, *182*, 594–611. [CrossRef]

40. Gallon, M.E.; Jaiyesimi, O.A.; Gobbo-Neto, L. LC–UV–HRMS dereplication of secondary metabolites from Brazilian Vernonieae (Asteraceae) species supported through in–house database. *Biochem. Syst. Ecol.* **2018**, *78*, 5–16. [CrossRef]

41. Mandel, J.R.; Barker, M.S.; Bayer, R.J.; Dikow, R.B.; Gao, T.G.; Jones, K.E.; Keeley, S.; Kilian, N.; Ma, H.; Siniscalchi, C.M. The Compositae tree of life in the age of phylogenomics. *J. Syst. Evol.* **2017**, *55*, 405–410. [CrossRef]

42. Katinas, L.; Funk, V.A. An updated classification of the basal grade of Asteraceae (=Compositae): From Cabrera's 1977 tribe Mutisieae to the present. *N. Z. J. Bot.* **2020**, *58*, 67–93. [CrossRef]

43. Morales, P.; Ferreira, I.C.; Carvalho, A.M.; Sánchez-Mata, M.C.; Cámara, M.; Fernández-Ruiz, V.; Pardo-de-Santayana, M.; Tardío, J. Mediterranean non–cultivated vegetables as dietary sources of compounds with antioxidant and biological activity. *LWT–Food Sci. Technol.* **2014**, *55*, 389–396. [CrossRef]

44. Zou, W.; Meng, J.; Lu, H.; Chen, G.; Shi, G.; Zhang, T.; Tan, R. Metabolites of Colletotrichum gloeosporioides, an endophytic fungus in *Artemisia mongolica*. *J. Nat. Prod.* **2000**, *63*, 1529–1530. [CrossRef] [PubMed]

45. Barbieri, R.; Coppo, E.; Marchese, A.; Daglia, M.; Sobarzo-Sánchez, E.; Nabavi, S.F.; Nabavi, S.M. Phytochemicals for human disease: An update on plant–derived compounds antibacterial activity. *Microbiol. Res.* **2017**, *196*, 44–68. [CrossRef]

46. Alvarenga, S.; Ferreira, M.; Emerenciano, V.d.P.; Cabrol-Bass, D. Chemosystematic studies of natural compounds isolated from Asteraceae: Characterization of tribes by principal component analysis. *Chemom. Intell. Lab. Syst.* **2001**, *56*, 27–37. [CrossRef]

47. Myers, N.; Mittermeier, R.A.; Mittermeier, C.G.; Da Fonseca, G.A.; Kent, J. Biodiversity hotspots for conservation priorities. *Nature* **2000**, *403*, 853. [CrossRef]

48. Rosa, L.H.; Gonçalves, V.N.; Caligiorne, R.B.; Alves, T.; Rabello, A.; Sales, P.A.; Romanha, A.J.; Sobral, M.E.; Rosa, C.A.; Zani, C.L. Leishmanicidal, trypanocidal, and cytotoxic activities of endophytic fungi associated with bioactive plants in Brazil. *Braz. J. Microbiol.* **2010**, *41*, 420–430. [CrossRef]

49. Rodríguez-Rodríguez, R.M.; Herrera, P.; Furrazola, E. Arbuscular mycorrhizal colonization in Asteraceae from white sand savannas, in Pinar del Río, Cuba. *Biota Neotrop.* **2013**, *13*, 136–140. [CrossRef]

50. Reis, I.M.A.; Ribeiro, F.P.C.; Almeida, P.R.M.; Costa, L.C.B.; Kamida, H.M.; Uetanabaro, A.P.T.; Branco, A. Characterization of the secondary metabolites from endophytic fungi *Nodulisporium* sp. isolated from the medicinal plant *Mikania laevigata* (Asteraceae) by reversed–phase high–performance liquid chromatography coupled with mass spectrometric multistage. *Pharmacogn. Mag.* **2018**, *14*, 495.

51. Zhao, K.; Penttinen, P.; Guan, T.; Xiao, J.; Chen, Q.; Xu, J.; Lindström, K.; Zhang, L.; Zhang, X.; Strobel, G.A. The diversity and anti–microbial activity of endophytic actinomycetes isolated from medicinal plants in Panxi plateau, China. *Curr. Microbiol.* **2011**, *62*, 182–190. [CrossRef]

52. Yuan, H.M.; Zhang, X.P.; Zhao, K.; Zhong, K.; Gu, Y.F.; Lindström, K. Genetic characterisation of endophytic actinobacteria isolated from the medicinal plants in Sichuan. *Ann. Microbiol.* **2008**, *58*, 597–604. [CrossRef]

53. Vieira, M.L.; Johann, S.; Hughes, F.M.; Rosa, C.A.; Rosa, L.H. The diversity and antimicrobial activity of endophytic fungi associated with medicinal plant *Baccharis trimera* (Asteraceae) from the Brazilian savannah. *Can. J. Microbiol.* **2014**, *60*, 847–856. [CrossRef] [PubMed]

54. Jeewon, R.; Liew, E.C.; Simpson, J.A.; Hodgkiss, I.J.; Hyde, K.D. Phylogenetic significance of morphological characters in the taxonomy of Pestalotiopsis species. *Mol. Phylogen. Evol.* **2003**, *27*, 372–383. [CrossRef]

55. Carvalho, C.R.; Gonçalves, V.N.; Pereira, C.B.; Johann, S.; Galliza, I.V.; Alves, T.M.; Rabello, A.; Sobral, M.E.; Zani, C.L.; Rosa, C.A. The diversity, antimicrobial and anticancer activity of endophytic fungi associated with the medicinal plant *Stryphnodendron adstringens* (Mart.) Coville (Fabaceae) from the Brazilian savannah. *Symbiosis* **2012**, *57*, 95–107. [CrossRef]

56. Hatamzadeh, S.; Rahnama, K.; Nasrollahnejad, S.; Fotouhifar, K.B.; Hemmati, K.; White, J.F.; Taliei, F. Isolation and identification of L–asparaginase–producing endophytic fungi from the Asteraceae family plant species of Iran. *PeerJ* **2020**, *8*, e8309. [CrossRef] [PubMed]

57. Cheng, D.; Tian, Z.; Feng, L.; Xu, L.; Wang, H. Diversity analysis of the rhizospheric and endophytic bacterial communities of *Senecio vulgaris* L.(Asteraceae) in an invasive range. *PeerJ* **2019**, *6*, e6162. [CrossRef] [PubMed]

58. Huang, W.-Y.; Cai, Y.-Z.; Xing, J.; Corke, H.; Sun, M. A potential antioxidant resource: Endophytic fungi from medicinal plants. *Econ. Bot.* **2007**, *61*, 14. [CrossRef]

59. Christian, N.; Sullivan, C.; Visser, N.D.; Clay, K. Plant host and geographic location drive endophyte community composition in the face of perturbation. *Microb. Ecol.* **2016**, *72*, 621–632. [CrossRef]

60. Jiang, H.; Shi, Y.-T.; Zhou, Z.-X.; Yang, C.; Chen, Y.-J.; Chen, L.-M.; Yang, M.-Z.; Zhang, H.-B. Leaf chemistry and co–occurring species interactions affecting the endophytic fungal composition of *Eupatorium adenophorum*. *Ann. Microbiol.* **2011**, *61*, 655–662. [CrossRef]

61. Mei, L.; Zhu, M.; Zhang, D.-Z.; Wang, Y.-Z.; Guo, J.; Zhang, H.-B. Geographical and temporal changes of foliar fungal endophytes associated with the invasive plant *Ageratina adenophora*. *Microb. Ecol.* **2014**, *67*, 402–409. [CrossRef]

62. Fang, K.; Miao, Y.-F.; Chen, L.; Zhou, J.; Yang, Z.-P.; Dong, X.-F.; Zhang, H.-B. Tissue–specific and geographical variation in endophytic fungi of *Ageratina adenophora* and fungal associations with the environment. *Front. Microbiol.* **2019**, *10*, 2919. [CrossRef]

63. Pereira, J.M.; Peixoto, V.; Teixeira, A.; Sousa, D.; Barros, L.; Ferreira, I.C.; Vasconcelos, M.H. *Achillea millefolium* L. hydroethanolic extract has phenolic compounds and inhibits the growth of human tumor cell lines. *Free Radic. Biol. Med.* **2018**, *120*, S145. [CrossRef]

64. Moghadam, M.H.; Mosaddegh, M.; Irani, M. Programmed cell death in breast adeno–carcinoma induced by *Achillea filipendulina*. *Med. Plants–Int. J. Phytomed. Relat. Ind.* **2019**, *11*, 435–439. [CrossRef]

65. Aminkhani, A.; Sharifi, S.; Ekhtiyari, S. *Achillea filipendulina* Lam.: Chemical constituents and antimicrobial activities of essential oil of stem, leaf, and flower. *Chem. Biodivers.* **2020**, *17*, e2000133. [CrossRef] [PubMed]

66. Boukhary, R.; Aboul-ElA, M.; El-Lakany, A. Review on chemical constituents and biological activities of genus *Anthemis*. *Pharm. J.* **2019**, *11*, 1155–1166. [CrossRef]

67. Buommino, E.; Baroni, A.; Canozo, N.; Petrazzuolo, M.; Nicoletti, R.; Vozza, A.; Tufano, M.A. Artemisinin reduces human melanoma cell migration by down–regulating αvβ3 integrin and reducing metalloproteinase 2 production. *Investig. New Drugs* **2009**, *27*, 412–418. [CrossRef] [PubMed]

68. Kontogianni, V.G.; Primikyri, A.; Sakka, M.; Gerothanassis, I.P. Simultaneous determination of artemisinin and its analogs and flavonoids in *Artemisia annua* crude extracts with the use of NMR spectroscopy. *Magn. Reson. Chem.* **2020**, *58*, 232–244. [CrossRef] [PubMed]

69. Song, Y.; Desta, K.T.; Kim, G.S.; Lee, S.J.; Lee, W.S.; Kim, Y.H.; Jin, J.S.; Abd El-Aty, A.; Shin, H.C.; Shim, J.H. Polyphenolic profile and antioxidant effects of various parts of *Artemisia annua* L. *Biomed. Chromatogr.* **2016**, *30*, 588–595. [CrossRef]

70. Lu, H.; Zou, W.X.; Meng, J.C.; Hu, J.; Tan, R.X. New bioactive metabolites produced by *Colletotrichum* sp., an endophytic fungus in *Artemisia annua*. *Plant Sci.* **2000**, *151*, 67–73. [CrossRef]

71. Abiri, R.; Silva, A.L.M.; de Mesquita, L.S.S.; de Mesquita, J.W.C.; Atabaki, N.; de Almeida, E.B., Jr.; Shaharuddin, N.A.; Malik, S. Towards a better understanding of *Artemisia vulgaris*: Botany, phytochemistry, pharmacological and biotechnological potential. *Food Res. Int.* **2018**, *109*, 403–415. [CrossRef]

72. Lösgen, S.; Magull, J.; Schulz, B.; Draeger, S.; Zeeck, A. Isofusidienols: Novel chromone-3-oxepines produced by the endophytic fungus *Chalara* sp. *Eur. J. Org. Chem.* **2008**, *2008*, 698–703. [CrossRef]

73. Pandey, A.K.; Singh, P. The genus *Artemisia*: A 2012–2017 literature review on chemical composition, antimicrobial, insecticidal and antioxidant activities of essential oils. *Medicines* **2017**, *4*, 68. [CrossRef] [PubMed]

74. Xu, K.; Jiang, J.-S.; Feng, Z.-M.; Yang, Y.-N.; Li, L.; Zang, C.-X.; Zhang, P.-C. Bioactive sesquiterpenoid and polyacetylene glycosides from *Atractylodes lancea*. *J. Nat. Prod.* **2016**, *79*, 1567–1575. [CrossRef] [PubMed]

75. Ren, C.-G.; Dai, C.-C. Jasmonic acid is involved in the signaling pathway for fungal endophyte–induced volatile oil accumulation of *Atractylodes lancea* plantlets. *BMC Plant Biol.* **2012**, *12*, 128. [CrossRef] [PubMed]

76. Cheriyan, B.V.; Joshi, S.; Mohamed, S. *Eupatorium triplinerve* (Vahl): An ethnobotanical review. *Asian J. Pharm. Res.* **2019**, *9*, 200–202. [CrossRef]

77. Da Silva Paes, L.; de Lucena, J.M.V.M.; da Silva Bentes, J.L.; de Oliveira Marques, J.D.; Casas, L.L.; Mendonca, M.S. Endophytic mycobiota of three Amazonian medicinal herbs: *Stachytarpheta cayennensis* (Verbenaceae), *Ayapana triplinervis* (Asteraceae) and *Costus spicatus* (Costaceae). *Int. J. Bot.* **2014**, *10*, 24.

78. Bonin, E.; Carvalho, V.M.; Avila, V.D.; dos Santos, N.C.A.; Benassi–Zanqueta, É.; Lancheros, C.A.C.; Previdelli, I.T.S.; Ueda–Nakamura, T.; de Abreu Filho, B.A.; do Prado, I.N. *Baccharis dracunculifolia*: Chemical constituents, cytotoxicity and antimicrobial activity. *LWT* **2020**, *120*, 108920. [CrossRef]

79. Cuzzi, C.; Link, S.; Vilani, A.; Sartori, C.; Onofre, S.B. Endophytic fungi of the "vassourinha" (*Baccharis dracunculifolia* DC, Asteraceae). *Rev. Bras. Bioc.* **2012**, *10*, 135.

80. Gonçalves, L.; Tavares, B.; Galdino, F.; Andrade, J.; Abrantes, L.; De Sousa, M.R.; Lira, P.G.; Pereira, L.R. Carqueja: Chemical composition and pharmacological effects. In Proceedings of the MOL2NET 2018, International Conference on Multidisciplinary Sciences, Paraiba, Brazil, 25 October 2018; p. 1.

81. Xuan, T.D.; Khanh, T.D. Chemistry and pharmacology of *Bidens pilosa*: An overview. *J. Pharm. Investig.* **2016**, *46*, 91–132. [CrossRef]

82. Abdou, R.; Scherlach, K.; Dahse, H.-M.; Sattler, I.; Hertweck, C. Botryorhodines A–D, antifungal and cytotoxic depsidones from *Botryosphaeria rhodina*, an endophyte of the medicinal plant *Bidens pilosa*. *Phytochemistry* **2010**, *71*, 110–116. [CrossRef]

83. Saeed, M.; Abd El-Hack, M.E.; Alagawany, M.; Arain, M.A.; Arif, M.; Mirza, M.A.; Naveed, M.; Chao, S.; Sarwar, M.; Sayab, M. Chicory (*Cichorium intybus*) herb: Chemical composition, pharmacology, nutritional and healthical applications. *Int. J. Pharm.* **2017**, *13*, 351–360.

84. Siriwatanametanon, N.; Heinrich, M. The Thai medicinal plant *Gynura pseudochina* var. *hispida*: Chemical composition and in vitro NF–κB inhibitory activity. *Nat. Prod. Commun.* **2011**, *6*, 627–630. [CrossRef] [PubMed]

85. Siriwach, R.; Kinoshita, H.; Kitani, S.; Igarashi, Y.; Pansuksan, K.; Panbangred, W.; Nihira, T. Bipolamides A and B, triene amides isolated from the endophytic fungus Bipolaris sp. MU34. *J. Antibiot.* **2014**, *67*, 167–170. [CrossRef]

86. Akpor, O.; Olaolu, T.; Rotimi, D. Antibacterial and antioxidant potentials of leave extracts of *Helianthus annuus*. *Potrav. Slovak J. Food Sci.* **2019**, *13*, 1026–1033. [CrossRef]

87. Al-Snafi, A.E. The pharmacological effects of *Helianthus annuus*—A review. *Indo Am. J. Pharm. Sci.* **2018**, *5*, 1745–1756.

88. Waqas, M.; Khan, A.L.; Hamayun, M.; Shahzad, R.; Kang, S.-M.; Kim, J.-G.; Lee, I.-J. Endophytic fungi promote plant growth and mitigate the adverse effects of stem rot: An example of *Penicillium citrinum* and *Aspergillus terreus*. *J. Plant Int.* **2015**, *10*, 280–287.

89. Waqas, M.; Khan, A.L.; Hamayun, M.; Shahzad, R.; Kim, Y.-H.; Choi, K.-S.; Lee, I.-J. Endophytic infection alleviates biotic stress in sunflower through regulation of defence hormones, antioxidants and functional amino acids. *Eur. J. Plant Pathol.* **2015**, *141*, 803–824. [CrossRef]

90. Getahun, T.; Sharma, V.; Gupta, N. The genus *Laggera* (Asteraceae)–Ethnobotanical and ethnopharmacological information, chemical composition as well as biological activities of its essential oils and extracts: A review. *Chem. Biodiver.* **2019**, *16*, e1900131. [CrossRef] [PubMed]

91. Matasyoh, J.C.; Dittrich, B.; Schueffler, A.; Laatsch, H. Larvicidal activity of metabolites from the endophytic *Podospora* sp. against the malaria vector *Anopheles gambiae*. *Parasitol. Res.* **2011**, *108*, 561–566. [CrossRef]

92. Gardiner, P. Complementary, holistic, and integrative medicine: Chamomile. *Pediatr. Rev.* **2007**, *28*, e16. [CrossRef]

93. Della Pasqua, C.; Iwamoto, R.; Antunes, E.; Borghi, A.; Sawaya, A.; Landucci, E. Pharmacological study of anti–inflammatory activity of aqueous extracts of *Mikania glomerata* (Spreng.) and *Mikania laevigata* (Sch. Bip. ex Baker). *J. Ethnopharm.* **2019**, *231*, 50–56. [CrossRef]

94. Polonio, J.; Almeida, T.; Garcia, A.; Mariucci, G.; Azevedo, J.; Rhoden, S.; Pamphile, J. Biotechnological prospecting of foliar endophytic fungi of guaco (*Mikania glomerata* Spreng.) with antibacterial and antagonistic activity against phytopathogens. *Gen. Mol. Res.* **2015**, *14*, 7297–7309. [CrossRef] [PubMed]

95. Azab, A. A facile method for testing antioxidant capacity and total phenolic content of *Notobasis syriaca* and *Scolymus maculatus* extracts and their antifungal activity. *Eur. Chem. Bull.* **2018**, *7*, 210–217. [CrossRef]

96. Hussain, H.; Tchimene, M.K.; Ahmed, I.; Meier, K.; Steinert, M.; Draeger, S.; Schulz, B.; Krohn, K. Antimicrobial chemical constituents from the endophytic fungus *Phomopsis* sp. from *Notobasis syriaca*. *Nat. Prod. Commun.* **2011**, *6*, 1905–1906. [CrossRef] [PubMed]

97. Saeed, M.; Xu, Y.; Rehman, Z.U.; Arain, M.A.; Soomro, R.N.; Abd El-Hack, M.E.; Bhutto, Z.A.; Abbasi, B.; Dhama, K.; Sarwar, M. Nutritional and healthical aspects of yacon (*Smallanthus sonchifolius*) for human, animals and poultry. *Int. J. Pharm.* **2017**, *13*, 361–369. [CrossRef]

98. Gallo, M.B.; Chagas, F.O.; Almeida, M.O.; Macedo, C.C.; Cavalcanti, B.C.; Barros, F.W.; de Moraes, M.O.; Costa-Lotufo, L.V.; Pessoa, C.; Bastos, J.K. Endophytic fungi found in association with *Smallanthus sonchifolius* (Asteraceae) as resourceful producers of cytotoxic bioactive natural products. *J. Basic Microbiol.* **2009**, *49*, 142–151. [CrossRef]

99. Gallo, M.B.; Cavalcanti, B.C.; Barros, F.W.; Odorico de Moraes, M.; Costa-Lotufo, L.V.; Pessoa, C.; Bastos, J.K.; Pupo, M.T. Chemical constituents of *Papulaspora immersa*, an endophyte from *Smallanthus sonchifolius* (Asteraceae), and their cytotoxic activity. *Chem. Biodivers.* **2010**, *7*, 2941–2950. [CrossRef]

100. Chagas, F.O.; Dias, L.G.; Pupo, M.T. A mixed culture of endophytic fungi increases production of antifungal polyketides. *J. Chem. Ecol.* **2013**, *39*, 1335–1342. [CrossRef]

101. Rosa, L.H.; Queiroz, S.C.; Moraes, R.M.; Wang, X.; Techen, N.; Pan, Z.; Cantrell, C.L.; Wedge, D.E. *Coniochaeta ligniaria*: Antifungal activity of the cryptic endophytic fungus associated with autotrophic tissue cultures of the medicinal plant *Smallanthus sonchifolius* (Asteraceae). *Symbiosis* **2013**, *60*, 133–142. [CrossRef]

102. Porwal, O.; Ameen, M.S.M.; Anwer, E.T.; Uthirapathy, S.; Ahamad, J.; Tahsin, A. *Silybum marianum* (Milk Thistle): Review on its chemistry, morphology, ethno medical uses, phytochemistry and pharmacological activities. *J. Drug Deliv. Ther.* **2019**, *9*, 199–206. [CrossRef]

103. El-Elimat, T.; Raja, H.A.; Graf, T.N.; Faeth, S.H.; Cech, N.B.; Oberlies, N.H. Flavonolignans from *Aspergillus iizukae*, a fungal endophyte of milk thistle (*Silybum marianum*). *J. Nat. Prod.* **2014**, *77*, 193–199. [CrossRef]

104. Sampaio, B.L.; Edrada-Ebel, R.; Da Costa, F.B. Effect of the environment on the secondary metabolic profile of *Tithonia diversifolia*: A model for environmental metabolomics of plants. *Sci. Rep.* **2016**, *6*, 29265. [CrossRef] [PubMed]

105. Guimarães, D.O.; Borges, W.S.; Kawano, C.Y.; Ribeiro, P.H.; Goldman, G.H.; Nomizo, A.; Thiemann, O.H.; Oliva, G.; Lopes, N.P.; Pupo, M.T. Biological activities from extracts of endophytic fungi isolated from *Viguiera arenaria* and *Tithonia diversifolia*. *FEMS Immunol. Med. Microbiol.* **2008**, *52*, 134–144. [CrossRef] [PubMed]

106. Borges, W.d.S.; Pupo, M.T. Novel anthraquinone derivatives produced by *Phoma sorghina*, an endophyte found in association with the medicinal plant *Tithonia diversifolia* (Asteraceae). *J. Braz. Chem. Soc.* **2006**, *17*, 929–934. [CrossRef]

107. Sales Junior, P.A.; Zani, C.L.; de Siqueira, E.P.; Kohlhoff, M.; Marques, F.R.; Caldeira, A.S.P.; Cota, B.B.; Maia, D.N.B.; Tunes, L.G.; Murta, S.M.F. Trypanocidal trixikingolides from *Trixis vauthieri*. *Nat. Prod. Res.* **2019**, *33*, 1–9. [CrossRef] [PubMed]

108. Cota, B.B.; Rosa, L.H.; Caligiorne, R.B.; Rabello, A.L.T.; Almeida Alves, T.M.; Rosa, C.A.; Zani, C.L. Altenusin, a biphenyl isolated from the endophytic fungus *Alternaria* sp., inhibits trypanothione reductase from *Trypanosoma cruzi*. *FEMS Microbiol. Lett.* **2008**, *285*, 177–182. [CrossRef] [PubMed]

109. Petropoulos, S.A.; Fernandes, Â.; Tzortzakis, N.; Sokovic, M.; Ciric, A.; Barros, L.; Ferreira, I.C. Bioactive compounds content and antimicrobial activities of wild edible Asteraceae species of the Mediterranean flora under commercial cultivation conditions. *Food Res. Int.* **2019**, *119*, 859–868. [CrossRef]

110. Aly, A.H.; Edrada-Ebel, R.; Wray, V.; Müller, W.E.; Kozytska, S.; Hentschel, U.; Proksch, P.; Ebel, R. Bioactive metabolites from the endophytic fungus *Ampelomyces* sp. isolated from the medicinal plant *Urospermum picroides*. *Phytochemistry* **2008**, *69*, 1716–1725. [CrossRef]

111. Marangoni, S.; Moraes, T.d.S.; Utrera, S.H.; Casemiro, L.A.; de Souza, M.G.; de Oliveira, P.F.; Veneziani, R.C.; Ambrósio, S.R.; Tavares, D.C.; Martins, C.H. Diterpenes of the pimarane type isolated from *Viguiera arenaria*: Promising in vitro biological potential as therapeutic agents for endodontics. *J. Pharm. Phytother.* **2018**, *10*, 34–44.

112. Verza, M.; Arakawa, N.S.; Lopes, N.P.; Kato, M.J.; Pupo, M.T.; Said, S.; Carvalho, I. Biotransformation of a tetrahydrofuran lignan by the endophytic fungus *Phomopsis* sp. *J. Braz. Chem. Soc.* **2009**, *20*, 195–200. [CrossRef]

113. Fattori, V.; Zarpelon, A.C.; Staurengo-Ferrari, L.; Borghi, S.M.; Zaninelli, T.H.; Da Costa, F.B.; Alves-Filho, J.C.; Cunha, T.M.; Cunha, F.Q.; Casagrande, R. Budlein a, a sesquiterpene lactone from *Viguiera robusta*, alleviates pain and inflammation in a model of acute gout arthritis in mice. *Front. Pharm.* **2018**, *9*, 1076. [CrossRef] [PubMed]

114. Momesso, L.d.S.; Kawano, C.Y.; Ribeiro, P.H.; Nomizo, A.; Goldman, G.H.; Pupo, M.T. Chaetoglobosinas produzidas por *Chaetomium globosum*, fungo endofítico associado a *Viguiera robusta* Gardn. (Asteraceae). *Química Nova* **2008**, *31*, 1680–1685. [CrossRef]

115. Khan, S.A.; Hamayun, M.; Yoon, H.; Kim, H.-Y.; Suh, S.-J.; Hwang, S.-K.; Kim, J.-M.; Lee, I.-J.; Choo, Y.-S.; Yoon, U.-H. Plant growth promotion and *Penicillium citrinum*. *BMC Microbiol.* **2008**, *8*, 231. [CrossRef] [PubMed]

116. Khan, A.L.; Hussain, J.; Al-Harrasi, A.; Al-Rawahi, A.; Lee, I.-J. Endophytic fungi: Resource for gibberellins and crop abiotic stress resistance. *Crit. Rev. Biotechnol.* **2015**, *35*, 62–74. [CrossRef] [PubMed]

117. Hamayun, M.; Khan, S.A.; Iqbal, I.; Ahmad, B.; Lee, I.-J. Isolation of a gibberellin–producing fungus (*Penicillium* sp. MH7) and growth promotion of crown daisy (*Chrysanthemum coronarium*). *J. Microbiol. Biotechnol.* **2010**, *20*, 202–207. [CrossRef]

118. Radić, N.; Štrukelj, B. Endophytic fungi—The treasure chest of antibacterial substances. *Phytomedicine* **2012**, *19*, 1270–1284. [CrossRef] [PubMed]

119. Fondi, M.; Fani, R. The horizontal flow of the plasmid resistome: Clues from inter-generic similarity networks. *Environ. Microbiol.* **2010**, *12*, 3228–3242. [CrossRef]

120. Dwivedi, G.R.; Sanchita; Singh, D.P.; Sharma, A.; Darokar, M.P.; Srivastava, S.K. Nano particles: Emerging warheads against bacterial superbugs. *Curr. Top. Med. Chem.* **2016**, *16*, 1963–1975. [CrossRef]

121. Chen, L.; Zhang, Q.-Y.; Jia, M.; Ming, Q.-L.; Yue, W.; Rahman, K.; Qin, L.-P.; Han, T. Endophytic fungi with antitumor activities: Their occurrence and anticancer compounds. *Crit. Rev. Microbiol.* **2016**, *42*, 454–473. [CrossRef]

122. Shishkoff, N.; McGrath, M. AQ10 biofungicide combined with chemical fungicides or AddQ spray adjuvant for control of cucurbit powdery mildew in detached leaf culture. *Plant Dis.* **2002**, *86*, 915–918. [CrossRef]

123. Chen, X.; Shi, Q.; Lin, G.; Guo, S.; Yang, J. Spirobisnaphthalene analogues from the endophytic fungus *Preussia* sp. *J. Nat. Prod.* **2009**, *72*, 1712–1715. [CrossRef]

124. Du, L.; King, J.B.; Morrow, B.H.; Shen, J.K.; Miller, A.N.; Cichewicz, R.H. Diarylcyclopentendione metabolite obtained from a *Preussia typharum* isolate procured using an unconventional cultivation approach. *J. Nat. Prod.* **2012**, *75*, 1819–1823. [CrossRef] [PubMed]

125. Klatt, S.; Simpson, L.; Maslov, D.A.; Konthur, Z. *Leishmania tarentolae*: Taxonomic classification and its application as a promising biotechnological expression host. *PLoS Negl. Trop. Dis.* **2019**, *13*, e0007424. [CrossRef] [PubMed]

126. Tanvir, R.; Sajid, I.; Hasnain, S. Larvicidal potential of Asteraceae family endophytic actinomycetes against *Culex quinquefasciatus* mosquito larvae. *Nat. Prod. Res.* **2014**, *28*, 2048–2052. [CrossRef] [PubMed]

127. Srivastava, V.; Negi, A.S.; Kumar, J.; Gupta, M.; Khanuja, S.P. Plant–based anticancer molecules: A chemical and biological profile of some important leads. *Bioorg. Med. Chem.* **2005**, *13*, 5892–5908. [CrossRef]

128. Global Biodiversity Information Facility. Available online: https://www.gbif.org/ (accessed on 12 June 2020).

129. MycoBank. Available online: http://www.mycobank.org/ (accessed on 12 June 2020).

Review

The Thin Line between Pathogenicity and Endophytism: The Case of *Lasiodiplodia theobromae*

Maria Michela Salvatore [1], **Anna Andolfi** [1,2,*] **and Rosario Nicoletti** [3,4]

[1] Department of Chemical Sciences, University of Naples 'Federico II', 80126 Naples, Italy; mariamichela.salvatore@unina.it

[2] BAT Center—Interuniversity Center for Studies on Bioinspired Agro-Environmental Technology, University of Naples 'Federico II', 80138 Naples, Italy

[3] Council for Agricultural Research and Economics, Research Centre for Olive, Fruit and Citrus Crops, 81100 Caserta, Italy; rosario.nicoletti@crea.gov.it

[4] Department of Agricultural Sciences, University of Naples 'Federico II', 80055 Portici, Italy

* Correspondence: andolfi@unina.it; Tel.: +39-081-2539179

Received: 18 September 2020; Accepted: 17 October 2020; Published: 21 October 2020

Abstract: Many fungi reported for endophytic occurrence are better known as plant pathogens on different crops, raising questions about their actual relationships with the hosts and other plants in the biocoenosis and about the factors underlying the lifestyle shift. This paper offers an overview of the endophytic occurrence of *Lasiodiplodia theobromae* (Dothideomycetes, Botryosphaeriaceae), a species known to be able to colonize many plants as both an endophyte and a pathogen. Prevalently spread in tropical and subtropical areas, there are concerns that it may propagate to the temperate region following global warming and the increasing trade of plant materials. The state of the art concerning the biochemical properties of endophytic strains of this species is also examined with reference to a range of biotechnological applications.

Keywords: endophytic fungi; mutualism; plant fitness; latent pathogens; *Botryosphaeria rhodina*; *Botryodiplodia theobromae*

---

1. Introduction

Endophytic fungi are plant-associated microorganisms that colonize the internal tissues of the host without inducing disease symptoms [1]. They represent a poorly understood endosymbiotic group of microbes that ought to be attentively considered by the scientific community, so as to provide comprehensive knowledge regarding their beneficial role and the actual extent of their interactions with plants.

A basic issue hindering studies on the ecological role of these microorganisms is represented by the reported endophytic occurrence of fungal pathogens. In fact, besides the cases where latency is a conspicuous phase of the disease cycle, there are more and more records of renowned pathogens found within asymptomatic hosts, for which an explanation is not immediately available [2–4]. Increasing and organizing the current knowledge on conditions associated with the occurrence of these ambiguous species is useful for a more conclusive assessment of their functions and impact on crops. This present paper offers an overview of a fungus which is mainly studied as a pathogen of tropical crops [4–6] but that is potentially able to spread as an endophytic associate of plants in the temperate zone.

2. Taxonomic and Phylogenetic Aspects

Lasiodiplodia theobromae (Pat.) Griffon & Maubl. (Dothideomycetes, Botryosphaeriaceae) is the accepted name of the species treated in this paper, prevailing over both the basionym *Botryodiplodia theobromae* Pat. and the teleomorphic name *Botryosphaeria rhodina* (Berk & M.A. Curtis) Arx, after the introduction in mycology of the principle "one species—one name" [7]. Isolated and morphologically identified from a wide range of plant hosts [5,8], it represents the type species of *Lasiodiplodia* which, for many years, was treated as a monotypic genus within the Botryosphaeriaceae [9,10]. However, such a simplified taxonomy was destined to dramatically change with the advent of DNA sequencing. In fact, starting from the year 2004, phylogenetic analyses carried out in the course of studies on *L. theobromae* in novel pathosystems showed the existence of several clades, even within the pool of strains stored in mycological collections [11–14]. Evidence of a higher complexity emerged gradually, to such an extent that more than 30 additional species have been described to date, with some of them, such as *L. endophytica*, *L. gonubiensis*, *L. pseudotheobromae*, *L. thailandica* and *L. venezuelensis*, reported as endophytes [15–23]. Hence, it is likely that several previous findings might be incorrectly classified and that some more recent records are going to be re-examined. The application of high-throughput DNA metabarcoding as a biomonitoring tool is expected to provide a notable contribution in investigations concerning the endophytic occurrence of *Lasiodiplodia* [24].

To further complicate the issue, the existence of hybrid strains has been ascertained [15,25], which is also considered to have affected species identification. As an example, the taxon *L. viticola* Úrbez-Torres, Peduto & Gubler [26] has been shown to be a hybrid between *L. theobromae* and *L. mediterranea*; both these taxa are known on grapevine (*Vitis vinifera*), which most likely represented the venue of the hybridization process [15]. An assumption in biology considers as a species an organism whose population is reproductively isolated from other phylogenetically related populations [27]; hence, the existence of hybrids between several *Lasiodiplodia* spp. may imply that the taxa described so far are not stable. Indeed, further reassessments are to be expected, particularly in consequence of new combinations possibly stimulated by the circulation of plant material hosting genotypes which are potentially capable of hybridizing with autochthonous strains. In order to avoid further misidentifications, the use of multiple genes is recommended when considering the phylogenetic relationships of novel strains, along with direct referencing to the type strains [15,20].

Apart from the variation characterizing the genus *Lasiodiplodia*, phylogenetic relationships have also been evaluated in the species under discussion. Low genotypic diversity was observed in a study considering three populations from different tree species in Venezuela, South Africa and Mexico. A few predominant genotypes were encountered in the first two countries, without evidence of host specificity and in the presence of a very high gene flow between populations from different hosts. The geographic isolation was substantiated by the finding of unique alleles fixed in the different populations. Moreover, the existence of some genotypes that were widely distributed throughout the three countries, coupled with the evidence that pseudothecia are rarely produced in nature, suggests that reproduction is predominantly clonal [8]. A similar conclusion was reached in another phylogeographic study carried out on coconut palm (*Cocos nucifera*) in Brazil, where higher genotypic variation was observed in the northeast in connection with the local higher host diversity and a conjectured repeated introduction from Central Africa, regarded as the possible center of radiation of the species. Differences between genotypes were mainly ascribed to mutations [28].

In Cameroon, cocoa (*Theobroma cacao*) and *Terminalia* spp. are frequently grown together in a peculiar agri-sylvicultural system. A comparison between strains from these two known hosts of *L. theobromae* showed high levels of gene diversity and low genotypic differentiation, in the presence of high gene flow between isolates. The absence of a geographic substructure in these populations across the region where the study was carried out is indicative of the symmetrical movement of the fungus between these hosts. Unlike the case documented on grapevine, no evidence of hybridization was found with the closely related *L. pseudotheobromae*, which also occurs on these plants [29].

Finally, quite a simple genetic structure was once more pointed out in a broader study including strains of more varied origin. In fact, one or two main haplotypes across all genes were identified, and these genotypes were unrelated to both the hosts and the geographic area. Such overall uniformity clearly indicates that large-scale dispersal of *L. theobromae* is essentially derived from commerce and human activities [4].

3. Endophytic Occurrence of *Lasiodiplodia theobromae*

After having basically been studied as a plant pathogen responsible for serious damages of crops, particularly in tropical and subtropical regions [5,6], in the last three decades, the literature regarding *L. theobromae* has been substantially enriched by many reports concerning its endophytic occurrence on plant species which are quite heterogeneous in botanical terms (Table 1).

Table 1. Plant hosts of endophytic *Lasiodiplodia theobromae*. Species where the fungus has been also reported as a pathogen are underlined.

	Source	Origin	Ref.
	Pinophyta		
Pinales, Pinaceae	*Pinus elliottii*	South Africa	[8]
	Pinus caribaea var. *hondurensis*	Venezuela	[8]
	Pinus pseudostrobus	Mexico	[8]
	Pinus tabulaeformis	China	[30]
Pinales, Taxaceae	*Cephalotaxus hainanensis*	China	[31]
	Taxus baccata	India	[32]
	Taxus chinensis	China	GenBank
	Magnoliids		
Magnoliales, Annonaceae	*Annona muricata*	Malaysia	GenBank
Piperales, Piperaceae	*Piper hispidum*	Brazil	[33,34]
	Piper nigrum	India	[35]
	Monocots		
Asparagales, Asparagaceae	*Dracaena draco*	Egypt	[36]
Asparagales, Orchidaceae	*Campylocentrum micranthum*	Costa Rica	[37]
	Cattleya sp.	Brazil	[38]
	Cymbidium aloifolium	India	[39]
	Dendrobium moschatum	India	[39]
	Encyclia fragrans	Costa Rica	[37]
	Epidendrum difforme	Costa Rica	[37]
	Epidendrum octomerioides	Costa Rica	[37]
	Epidendrum radicans	India	GenBank
	Eria flava	India	[39]
	Nidema boothii	Costa Rica	[37]
	Oncidium sp.	Brazil	[38]
	Paphiopedilum fairrieanum	India	[39]
	Phalaenopsis sp.	Brazil	[38]
	Pholidota imbricata	India	[39]
	Pholidota pallida	India	[40]
	Pleurothallis guanacastensis	Costa Rica	[37]
	Pleurothallis phyllocardioides	Costa Rica	[37]
	Sobralia mucronata	Costa Rica	[37]

Table 1. *Cont.*

	Source	Origin	Ref.
Asparagales, Orchidaceae	*Sobralia* sp.	Costa Rica	[37]
	Trichosalpinx blasdellii	Costa Rica	[37]
	Vanilla planifolia	India	[39]
Pandanales, Pandanaceae	*Pandanus* sp.	Thailand	[41]
Arecales, Arecaceae	*Calamus thwaitesii*	Sri Lanka	[42]
	Cocos nucifera	Brazil	[28]
	Cocos nucifera	India	[43]
	Cocos nucifera	Philippines	[44]
	Euterpe oleracea	Brazil	[45]
	Nypa fruticans	Malaysia	[46]
Poales, Cyperaceae	*Mapania kurzii*	Malaysia	[47]
Poales, Poaceae	*Cynodon dactylon*	India	GenBank
Zingiberales, Costaceae	*Costus igneus*	India	[48]
Zingiberales, Musaceae	*Musa* spp.	Malaysia	[49]
Eudicots			
Proteales, Proteaceae	*Grevillea agrifolia*	Australia	[50]
Ranunculales, Menispermaceae	*Tinospora cordifolia*	India	[51]
Santalales, Santalaceae	*Viscum coloratum*	China	[52]
Saxifragales, Hamamelidaceae	*Distilium chinense*	China	[53]
Vitales, Vitaceae	*Vitis vinifera*	China	[54]
	Vitis vinifera	Italy	[55]
Celastrales, Celastraceae	*Elaeodendrum glaucum*	India	[56]
	Salacia oblonga	India	[57]
Fabales, Fabaceae	*Acacia karroo*	South Africa	[58]
	Acacia mangium	Venezuela	[8]
	Acacia synchronicia	Australia	[50]
	Albizzia lebbeck	India	Genbank
	Arachis hypogaea	India	[56]
	Bauhinia racemosa	India	[56]
	Butea monosperma	India	[59,60]
	Cassia fistula	India	[56]
	Crotalaria medicaginea	Australia	[50]
	Dalbergia lanceolaria	India	[60]
	Dalbergia latifolia	India	[56]
	Glycyrrhiza glabra	India	[61]
	Humboldtia brunonis	India	[62]
	Indigofera suffruticosa	Brazil	[63]

Table 1. *Cont.*

	Source	Origin	Ref.
Fabales, Fabaceae	*Libidibia (Caesalpinia) ferrea*	Brazil	[64]
	Lysiphyllum cunninghamii	Australia	[50]
	Mimosa caesalpinifolia	Brazil	[64]
	Ougeinia oojeinensis	India	[60]
	Phaseolus lunatus	Mexico	[65]
	Pongamia pinnata	India	[43]
	Saraca asoca	India	[66,67]
	Sophora tonkinensis	China	[68]
Malpighiales, Chrysobalanaceae	*Licania rigida*	Brazil	[64]
Malpighiales, Clusiaceae	*Garcinia mangostana*	Thailand	[69]
Malpighiales, Euphorbiaceae	*Croton campestris*	Brazil	[64]
	Croton sonderianus	Brazil	[64]
	Givotia rottleriformis	India	[60]
	Hevea brasiliensis	Malaysia	GenBank
		Peru	[70]
Malpighiales, Hypericaceae	*Hypericum mysorense*	India	[71]
Malpighiales, Rhizophoraceae	*Bruguiera cylindrica*	Philippines	[72]
	Ceriops tagal	China	GenBank
	Rhizophora mucronata	China	[73]
Malpighiales, Salicaceae	*Populus* sp.	China	[74]
Oxalidales, Elaeocarpaceae	*Elaeocarpus ganitrus*	India	GenBank
	Elaeocarpus tuberculatus	India	[56]
Rosales, Moraceae	*Artocarpus altilis*	Ecuador	Genbank
	Ficus opposita	Australia	[50]
	Ficus racemosa	India	GenBank
	Ficus trigona	Ecuador	GenBank
Rosales, Rhamnaceae	*Ziziphus xylopyrus*	India	[60]
Rosales, Ulmaceae	*Zelkova carpinifolia*	Iran	GenBank
Cucurbitales, Cucurbitaceae	*Momordica charantia*	China	[75]
Fagales, Fagaceae	*Quercus castaneifolia*	Iran	GenBank
Fagales, Juglandaceae	*Pterocarya fraxinifolia*	Iran	GenBank
Brassicales, Moringaceae	*Moringa oleifera*	Brazil	[64]
Malvales, Malvaceae	*Adansonia digitata*	Australia	[50]
		Cameroon	[15]
	Adansonia gregorii	Australia	[50]
	Adansonia za	Australia	[50]
	Gossypium hirsutum	India	[76]

Table 1. *Cont.*

	Source	Origin	Ref.
Malvales, Malvaceae	*Grewia tiliaefolia*	India	[56]
	Helicteres isora	India	[60]
	Kydia calycina	India	[60]
	Theobroma cacao	Brazil	[77]
		India	[78]
	Theobroma gileri	Ecuador	[79]
Malvales, Thymelaeaceae	*Aquilaria malaccensis*	India	[80]
	Aquilaria sinensis	China	[81,82]
		Taiwan	GenBank
Myrtales, Combretaceae	*Anogeissus latifolia*	India	[60]
	Combretum leprosum	Brazil	[64]
	Lumnitzera littorea	Philippines	[72]
	Terminalia arjuna	India	[83,84]
	Terminalia bellerica	India	[56]
	Terminalia catappa	Cameroon	[85,86]
	Terminalia crenulata	India	[60]
	Terminalia ivorensis	Cameroon	[87]
	Terminalia mantaly	Cameroon	[86,87]
	Terminalia pterocarya	Australia	[50]
	Terminalia superba	Cameroon	[87]
	Terminalia tomentosa	India	[56]
Myrtales, Lythraceae	*Lagerstroemia microcarpa*	India	[60]
	Lagerstroemia parviflora	India	[60]
Myrtales, Melastomataceae	*Memecylon umbellatum*	India	[88]
Myrtales, Myrtaceae	*Calytrix* sp.	Australia	[50]
	Corymbia sp.	Australia	[50]
	Eucalyptus sp.	Australia	[50]
	Eucalyptus urophylla	Venezuela	[8]
	Eugenia uniflora	Brazil	[64]
	Psidium guajava	Venezuela	[89]
		Brazil	[64]
		India	[90]
		Nigeria	GenBank
	Psidium rufum	Brazil	[64]
	Syzygium cordatum	South Africa	[11]
	Syzygium cumini	India	[60]

Table 1. *Cont.*

	Source	Origin	Ref.
Sapindales, Anacardiaceae	*Anacardium occidentale*	Brazil	[91,92]
	Astronium fraxinifolium	Brazil	[64]
	Mangifera indica	Australia	[93]
		Brazil	[91]
		Venezuela	[94]
		Costa Rica	[95]
	Myracrodruon urundeuva	Brazil	[64]
	Spondias mombin	Brazil	[64]
	Spondias sp.	Brazil	[64]
Sapindales, Burseraceae	*Boswellia ovalifoliata*	India	[96]
	Boswellia sacra	Oman	[97]
	Protium heptaphyllum	Brazil	[64]
Sapindales, Meliaceae	*Azadirachta indica*	India	[43]
	Khaya anthotheca	Ghana	[98]
Sapindales, Rutaceae	*Citrus sinensis*	USA	[99]
Sapindales, Sapindaceae	*Nephelium lappaceum*	Malaysia	GenBank
	Paullinia cupana	Brazil	GenBank
Sapindales, Simaroubaceae	*Ailanthus excelsa*	India	[100]
	Simarouba amara	Brazil	[64]
Ericales, Ebenaceae	*Diospyros montana*	India	[60]
Ericales, Lecythidaceae	*Barringtonia racemosa*	South Africa	[101]
	Careya arborea	India	[60]
Ericales, Sapotaceae	*Madhuca indica*	India	[102]
Icacinales, Icacinaceae	*Nothapodytes nimmoniana*	India	[103]
	Pyrenacantha sp.	India	GenBank
Boraginales, Boraginaceae	*Auxemma oncocalyx*	Brazil	[64]
	Cordia obliqua	India	[60]
	Cordia trichotoma	Brazil	[64]
	Cordia wallichi	India	[60]
Gentianales, Apocynaceae	*Alstonia scholaris*	India	[56]
	Catharanthus roseus	India	[90,104,105]
	Hancornia speciosa	Brazil	[106]
	Holarrhena antidysenterica	India	[59]
	Plumeria rubra	India	[107]
	Rauwolfia serpentina	India	[108]
Gentianales, Loganiaceae	*Strychnos potatorum*	India	[60]
Gentianales, Rubiaceae	*Coffea arabica*	Puerto Rico	[109]
	Ixora nigricans	India	[60]

Table 1. *Cont.*

	Source	Origin	Ref.
Gentianales, Rubiaceae	*Morinda citrifolia*	India	[110]
	Psychotria flavida	India	[62,111]
	Psychotria sp.	Brazil	[64]
Lamiales, Acanthaceae	*Acanthus ilicifolius*	China	[112,113]
	Avicennia lanata	Philippines	[114]
		Malaysia	[115]
Lamiales, Bignoniaceae	*Jacaranda* sp.	Guyana	[116]
	Kigelia pinnata	India	[117]
	Radermachera xylocarpa	India	[56]
	Stereospermum angustifolium	India	[60]
Lamiales, Lamiaceae	*Gmelina arborea*	India	[60]
	Plectranthus amboinicus	India	[118]
	Pogostemon cablin	China	GenBank
	Premna tomentosa	India	[60]
	Tectona grandis	India	[60,119]
	Teucrium polium	Egypt	[120]
	Vitex negundo	India	[121]
	Vitex pinnata	Malaysia	[122]
Lamiales, Oleaceae	*Ligustrum lucidum*	Argentina	[123]
	Olea dioica	India	[56]
Solanales, Solanaceae	*Solanum melongena*	Brazil	GenBank
	Solanum nigrum	Egypt	[124]
	Solanum surratense	India	[125]
	Solanum torvum	India	[125]
	Withania somnifera	India	[125]
Apiales, Araliaceae	*Dendropanax laurifolius*	Malaysia	GenBank
Asterales, Asteraceae	*Bidens pilosa*	Egypt	[126]

The total number of 203 findings summarized in Table 1 is indicative of the widespread adaptation of *L. theobromae* to an endophytic lifestyle. They refer to as many as 189 plant species from 60 families, including representatives of the Pinophyta (seven species) along with the more numerous angiosperms. Among the latter, there are just *Annona muricata* and two *Piper* species in the Magnoliids, while Monocots and Eudicots are more common—particularly the families Orchidaceae (21 species) within the former, and Fabaceae (22 species), Combretaceae (12 species), Myrtaceae and Malvaceae (9 species each) within the latter grouping. Most of these plants are trees, which likely depends on both a preference of the fungus for lignified tissues and on the higher number of investigations on endophytes which have been carried out in forests and on woody hosts.

In geographical terms, a greater diffusion of *L. theobromae* is evident in tropical and subtropical countries (Figure 1), which is related to both the known prevalence of the fungus in this climatic zone and to the more consistent investigational activity in these countries, particularly India and Brazil, with, respectively, 81 and 32 records (ca. 40 and 16% of the total). Some reports are inaccurate and do not allow us to match the endophytic finding of *L. theobromae* with a definite host [127,128].

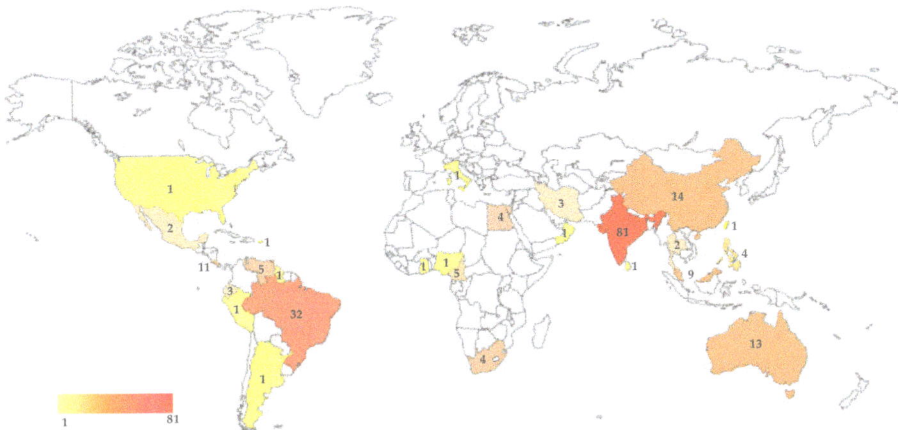

Figure 1. Geographical distribution of endophytic *Lasiodiplodia theobromae* as resulting from entries of Table 1. The color scale ranging from yellow to red is representative of the number of findings for each country.

4. Biological and Ecological Traits

As introduced above, endophytes are basically defined by their ability to spread in host tissues without inducing disease symptoms. However, the contraposition with pathogens is not so obvious, considering that many pathogens have a latent stage in their life cycle during which they are characteristically asymptomatic. The duration of this stage is very variable, and the pathogenic shift often depends on changes in the host susceptibility induced by several kinds of stress, which may reduce their tolerance or trigger a more aggressive behavior by the latent pathogen. For instance, plant stress is presented as a fundamental factor stimulating the pathogenic behavior of *L. theobromae* on dogwoods (*Cornus florida*), also considering the occasional failure of artificial inoculations during pathogenicity trials [129]. Genetic factors also actively influence the lifestyle shift of plant-associated fungi, as documented in a dedicated study disclosing repeated conversions during the evolutionary history of several species [130].

Members of the Botryosphaeriaceae are renowned as latent pathogens with a wide host range and geographical distribution [50,131]. Confirming this general feature, *L. theobromae* exerts such an ecological adaptability, particularly in tropical and subtropical regions [4,5]. However, the recent increasing trend in temperature may result in a major range expansion, placing more known and unknown hosts at risk.

Until recently, the incidence of latent pathogens has been underestimated, particularly in the trade of forest and horticulture plants and products; indeed, endophytes have been long disregarded in quarantine measures [131], which has enabled fungi to spread in plant germplasm circulating around the world [2,132]. With specific reference to *L. theobromae*, it has been conjectured that this fungus might have spread from Mexico to other subtropical countries through the trade of pine seeds [8]. Another hypothesis based on a phylogeographic approach considers the possible spread to South America from Africa to have repeatedly occurred as a consequence of human activities [28]. The availability of molecular techniques for the routine screening of plant material has increased the awareness that this risk has to be monitored [133]. In fact, besides considering pathogenic fungi of crops with an undefined latent stage [134,135], the European Food Safety Agency (EFSA) has recently started to consider the potential presence of disease agents occurring as endophytes in traded ornamental plants [136]. This concern is further supported by data gathered in this review, also considering that several hosts belong to widespread tree genera in boreal forests (e.g., *Pinus*, *Populus*, *Quercus*, *Taxus* and *Zelkova*). On the other hand, the accumulation of data on the occurrence of endophytes also provides an indication that some plants could be exempt. As an example, a recent review on the endophytic fungi of olive tree (*Olea europaea*), gathering all the available data concerning this important crop, has disclosed that, to date, there are no citations concerning *L. theobromae*, not only in the Mediterranean area but also in several tropical and subtropical countries where the plant has been introduced [3].

Many fungi reported for their endophytic occurrence are better known as plant pathogens. This is to be interpreted not only considering a more or less enduring latent stage within the disease cycle, as introduced above, but also with reference to a variable capacity by plant species to host certain fungal pathogens without showing symptoms of infection. Besides the more established concepts considering an improvement of host fitness in terms of growth promotion and protection against pests and pathogens, in the case of renowned disease agents, it has been conjectured that the capacity of a plant to host and promote their horizontal spread in the biocoenosis reflects a competitive advantage against other susceptible species [137]. This concept is quite appropriate for *L. theobromae*, which has such a high number of hosts as both a pathogen and an endophyte.

The problematic discernment of the real relationships with the host plant particularly emerged in our overview of the endophytic occurrence of *L. theobromae*. Indeed, defining this fungus as an endophyte in crops where it is known to cause disease (at least 46 plant species listed in Table 1, underlined) arouses a certain perplexity and raises the question of how to consider observations in the wild. The subject of plant pathology basically consists of diseases affecting crops or forest plants, and thorough assessments concerning fungal diseases of non-crop species are infrequent. In the absence of previous records and symptom descriptions, how can we be sure that a fungus isolated from "asymptomatic" tissues of a plant growing in whatever natural context is not exerting pathogenicity? It is worth observing that most of the plant species listed in Table 1 are not crops, and that for the majority of them, there is just a single finding, which is not at all sufficient for making a decision in this respect. Moreover, rather than being focused on the moment and circumstances of isolation, the issue should be considered with reference to the entire life cycle of the host plant: in this respect, how to consider reports of endophytic occurrence in centuries-old trees such as baobabs? [15,50].

Besides baobabs, there are more plants where it has been clearly demonstrated that the presumed endophytic occurrence is rather considered to refer to isolations carried out during the latent stage of the disease cycle. This is the case of cashew (*Anacardium occidentale*), where *L. theobromae* was recovered from healthy tissues at a distance of up to 80 cm from cankers caused by the same, and it was found to transmit through apparently healthy propagation material [92]. In other cases, the issue may be considered to have a "topographical" connotation, basically when the fungus exerts its pathogenic aptitude in some plant parts only. In fact, endophytic asymptomatic colonization of mango (*Mangifera indica*) shoots and branches has been shown to be prodromal to postharvest fruit rot [93,138]. In the case of *Aquilaria* spp. used for the production of agarwood, designating *L. theobromae* as an endophyte seems inappropriate too; in fact, resin formation is promoted as a reaction to an infection process which rather qualifies the fungus as a pathogen [139]. Likewise, internal infections by *L. theobromae* are reported to cause blue stain of wood after felling in *Pinus elliottii* [8], as well as in *Terminalia* spp. [87] and rubberwood (*Hevea brasiliensis*) [140]. It is worth considering that in similar cases observed on neem (*Azadirachta indica*) [141] and *Ficus insipida* [142], the occurrence of the fungus is merely referred to as a pathogenic association.

5. Bioactivities of Endophytic Isolates of *Lasiodiplodia theobromae*

Endophytes present potential for the exploitation of metabolites and enzymes. The biosynthesis of many secondary metabolites is often a response to environmental factors and fulfils different functions, such as defense, signaling and nutrient acquisition. Moreover, endophytes can influence the metabolism of the host and modify secondary metabolites by enzymatic steps of biochemical transformation [143].

Many studies have shown that endophytic fungi can synthesize bioactive products identical or similar to those produced by plants, representing an alternative source of some drugs and new useful medicinal compounds [144,145]. For this reason, many researchers have focused their attention on endophytes of medicinal plants, and many strains have been isolated which could be used for producing plant-derived drugs through fermentation. Among these fungi, *L. theobromae* particularly stands out for its ability to synthesize a high number of bioactive compounds [146]. The current panel of products is expected to further increase with reference to the many studies disclosing bioactive properties by endophytic strains of this species. Table 2 refers to investigations concerning endophytic strains of *L. theobromae* as a possible source of bioactive products, which sometimes are limited to assays carried out with culture filtrates.

Table 2. Bioactivities of endophytic isolates of *Lasiodiplodia theobromae*.

Bioactivity	Source	Sample tested	Ref.
Antibacterial	*Acanthus ilicifolius*	Secondary metabolites	[112]
	Aquilaria sinensis	Culture filtrate extract	[81]
	Calamus thwaitesii	Culture filtrate extract	[42]
	Dracaena draco	Culture filtrate extract	[36]
	Garcinia mangostana	Secondary metabolites	[69]
	Hancornia speciosa	Culture filtrate extract	[106]
	Humboldtia brunonis	Culture filtrate extract	[62]
	Madhuca indica	Culture filtrate extract	[102]
	Piper hispidum	Culture filtrate extract	[33]
	Terminalia arjuna	Culture filtrate extract	[84]
Antifungal	*A. sinensis*	Culture filtrate extract	[81]
	Avicennia lanata	Culture filtrate extract	[114]
	Bidens pilosa	Culture filtrate extract and secondary metabolites	[126]
	H. speciosa	Culture filtrate extract	[106]
	H. brunonis	Culture filtrate extract	[62]
	T. arjuna	Culture filtrate extract	[84]
Anti-inflammatory	*Acanthus ilicifolius*	Secondary metabolites	[113]
Antioxidant	*Catharanthus roseus*	Culture filtrate and mycelial extracts	[104]
	C. roseus	Silver nanoparticles	[105]
	T. arjuna	Culture filtrate extract	[84]
Antiprotozoal	*A. lanata*	Culture filtrate extract and chromatographic fraction	[115]
	Vitex pinnata	Secondary metabolites	[122]
Cytotoxic	*Acanthus ilicifolius*	Secondary metabolites	[112]
	A. sinensis	Culture filtrate extract	[81]
	B. pilosa	Culture filtrate extract and secondary metabolites	[126]
	C. roseus	Silver nanoparticles	[90]
	C. roseus	Culture filtrate and mycelial extracts	[104]
	Morinda citrifolia	Secondary metabolite	[110]
	Plectranthus amboinicus	Secondary metabolite	[118]
Enzymatic	*Azadirachta indica*	Isolate	[43]
	Cocos nucifera	Isolate	[43]
	Pongamia pinnata	Isolate	[43]
	Psychotria flavida	Isolate	[111]
	Terminalia catappa	Isolate	[86]
	Terminalia mantaly	Isolate	[86]
Heavy metal tolerance	*Boswellia ovalifoliata*	Isolate	[96]

Concerning the antibacterial activity, extracts produced by endophytic strains from the medicinal plant *Piper hispidum* were effective against four human pathogenic bacteria (i.e., *Enterococcus hirae*, *Escherichia coli*, *Micrococcus luteus* and *Staphylococcus aureus*) and showed good activity against *Salmonella tiphy* [33]. Antimicrobial activity was again displayed by endophytic strains from *Hancornia speciosa*, a plant native to Brazil, used to treat various pathologies [106].

Strains isolated from leaves, twigs and bark of *Terminalia arjuna* showed antimicrobial activity against *Bacillus subtilis* and *Aspergillus niger*, along with significant antioxidant properties [84]. The culture extract from an endophytic strain isolated from the mangrove *Avicennia lanata* in the Philippines was very active against the yeast *Saccharomyces cereviseae* but inactive against several Gram-negative and Gram-positive bacteria [114].

The culture extracts of endophytic strains from leaf and stem segments of *Humboldtia brunonis* were inhibitory against *Bacillus subtilis*, *S. aureus*, *Klebsiella pneumoniae*, *Proteus volgaris* and *Candida albicans* [62]. The crude extract from another endophytic strain isolated from *Madhuca indica* in India was found to be active against several common bacteria [102]. A strain isolated from *A. sinensis* showed low antimicrobial activity against microbial pathogens, particularly *Aspergillus famigatus*. This strain also displayed cytotoxic activity against some cancer cell lines [81]. Likewise, the culture extract of a strain from *Catharanthus roseus* exhibited cytotoxicity against the human cervical adenocarcinoma (HeLa) cell line [104].

The anticancer activity was particularly prominent when metal nanoparticles were prepared by exposing the endophytic fungus to metal salt solution. In fact, *L. theobromae* from leaves of *Psidium guajava* was used for the biological synthesis of silver nanoparticles, which provided powerful antitumor activity against human breast and lung cancer cells [90]. Silver nanoparticles were also prepared using an endophytic strain of *L. theobromae* isolated from *C. roseus*, inducing apoptosis in various types of cancer cells and promoting free radical scavenging [105]. These findings suggest that natural compounds produced by these isolates and incorporated into the nanoparticles have potential as a novel chemotherapeutic agent.

Finally, an endophytic strain of *Boswellia ovalifoliolata* is capable of growing in the presence of heavy metals (i.e., Co, Cd, Cu and Zn) in concentrations up to 600 ppm, showing that it may be used to remove heavy metals from solid substrates [96].

6. Secondary Metabolites and Enzymes of Endophytic *Lasiodiplodia theobromae*

As introduced above, the biological properties of culture extracts of endophytic *L. theobromae* might be linked to the capacity of the fungus to produce bioactive compounds (Figure 2). In fact, *L. theobromae* is a proficient producer of compounds belonging to different classes of secondary metabolites, such as diketopiperazines, indoles, jasmonates, melleins, lactones and phenols [146].

Biotic and abiotic stimuli influence the capacity of *L. theobromae* to grow and produce secondary metabolites, with implications for its physiology, lifestyle and pathogenic aptitude [146–148]. Studies on fungal genomes have shown that the capability of fungi to produce secondary metabolites has been underestimated, because many secondary metabolite biosynthetic gene clusters are silent under standard cultivation conditions [149,150]. In fact, different metabolomic profiles have been reported for *L. theobromae* strains according to variation in growth conditions, with reference to temperature [147,148], nutrient availability [151,152], presence of signal molecules [153] and incubation period [122].

Metabolomic investigations of *L. theobromae* have pointed out that some compounds are produced by endophytic strains only. This is the case of preussomerins and cloropreussomerins, compounds with an unusual structure isolated from the culture extract of a strain from leaves of the mangrove *Acanthus ilicifolius* and characterized for their cytotoxicity against five human cancer cell lines [112]. Moreover, endophytic strains from *Aquilaria sinensis* have been reported to produce 2-(2-phenylethyl)chromones, which are among the most abundant constituents of agarwood [154]. The coumarins meranzine and monocerin could be responsible for the antimicrobial activity of the culture extract of an endophytic strain from *Dracaena draco*, displaying characteristic inhibition zones against Gram-positive and Gram-negative bacteria [36].

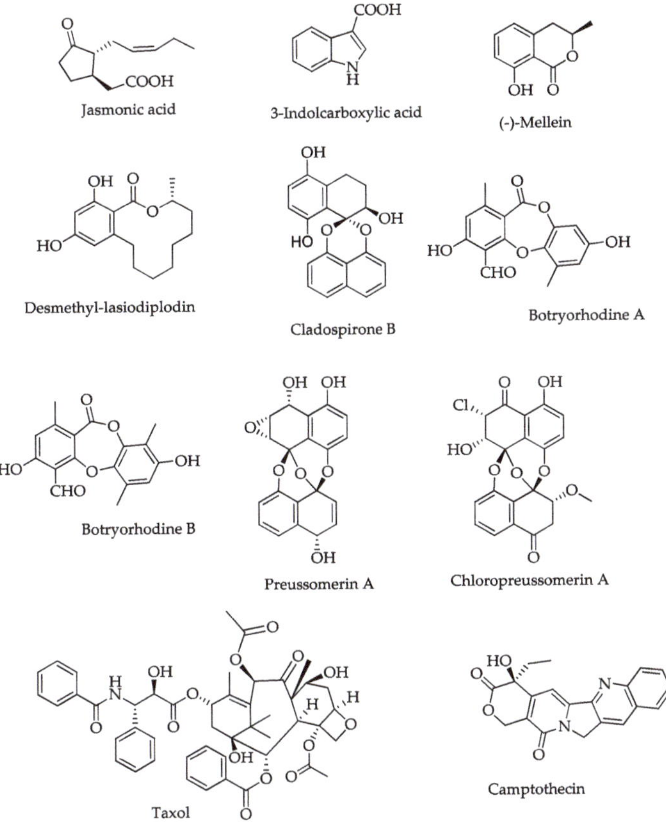

Figure 2. Representative secondary metabolites produced by endophytic *Lasiodiplodia theobromae*.

Lasiodiplodins were frequently, although not exclusively, reported as products of endophytic strains of *L. theobromae* [47,69,113]. These macrolides are relevant for a variety of biological properties including cytotoxic, antimicrobial and anti-inflammatory activities [69,155]. Within this class, lasiodiplactone A was obtained from a mangrove endophytic strain showing anti-inflammatory activity [113]. Furthermore, desmethyl-lasiodiplodin was isolated, together with cladospirone B and (-)-mellein, from the crude extract of a strain from leaves of *Vitex pinnata*. Interestingly, cladospirone B and desmethyl-lasiodiplodin showed good activity against *Trypanosoma brucei* [122].

An endophytic strain from the medicinal plant *Bidens pilosa* yielded four depsidones, botryorhodines A-D, and the auxin 3-indolecarboxylic acid, which are not exclusively produced by endophytic strains. Botryorhodines A and B show moderate cytotoxic activity against cervical cancer cells (i.e., HeLa) and antifungal activity against pathogenic fungi, such as *Aspergillus terreus* and *Fusarium oxysporum* [126].

The fact that two of the leading natural products, namely camptothecin and taxol, in cancer chemotherapy were originally extracted from plants is quite interesting from an applicative perspective [144]. The first compound has been detected as a secondary metabolite of strains isolated from the leaves and stem of *Nothapodytes nimmoniana* in the Western Ghats, India [103]. One of these strains (L-6) was investigated in depth with reference to the common phenomenon of attenuation of bioactive metabolite production in axenic cultures. It was found that its re-inoculation in the host promoted higher production of camptothecin, indicating that the fungus receives eliciting signals from the host tissues, or some factors which prevent silencing of the genes responsible for biosynthesis [156].

Taxol, the first billion-dollar natural antitumor product [157], has been reported as a secondary metabolite of several endophytic strains of *L. theobromae*, from *Taxus baccata*, *Morinda citrifolia*, *Salacia oblonga* and *Piper nigrum* [32,35,57,110]. Investigational activity carried out on the product extracted from these strains pointed out its ability to counteract the carcinogenic effects of dimethylbenzanthracene [158]. Moreover, valuable studies have disclosed the capacity by non-*Taxus* endophytic strains to produce the compound through a similar biosynthetic pathway as the one reported from the plant. In fact, the gene encoding 10-deacetylbaccatin-III-*O*-acetyltransferase, as well as the open reading frame of WRKY1 transcription factor, were cloned and sequenced and found to share high similarity with deposited sequences from *Taxus chinensis*, *T. cuspidata* and *T. celebica* [35].

Of great interest in endophytic *L. theobromae* is the production of phytohormones, such as indole derivatives and jasmonic acid analogues [146]. It is known that 3-indoleacetic acid and 3-indolecarboxylic acid are the most studied auxins regulating plant growth and development. These compounds have been frequently reported as fungal metabolites [144] and have also been documented as being produced by *L. theobromae* strains. The biological role of 3-indolecarboxylic acid has not been fully investigated, but some studies address its biosynthesis [159–161] and toxicity [147]. Several *L. theobromae* strains with different lifestyles are in vitro producers of jasmonic acid and analogues. Jasmonic acid is one of the most important signal molecules involved in several plant processes including seed germination, senescence and blooming. Hence, investigations of the bioactive properties of jasmonic acid and related compounds are essentially focused on their role in the interaction between host and pathogen.

The great ability of adaptation to different environments, the capacity to colonize a high number of hosts and the expression of high amounts of extracellular enzymes make *L. theobromae* a producer of relevant enzymes (Table 2) to be considered for biotechnological applications [162]. The most recognized extracellular enzymes used to penetrate the plant host include cellulases, proteases and lipases. Endophytic strains colonizing *C. nucifera*, *Pongamia pinnata* and *A. indica* exhibited great lipase activity [43]. Moreover, endophytic strains from *Terminalia catappa* and *T. mantaly* were found to produce amylases and cellulases [86]. Finally, *L. theobromae* isolated from *Psychotria flavida* turned out to be able to degrade irradiated polypropylene thanks to the production of laccases [62].

7. Conclusions

This overview of the endophytism of *L. theobromae* based on the literature published in the last three decades has pointed out its widespread occurrence in tropical and subtropical areas and the likeliness of further spread to regions with a temperate climate following the increasing trade of plant material. Hints concerning the biochemical properties are indicative of a certain degree of adaptation to the endophytic lifestyle, particularly deriving from the ability to synthesize bioactive products which may contribute to protection against biological adversities and improve plant fitness. However, the analysis of the available information also raises questions on whether the ability of *L. theobromae* to colonize such a high number of hosts is rather to be referred to as a fundamental pathogenic aptitude and whether a number of reports are actually referable to its interception during the latency phase of the disease cycle. Finding reasonable answers is clearly dependent on the analysis of additional data resulting from dedicated investigations in both natural and agricultural contexts.

Author Contributions: Conceptualization, R.N. and A.A.; writing—original draft preparation, M.M.S.; writing—review and editing, M.M.S., R.N. and A.A. All authors have read and agreed to the published version of the manuscript.

Funding: This research received no external funding.

Conflicts of Interest: The authors declare no conflict of interest.

References

1. Stone, J.K.; Bacon, C.W.; White, J.F. An overview of endophytic microbes: Endophytism defined. *Microb. Endophytes* **2000**, *3*, 29–33.
2. Nicoletti, R. Endophytic Fungi of Citrus Plants. *Agriculture* **2019**, *9*, 247. [CrossRef]
3. Nicoletti, R.; Di Vaio, C.; Cirillo, C. Endophytic fungi of olive tree. *Microorganisms* **2020**, *8*, 1321. [CrossRef]
4. Mehl, J.; Wingfield, M.J.; Roux, J.; Slippers, B. Invasive everywhere? Phylogeographic analysis of the globally distributed tree pathogen *Lasiodiplodia theobromae*. *Forests* **2017**, *8*, 145. [CrossRef]
5. Punithalingam, E. *Plant Diseases Attributed to Botryodiplodia theobromae Pat*; J. Cramer: Vaduz, Lichtenstein, 1980; ISBN 9783768212564.
6. Cilliers, A.J.; Swart, W.J.; Wingfield, M.J. A review of *Lasiodiplodia theobromae* with particular reference to its occurrence on coniferous seeds. *S. Afr. For. J.* **1993**, *166*, 47–52.
7. Hawksworth, D.L.; Crous, P.W.; Redhead, S.A.; Reynolds, D.R.; Samson, R.A.; Seifert, K.A.; Taylor, J.W.; Wingfield, M.J.; Abaci, Ö.; Aime, C.; et al. The Amsterdam declaration on fungal nomenclature. *IMA Fungus* **2011**, *2*, 105–111. [CrossRef] [PubMed]
8. Mohali, S.; Burgess, T.I.; Wingfield, M.J. Diversity and host association of the tropical tree endophyte *Lasiodiplodia theobromae* revealed using simple sequence repeat markers. *For. Pathol.* **2005**, *35*, 385–396. [CrossRef]
9. Larignon, P.; Fulchic, R.; Cere, L.; Dubos, B. Observation on black dead arm in French vineyards. *Phytopathol. Mediterr.* **2001**, *40*, 336–342.
10. Slippers, B.; Boissin, E.; Phillips, A.J.L.; Groenewald, J.Z.; Lombard, L.; Wingfield, M.J.; Postma, A.; Burgess, T.; Crous, P.W. Phylogenetic lineages in the botryosphaeriales: A systematic and evolutionary framework. *Stud. Mycol.* **2013**, *76*, 32–49. [CrossRef]
11. Pavlic, D.; Slippers, B.; Coutinho, T.A.; Gryzenhout, M.; Wingfield, M.J. *Lasiodiplodia gonubiensis* sp. nov., a new *Botryosphaeria* anamorph from native *Syzygium cordatum* in South Africa. *Stud. Mycol.* **2004**, *50*, 313–322.
12. Burgess, T.I.; Barber, P.A.; Mohali, S.; Pegg, G.; De Beer, W.; Wingfield, M.J. Three new *Lasiodiplodia* spp. from the tropics, recognized based on DNA sequence comparisons and morphology. *Mycologia* **2006**, *98*, 423–435. [CrossRef] [PubMed]
13. Damm, U.; Crous, P.W.; Fourie, P.H. Botryosphaeriaceae as potential pathogens of Prunus species in South Africa, with descriptions of *Diplodia africana* and *Lasiodiplodia plurivora* sp. nov. *Mycologia* **2007**, *99*, 664–680. [CrossRef] [PubMed]
14. Alves, A.; Crous, P.W.; Correia, A.; Phillips, A.J.L. Morphological and molecular data reveal cryptic speciation in *Lasiodiplodia theobromae*. *Fungal Divers.* **2008**, *28*, 1–13.
15. Cruywagen, E.M.; Slippers, B.; Roux, J.; Wingfield, M.J. Phylogenetic species recognition and hybridisation in *Lasiodiplodia*: A case study on species from baobabs. *Fungal Biol.* **2017**, *121*, 420–436. [CrossRef]
16. Coutinho, I.B.L.; Freire, F.C.O.; Lima, C.S.; Lima, J.S.; Gonçalves, F.J.T.; Machado, A.R.; Silva, A.M.S.; Cardoso, J.E. Diversity of genus *Lasiodiplodia* associated with perennial tropical fruit plants in northeastern Brazil. *Plant Pathol.* **2017**, *66*, 90–104. [CrossRef]
17. Netto, M.S.B.; Lima, W.G.; Correia, K.C.; da Silva, C.F.B.; Thon, M.; Martins, R.B.; Miller, R.N.G.; Michereff, S.J.; Câmara, M.P.S. Analysis of phylogeny, distribution, and pathogenicity of *Botryosphaeriaceae* species associated with gummosis of *Anacardium* in Brazil, with a new species of *Lasiodiplodia*. *Fungal Biol.* **2017**, *121*, 437–451. [CrossRef]
18. Jiang, N.; Wang, X.W.; Liang, Y.M.; Tian, C.M. *Lasiodiplodia cinnamomi* sp. nov. from *Cinnamomum camphora* in China. *Mycotaxon* **2018**, *133*, 249–259. [CrossRef]
19. Budiono, B.; Elfita, E.; Muharni, M.; Yohandini, H.; Widjajanti, H. Antioxidant activity of *Syzygium samarangense* L. and their endophytic fungi. *Molekul* **2019**, *14*, 48–55. [CrossRef]
20. de Silva, N.I.; Phillips, A.J.L.; Liu, J.K.; Lumyong, S.; Hyde, K.D. Phylogeny and morphology of *Lasiodiplodia* species associated with *Magnolia* forest plants. *Sci. Rep.* **2019**, *9*, 14355. [CrossRef]
21. Wang, Y.; Lin, S.; Zhao, L.; Sun, X.; He, W.; Zhang, Y.; Dai, Y.C. *Lasiodiplodia* spp. associated with *Aquilaria crassna* in Laos. *Mycol. Prog.* **2019**, *18*, 683–701. [CrossRef]
22. Zhao, L.; Wang, Y.; He, W.; Zhang, Y. Stem blight of blueberry caused by *Lasiodiplodia vaccinii* sp. nov. in China. *Plant Dis.* **2019**, *103*, 2041–2050. [CrossRef] [PubMed]

23. Berraf-Tebbal, A.; Mahamedi, A.E.; Aigoun-Mouhous, W.; Špetík, M.; Čechová, J.; Pokluda, R.; Baránek, M.; Eichmeier, A.; Alves, A. *Lasiodiplodia mitidjana* sp. nov. and other *Botryosphaeriaceae* species causing branch canker and dieback of *Citrus sinensis* in Algeria. *PLoS ONE* **2020**, *15*, e0232448. [CrossRef] [PubMed]

24. Laurent, B.; Marchand, M.; Chancerel, E.; Saint-Jean, G.; Capdevielle, X.; Poeydebat, C.; Bellée, A.; Comont, G.; Villate, L.; Desprez-Loustau, M.-L. A richer community of Botryosphaeriaceae within a less diverse community of fungal endophytes in grapevines than in adjacent forest trees revealed by a mixed metabarcoding strategy. *Phytobiomes J.* **2020**, *4*, 252–267. [CrossRef]

25. Rodríguez-Gálvez, E.; Guerrero, P.; Barradas, C.; Crous, P.W.; Alves, A. Phylogeny and pathogenicity of *Lasiodiplodia* species associated with dieback of mango in Peru. *Fungal Biol.* **2017**, *121*, 452–465. [CrossRef]

26. Úrbez-Torres, J.R. The status of Botryosphaeriaceae species infecting grapevines. *Phytopathol. Mediterr.* **2011**, *50*, S5–S45.

27. Taylor, J.W.; Jacobson, D.J.; Kroken, S.; Kasuga, T.; Geiser, D.M.; Hibbett, D.S.; Fisher, M.C. Phylogenetic species recognition and species concepts in fungi. *Fungal Genet. Biol.* **2000**, *31*, 21–32. [CrossRef]

28. Santos, P.H.D.; Carvalho, B.M.; Aguiar, K.P.; Aredes, F.A.S.; Poltronieri, T.P.S.; Vivas, J.M.S.; Mussi Dias, V.; Bezerra, G.A.; Pinho, D.B.; Pereira, M.G.; et al. Phylogeography and population structure analysis reveals diversity by mutations in *Lasiodiplodia theobromae* with distinct sources of selection. *Genet. Mol. Res.* **2017**, *16*. [CrossRef]

29. Boyogueno, A.D.B.; Slippers, B.; Perez, G.; Wingfield, M.J.; Roux, J. High gene flow and outcrossing within populations of two cryptic fungal pathogens on a native and non-native host in Cameroon. *Fungal Biol.* **2012**, *116*, 343–353. [CrossRef]

30. Wang, Y.; Guo, L.D.; Hyde, K.D. Taxonomic placement of sterile morphotypes of endophytic fungi from *Pinus tabulaeformis* (Pinaceae) in northeast China based on rDNA sequences. *Fungal Divers.* **2005**, *20*, 235–260.

31. Ma, Y.M.; Ma, C.C.; Li, T.; Wang, J. A new furan derivative from an endophytic *Aspergillus flavus* of *Cephalotaxus fortunei*. *Nat. Prod. Res.* **2016**, *30*, 79–84. [CrossRef]

32. Venkatachalam, R.; Subban, K.; Paul, M.J. Taxol from *Botryodiplodia theobromae* (BT 115)—An endophytic fungus of *Taxus baccata*. In Proceedings of the 13th International Biotechnology Symposium and Exhibition, Dalian, China, 12–17 October 2008; pp. 189–190.

33. Orlandelli, R.C.; Alberto, R.N.; Almeida, T.T.; Azevedo, J.L.; Pamphile, J.A. In vitro antibacterial activity of crude extracts produced by endophytic fungi isolated from *Piper hispidum* sw. *J. Appl. Pharm. Sci.* **2012**, *2*, 137–141. [CrossRef]

34. Orlandelli, R.C.; Alberto, R.N.; Rubin Filho, C.J.; Pamphile, J.A. Diversity of endophytic fungal community associated with *Piper hispidum* (Piperaceae) leaves. *Genet. Mol. Res.* **2012**, *11*, 1575–1585. [CrossRef] [PubMed]

35. Sah, B.; Subban, K.; Chelliah, J. Cloning and sequence analysis of 10-deacetylbaccatin III-10-O-acetyl transferase gene and WRKY1 transcription factor from taxol-producing endophytic fungus *Lasiodiplodia theobromea*. *FEMS Microbiol. Lett.* **2017**, *364*, fnx253. [CrossRef] [PubMed]

36. Zaher, A.M.; Moharram, A.M.; Davis, R.; Panizzi, P.; Makboul, M.A.; Calderón, A.I. Characterisation of the metabolites of an antibacterial endophyte *Botryodiplodia theobromae* Pat. of *Dracaena draco* L. by LC-MS/MS. *Nat. Prod. Res.* **2015**, *29*, 2275–2281. [CrossRef]

37. Richardson, K.A.; Currah, R.S. The fungal community associated with the roots of some rainforest epiphytes of Costa Rica. *Selbyana* **1995**, *16*, 49–73.

38. Duarte dos Santos, C.; Rocha Sousa, E.M.; Leal Candeias, E.; Santos Vitória, N.; Bezerra, J.L.; Newman, M.; Luz, E.D. *Lasiodiplodia theobromae* as an endophyte on Orchidaceae in Bahia. *Agrotrópica* **2012**, *24*, 179–182. [CrossRef]

39. Govinda Rajulu, M. Endophytic fungi of orchids of Arunachal Pradesh, North Eastern India. *Curr. Res. Environ. Appl. Mycol.* **2016**, *6*, 293–299. [CrossRef]

40. Sawmya, K.; Vasudevan, T.G.; Murali, T.S. Fungal endophytes from two orchid species—Pointer towards organ specificity. *Czech Mycol.* **2013**, *65*, 89–101. [CrossRef]

41. Tibpromma, S.; Hyde, K.D.; Bhat, J.D.; Mortimer, P.E.; Xu, J.; Promputtha, I.; Doilom, M.; Yang, J.-B.; Tang, A.M.C.; Karunarathna, S.C. Identification of endophytic fungi from leaves of Pandanaceae based on their morphotypes and DNA sequence data from southern Thailand. *MycoKeys* **2018**, *33*, 25–67. [CrossRef]

42. Dissanayake, R.K.; Ratnaweera, P.B.; Williams, D.E.; Wijayarathne, C.D.; Wijesundera, R.L.C.; Andersen, R.J.; de Silva, E.D. Antimicrobial activities of mycoleptodiscin B isolated from endophytic fungus *Mycoleptodiscus* sp. of *Calamus thwaitesii* Becc. *J. Appl. Pharm. Sci.* **2016**, *6*, 1–6. [CrossRef]

43. Venkatesagowda, B.; Ponugupaty, E. Diversity of plant oil seed-associated fungi isolated from seven oil-bearing seeds and their potential for the production of lipolytic enzymes. *World J. Microbiol. Biotechnol.* **2012**, *28*, 71–80. [CrossRef] [PubMed]

44. Matsumoto, M.; Nago, H. (R)-2-octeno-δ-lactone and other volatiles produced by *Lasiodiplodia theobromae*. *Biosci. Biotechnol. Biochem.* **1994**, *58*, 1262–1266. [CrossRef]

45. Rodrigues, K.F. The foliar fungal endophytes of the Amazonian palm *Euterpe oleracea*. *Mycologia* **1994**, *86*, 376–385. [CrossRef]

46. Aun, E.S.S.; Hui, J.Y.S.; Ming, J.W.W.; Yi, J.C.C.; Wong, C.; Mujahid, A.; Müller, M. Screening of endophytic fungi for biofuel feedstock production using palm oil mill effluent as a carbon source. *Malays. J. Microbiol.* **2017**, *13*, 203–209.

47. Sultan, S.; Sun, L.; Blunt, J.W.; Cole, A.L.J.; Munro, M.H.G.; Ramasamy, K.; Weber, J.-F.F. Evolving trends in the dereplication of natural product extracts. 3: Further lasiodiplodins from *Lasiodiplodia theobromae*, an endophyte from *Mapania kurzii*. *Tetrahedron Lett.* **2014**, *55*, 453–455. [CrossRef]

48. Amirita, A.; Sindhu, P.; Swetha, J.; Vasanthi, N.S.; Kannan, K.P. Enumeration of endophytic fungi from medicinal plants and screeening of extracellular enzymes. *World J. Sci. Technol.* **2012**, *2*, 13–19.

49. Zakaria, L.; Nuraini, W.; Aziz, W.; Pisang, D. Molecular identification of endophytic fungi from banana leaves (*Musa* spp.). *Trop. Life Sci. Res.* **2018**, *29*, 201–211. [CrossRef]

50. Sakalidis, M.L.; Hardy, G.E.S.J.; Burgess, T.I. Endophytes as potential pathogens of the baobab species *Adansonia gregorii*: A focus on the Botryosphaeriaceae. *Fungal Ecol.* **2011**, *4*, 1–14. [CrossRef]

51. Mishra, A.; Gond, S.K.; Kumar, A.; Sharma, V.K.; Verma, S.K.; Kharwar, R.N.; Sieber, T.N. Season and tissue type affect fungal endophyte communities of the indian medicinal plant *Tinospora cordifolia* more strongly than geographic location. *Microb. Ecol.* **2012**, *64*, 388–398. [CrossRef]

52. Jin, B.; Jiang, F.; Xu, F.; Ding, Z. An antitumor activity endophytic fungus A33 isolated from *Viscum Coloratum* of Chinese. In Proceedings of the 2011 International Conference on Remote Sensing, Environment and Transportation Engineering, RSETE 2011, Nanjing, China, 24–26 June 2011; pp. 7368–7371.

53. Duan, X.; Xu, F.; Qin, D.; Gao, T.; Shen, W.; Zuo, S.; Yu, B.; Xu, J.; Peng, Y.; Dong, J. Diversity and bioactivities of fungal endophytes from *Distylium chinense*, a rare waterlogging tolerant plant endemic to the Three Gorges Reservoir. *BMC Microbiol.* **2019**, *19*, 278. [CrossRef]

54. Dissanayake, A.J.; Purahong, W.; Wubet, T.; Hyde, K.D.; Zhang, W.; Xu, H.; Zhang, G.; Fu, C.; Liu, M.; Xing, Q.; et al. Direct comparison of culture-dependent and culture-independent molecular approaches reveal the diversity of fungal endophytic communities in stems of grapevine (*Vitis vinifera*). *Fungal Divers.* **2018**, *90*, 85–107. [CrossRef]

55. Burruano, S.; Alfonzo, A.; Conigliaro, G.; Mondello, V.; Torta, L. First observations on the interaction between *Lasiodiplodia theobromae* and *Epicoccum purpurascens*, both endophytes in grapevine buds. In Proceedings of the XV Convegno Nazionale della Società Italiana di Patologia Vegetale (SIPaV), Locorotondo, Italy, 28 September–1 October 2009.

56. Suryanarayanan, T.S.; Murali, T.S.; Venkatesan, G. Occurrence and distribution of fungal endophytes in tropical forests across a rainfall gradient. *Can. J. Bot.* **2002**, *80*, 818–826. [CrossRef]

57. Roopa, G.; Madhusudhan, M.C.; Sunil, K.C.R.; Lisa, N.; Calvin, R.; Poornima, R.; Zeinab, N.; Kini, K.R.; Prakash, H.S.; Geetha, N. Identification of taxol-producing endophytic fungi isolated from *Salacia oblonga* through genomic mining approach. *J. Genet. Eng. Biotechnol.* **2015**, *13*, 119–127. [CrossRef] [PubMed]

58. Jami, F.; Slippers, B.; Wingfield, M.J.; Loots, M.T.; Gryzenhout, M. Temporal and spatial variation of Botryosphaeriaceae associated with *Acacia karroo* in South Africa. *Fungal Ecol.* **2015**, *15*, 51–62. [CrossRef]

59. Tejesvi, M.V.; Mahesh, B.; Nalini, M.S.; Prakash, H.S.; Kini, K.R.; Subbiah, V.; Shetty, H.S. Fungal endophyte assemblages from ethnopharmaceutically important medicinal trees. *Can. J. Microbiol.* **2006**, *52*, 427–435. [CrossRef]

60. Murali, T.S.; Suryanarayanan, T.S.; Venkatesan, G. Fungal endophyte communities in two tropical forests of southern India: Diversity and host affiliation. *Mycol. Prog.* **2007**, *6*, 191–199. [CrossRef]

61. Arora, P.; Wani, Z.A.; Ahmad, T.; Sultan, P.; Gupta, S.; Riyaz-ul-hassan, S. Community structure, spatial distribution, diversity and functional characterization of culturable endophytic fungi associated with *Glycyrrhiza glabra* L. *Fungal Biol.* **2019**, *123*, 373–383. [CrossRef]

62. Sheik, S.; Chandrashekar, K.R. Fungal endophytes of an endemic plant *Humboldtia brunonis* wall of western Ghats (India) and their antimicrobial and DPPH radical scavenging potentiality. *Orient. Pharm. Exp. Med.* **2018**, *18*, 115–125. [CrossRef]

63. Prazeres, I.; Diogo, J.; Bezerra, P.; De Souza, C.M.; Cavalcanti, S.; Lucia, V.; Lima, D.M. Endophytic mycobiota from leaves of *Indigofera suffruticosa* Miller (Fabaceae): The relationship between seasonal change in Atlantic Coastal Forest and tropical dry forest (Caatinga), Brazil. *Afr. J. Microbiol. Res.* **2015**, *9*, 1227–1235.

64. Gonçalves, F.J.T.; Freire, F.D.C.O.; Lima, J.S.; Melo, J.G.M.; Câmara, M.P.S. Patogenicidade de espécies de *Botryosphaeriaceae* endofíticas de plantas da Caatinga do estado do Ceará em manga e umbu-cajá. *Summa Phytopath.* **2016**, *42*, 43–52. [CrossRef]

65. López-González, R.C.; Gómez-Cornelio, S.; Susana, C.; Garrido, E.; Oropeza-Mariano, O.; Heil, M.; Partida-Martínez, L.P. The age of lima bean leaves influences the richness and diversity of the endophytic fungal community, but not the antagonistic effect of endophytes against *Colletotrichum lindemuthianum*. *Fungal Ecol.* **2017**, *26*, 1–10. [CrossRef]

66. Jinu, M.V.; Gini, C.K.; Jayabaskaran, C. In vitro antioxidant activity of cholestanol glucoside from an endophytic fungus *Lasiodiplodia theobromae* isolated from *Saraca asoca*. *J. Chem. Pharm. Res.* **2015**, *7*, 952–962.

67. Jinu, M.V.; Jayabaskaran, C. Diversity and anticancer activity of endophytic fungi associated with the medicinal plant *Saraca asoca*. *Curr. Res. Environ. Appl. Mycol.* **2015**, *5*, 169–179. [CrossRef]

68. Yao, Y.Q.; Lan, F.; Ming, Y.; Ji, Q.; Wei, G.; Shao, R.; Liang, H.; Li, B. Endophytic fungi harbored in the root of *Sophora tonkinensis* Gapnep: Diversity and biocontrol potential against phytopathogens. *Microbiol. Open* **2017**, *6*, e00437. [CrossRef] [PubMed]

69. Rukachaisirikul, V.; Arunpanichlert, J.; Sukpondma, Y. Metabolites from the endophytic fungi *Botryosphaeria rhodina* PSU-M35 and PSU-M114. *Tetrahedron* **2009**, *65*, 10590–10595. [CrossRef]

70. Gazis, R.; Chaverri, P. Diversity of fungal endophytes in leaves and stems of wild rubber trees (*Hevea brasiliensis*) in Peru. *Fungal Ecol.* **2010**, *3*, 240–254. [CrossRef]

71. Samaga, P.V.; Vittal, R.R. Diversity and bioactive potential of endophytic fungi from *Nothapodytes foetida*, *Hypericum mysorense* and *Hypericum japonicum* collected from Western Ghats of India. *Ann. Microbiol.* **2015**, *66*, 229–244. [CrossRef]

72. Guerrero, J.J.G.; General, M.A.; Serrano, J.E. Culturable foliar fungal endophytes of mangrove epecies in Bicol region, Philippines. *Philipp. J. Sci.* **2018**, *147*, 563–574.

73. Zhou, J.; Li, G.; Deng, Q.; Zheng, D.; Yang, X.; Xu, J. Cytotoxic constituents from the mangrove endophytic *Pestalotiopsis* sp. induce G 0 /G 1 cell cycle arrest and apoptosis in human cancer cells. *Nat. Prod. Res.* **2018**, *32*, 2968–2972. [CrossRef]

74. Li, R.; Yan, D.; Feng, X.; Sun, X. Diversity of *Botryosphaeria* spp., as endophytes in poplars in Beijing, based on molecular operational taxonomic units. *Sci. Silvae Sin.* **2014**, *50*, 109–115.

75. Huang, J.H.; Xiang, M.M.; Jiang, Z.D. Endophytic fungi of bitter melon (*Momordica Charantia*) in Guangdong province, China. *Gt. Lakes Entomol.* **2012**, *45*, 2.

76. Suryanarayanan, T.S.; Venkatachalam, A.; Rajulu, M.G. A comparison of endophyte assemblages in transgenic and non-transgenic cotton plant tissues. *Curr. Sci.* **2011**, *101*, 1472–1474.

77. Rubini, M.R.; Silva-ribeiro, R.T.; Pomella, A.W.V.; Maki, C.S. Diversity of endophytic fungal community of cacao (*Theobroma cacao* L.) and biological control of *Crinipellis perniciosa*, causal agent of Witches' Broom Disease. *Int. J. Mol. Sci.* **2005**, *1*, 24–33. [CrossRef] [PubMed]

78. Chaithra, M.; Vanitha, S.; Ramanathan, A.; Jegadeeshwari, V.; Rajesh, V. Morphological and molecular characterization of endophytic fungi associated with Cocoa (*Theobroma cacao* L.) in India. *Curr. Appl. Sci. Technol.* **2020**, *38*, 1–8. [CrossRef]

79. Evans, H.C.; Holmes, K.A.; Thomas, S.E. Endophytes and mycoparasites associated with an indigenous forest tree. *Mycol. Prog.* **2003**, *2*, 149–160. [CrossRef]

80. Chhipa, H.; Kaushik, N. Fungal and bacterial diversity isolated from *Aquilaria malaccensis* tree and soil, induces agarospirol formation within 3 months after artificial infection. *Front. Microbiol.* **2017**, *8*, 1286. [CrossRef]

81. Cui, J.; Guo, S.; Xiao, P. Antitumor and antimicrobial activities of endophytic fungi from medicinal parts of *Aquilaria sinensis*. *J. Zhejiang Univ. Sci. B* **2011**, *12*, 385–392. [CrossRef]

82. Wang, L.; Zhang, W.M.; Pan, Q.L.; Li, H.H.; Tao, M.H.; Gao, X.X. Isolation and molecular identifcation of endophytic fungi from *Aquilaria sinensis*. *J. Fungal Res.* **2009**, *7*, 37–42.

83. Tejesvi, M.V.; Mahesh, B.; Nalini, M.S.; Prakash, H.S.; Kini, K.R.; Subbiah, V.; Shetty, H.S. Endophytic fungal assemblages from inner bark and twig of *Terminalia arjuna* W. & A. (Combretaceae). *World J. Microbiol. Biotechnol.* **2005**, *21*, 1535–1540.

84. Patil, M.P.; Patil, R.H.; Patil, S.G.; Maheshwari, V.L. Endophytic mycoflora of indian medicinal plant, *Terminalia arjuna* and their biological activities. *Int. J. Biotechnol. Wellness Ind.* **2014**, *3*, 53–61.

85. Begoude, B.A.D.; Slippers, B. Botryosphaeriaceae associated with *Terminalia catappa* in Cameroon, South Africa and Madagascar. *Mycol. Prog.* **2010**, *9*, 101–123. [CrossRef]

86. Toghueo, R.M.K.; Zabalgogeazcoa, I.; De Aldana, B.R.V.; Boyom, F.F. Enzymatic activity of endophytic fungi from the medicinal plants *Terminalia catappa*, *Terminalia mantaly* and *Cananga odorata*. *S. Afr. J. Bot.* **2017**, *109*, 146–153. [CrossRef]

87. Begoude, B.A.D.; Slippers, B.; Wingfield, M.J.; Roux, J. The pathogenic potential of endophytic Botryosphaeriaceous fungi on *Terminalia* species in Cameroon. *For. Pathol.* **2011**, *41*, 281–292. [CrossRef]

88. Suryavamshi, G.; Shivanna, M.B. Diversity and antibacterial activity of endophytic fungi in *Memecylon umbellatum* Burm. F.—A medicinal plant in the Western Ghats of Karnataka, India. *Indian J. Ecol.* **2020**, *47*, 171–180.

89. Urdaneta, L.; Araujo, D.; Quirós, M.; Rodríguez, D.; Poleo, N.; Petit, Y. Endophytic mycobiota associated to guava floral buds (*Psidium guajava* L.) and to flat mite (*Brevipalpus phoenicis*)(Geijskes)(Acari: Tenuipalpidae). *UDO Agríc.* **2009**, *9*, 166–174.

90. Janakiraman, V.; Govindarajan, K.; Magesh, C.R. Biosynthesis of silver nanoparticles from endophytic fungi, and its cytotoxic activity. *Bionanoscience* **2019**, *9*, 573–579. [CrossRef]

91. Freire, F.C.; Bezerra, J.L. Foliar endophytic fungi of Ceará State (Brazil): A preliminary study. *Summa Phytopath.* **2001**, *27*, 304–308.

92. Cardoso, J.E.; Bezerra, M.A.; Viana, F.M.P.; de Sousa, T.R.M.; Cysne, A.Q.; Farias, F. Endophyte occurrence of *Lasiodiplodia theobromae* in cashew tissues and its transmission by vegetative propagules. *Summa Phytopath.* **2009**, *35*, 262–266. [CrossRef]

93. Johnson, G.I.; Mead, A.J.; Cooke, A.W.; Dean, J.R. Mango stem end rot pathogens—Fruit infection by endophytic colonisation of the inflorescence and pedicel. *Ann. Appl. Biol.* **1992**, *120*, 225–234. [CrossRef]

94. Morales, R.V.; Rodríguez, G.M. Micobiota endofítica asociada al cultivo Del mango 'Haden'(*Mangifera indica* L.) en el oriente de Venezuela. *Rev. Científica UDO Agríc.* **2009**, *9*, 393–402.

95. González, E.; Umana, G.; Arauz, L.F. Population fluctuation of *Botryodiplodia theobromae* Pat. in mango. *Agron. Costarric.* **1999**, *23*, 21–29.

96. Aishwarya, S.; Nagam, N.; Vijaya, T.; Netala, R.V. Screening and identification of heavy metal-tolerant endophytic fungi *Lasiodiplodia theobromae* from *Boswellia ovalifoliolata* an endemic plant of tirumala hills. *Asian J. Pharm. Clin. Res.* **2017**, *10*, 488–491.

97. El-Nagerabi, S.A.F.; Elshafie, A.E.; Alkhanjari, S.S. Endophytic fungi associated with endogenous *Boswellia sacra*. *Biodiversitas* **2014**, *15*, 24–30. [CrossRef]

98. Linnakoski, R.; Puhakka-tarvainen, H.; Pappinen, A. Endophytic fungi isolated from *Khaya anthotheca* in Ghana. *Fungal Ecol.* **2012**, *5*, 298–308. [CrossRef]

99. Zhao, W.; Bai, J.; McCollum, G.; Baldwin, E. High incidence of preharvest colonization of huanglongbing-symptomatic *Citrus sinensis* fruit by *Lasiodiplodia theobromae* (*Diplodia natalensis*) and exacerbation of postharvest fruit decay by that fungus. *Appl. Environ. Microbiol.* **2015**, *81*, 364–372. [CrossRef]

100. Senthilkumar, N.; Mohan, V.; Murugesan, S. Studies on endophytic fungi of commercially important tropical tree species in India. *Kavaka* **2014**, *31*, 22–31.

101. Osorio, J.A.; Crous, C.J.; de Beer, Z.W.; Wingfield, M.J.; Roux, J. Endophytic Botryosphaeriaceae, including five new species, associated with mangrove trees in South Africa. *Fungal Biol.* **2017**, *121*, 361–393. [CrossRef]

102. Verma, S.K.; Gond, S.K.; Mishra, A.; Sharma, V.K.; Kumar, J.; Singh, D.K.; Kumar, A.; Goutam, J.; Kharwar, R.N. Impact of environmental variables on the isolation, diversity and antibacterial activity of endophytic fungal communities from *Madhuca indica* Gmel. at different locations in India. *Ann. Microbiol.* **2014**, *64*, 721–734. [CrossRef]

103. Shweta, S.; Gurumurthy, B.R.; Vasanthakumari, M.M.; Ravikanth, G.; Dayanandan, S.; Storms, R.; Shivanna, M.B.; Uma Shaanker, R. Endophyte fungal diversity in *Nothapodytes nimmoniana* along its distributional gradient in the Western Ghats, India: Are camptothecine (anticancer alkaloid) producing endophytes restricted to specific clades? *Curr. Sci.* **2015**, *109*, 127–138.

104. Dhayanithy, G.; Subban, K.; Chelliah, J. Diversity and biological activities of endophytic fungi associated with *Catharanthus roseus*. *BMC Microbiol.* **2019**, *19*, 22. [CrossRef]

105. Akther, T.; Mathipi, V.; Kumar, N.S.; Davoodbasha, M.; Srinivasan, H. Fungal-mediated synthesis of pharmaceutically active silver nanoparticles and anticancer property against A549 cells through apoptosis. *Environ. Sci. Pollut. Res.* **2019**, *26*, 13649–13657. [CrossRef] [PubMed]

106. de Oliveira Chagas, M.B.; dos Santos, I.P.; da Silva, L.C.N.; dos Santos Correia, M.T.; de Araújo, J.M.; da Silva Cavalcanti, M.; de Menezes Lima, V.L. Antimicrobial activity of cultivable endophytic fungi associated with *Hancornia speciosa* gomes bark. *Open Microbiol. J.* **2017**, *11*, 179–188. [CrossRef]

107. Suryanarayanan, T.S.; Thennarasan, S. Temporal variation in endophyte assemblages of *Plumeria rubra* leaves. *Fungal Divers.* **2004**, *15*, 197–204.

108. Qadri, M.; Johri, S.; Shah, B.A.; Khajuria, A.; Sidiq, T.; Lattoo, S.K.; Abdin, M.Z.; Riyaz-Ul-Hassan, S. Identification and bioactive potential of endophytic fungi isolated from selected plants of the Western Himalayas. *SpringerPlus* **2013**, *2*, 8. [CrossRef] [PubMed]

109. Santamaría, J.; Bayman, P. Fungal epiphytes and endophytes of coffee leaves (*Coffea arabica*). *Microb. Ecol.* **2005**, *50*, 1–8. [CrossRef]

110. Pandi, M.; Kumaran, R.S.; Choi, Y.K.; Kim, H.J.; Muthumary, J. Isolation and detection of taxol, an anticancer drug produced from *Lasiodiplodia theobromae*, an endophytic fungus of the medicinal plant *Morinda citrifolia*. *Afr. J. Biotechnol.* **2011**, *10*, 1428–1435.

111. Sheik, S.; Chandrashekar, K.R.; Swaroop, K.; Somashekarappa, H.M. Lasiodiplactone A, a novel lactone from the mangrove endophytic fungus *Lasiodiplodia theobromae* ZJ-HQ1. *Int. Biodeterior. Biodegrad.* **2015**, *105*, 21–29. [CrossRef]

112. Chen, S.; Chen, D.; Cai, R.; Cui, H.; Long, Y.; Lu, Y.; Li, C.; She, Z. Cytotoxic and antibacterial preussomerins from the mangrove endophytic fungus *Lasiodiplodia theobromae* ZJ-HQ1. *J. Nat. Prod.* **2016**, *79*, 2397–2402. [CrossRef]

113. Chen, S.; Liu, Z.; Liu, H.; Long, Y.; Chen, D.; Lu, Y.; She, Z. Lasiodiplactone A, a novel lactone from the mangrove endophytic fungus *Lasiodiplodia*. *Org. Biomol. Chem.* **2017**, *73*, 6338–6341. [CrossRef]

114. Moron, L.S.; Lim, Y. Antimicrobial activities of crude culture extracts from mangrove fungal endophytes collected in Luzon Island, Philippines. *Philipp. Sci. Lett.* **2018**, *11*, 28–36.

115. Mazlan, N.; Tate, R.; Clements, C.; Edrada-Ebel, R. Metabolomics studies of endophytic metabolites from Malaysian mangrove plant in the search for new potential antibiotics. *Planta Med.* **2013**, *79*, PA22. [CrossRef]

116. Cannon, P.F.; Simmons, C.M. Diversity and host preference of leaf endophytic fungi in the Iwokrama forest reserve, Guyana diversity and host preference of leaf endophytic fungi in the. *Mycologia* **2017**, *5514*, 210–220.

117. Maheswari, S.; Rajagopal, K. Biodiversity of endophytic fungi in *Kigelia pinnata* during two different seasons. *Curr. Sci.* **2013**, 515–518.

118. Rajendran, L.; Rajagopal, K.; Subbarayan, K. Efficiency of fungal taxol on human liver carcinoma cell lines. *Am. J. Res. Commun.* **2013**, *1*, 112–121.

119. Singh, D.K.; Sharma, V.K.; Kumar, J.; Mishra, A.; Verma, S.K.; Sieber, T.N.; Kharwar, R.N. Diversity of endophytic mycobiota of tropical tree *Tectona grandis* Linn.f.: Spatiotemporal and tissue type effects. *Sci. Rep.* **2017**, *7*, 3745. [CrossRef] [PubMed]

120. Balbool, B.A.; Ahmed, A.A.; Moubasher, M.H.; Helmy, E.A. Production of L-asparaginase (L-ASN) from endophytic *Lasiodiplodia theobromae* hosted *Teucrium polium* in Egypt. *Microb. Biosyst.* **2018**, *3*, 46–55.

121. Sunayana, N.; Nalini, M.S.; Kumara Sampath, K.K.; Prakash, H.S. Diversity studies on the endophytic fungi of *Vitex negundo* L. *Miycosphere* **2014**, *5*, 578–590. [CrossRef]

122. Kamal, N.; Viegelmann, C.V.; Clements, C.J.; Edrada-Ebel, R.A. Metabolomics-guided isolation of anti-trypanosomal metabolites from the endophytic fungus *Lasiodiplodia theobromae*. *Planta Med.* **2017**, *83*, 565–573. [CrossRef]

123. De Errasti, A.; Carmarán, C.C.; Novas, M.V. Diversity and significance of fungal endophytes from living stems of naturalized trees from Argentina. *Fungal Divers.* **2010**, *41*, 29–40. [CrossRef]

124. El-hawary, S.S.; Sayed, A.M.; Rateb, M.E.; Bakeer, W.; Abouzid, S.F.; Mohammed, R.; Sayed, A.M.; Rateb, M.E. Secondary metabolites from fungal endophytes of *Solanum nigrum*. *Nat. Prod. Res.* **2017**, *31*, 2568–2571. [CrossRef]

125. Kannan, K.P.; Muthumary, J. Comparative analysis of endophytic mycobiota in different tissues of medicinal plants. *Afr. J. Microbiol. Res.* **2012**, *6*, 4219–4225.

126. Abdou, R.; Scherlach, K.; Dahse, H.; Sattler, I.; Hertweck, C. Phytochemistry Botryorhodines A–D, antifungal and cytotoxic depsidones from *Botryosphaeria rhodina*, an endophyte of the medicinal plant Bidens pilosa. *Phytochemistry* **2010**, *71*, 110–116. [CrossRef] [PubMed]

127. George, T.S.; Samy, K.; Guru, S.; Sankaranarayanan, N. Extraction, purification and characterization of chitosan from endophytic fungi isolated from medicinal plants. *World J. Sci. Technol.* **2015**, *1*, 43–48.

128. Vardhana, J.; Kathiravan, G.; Dhivya, R. Biodiversity of endophytic fungi and its seasonal recurrence from some plants. *Res. J. Pharm. Technol.* **2017**, *10*, 490–496. [CrossRef]

129. Mullen, J.M.; Gilliam, C.H.; Hagan, A.K.; Morgan-Jones, G. Canker of dogwood caused by *Lasiodiplodia theobromae*, a disease influenced by drought stress or cultivar selection. *Plant Dis.* **1991**, *75*, 886–889. [CrossRef]

130. Delaye, L.; García-guzmán, G.; Heil, M. Endophytes versus biotrophic and necrotrophic pathogens—Are fungal lifestyles evolutionarily stable traits? *Fungal Divers.* **2013**, *1*, 125–135. [CrossRef]

131. Slippers, B.; Wingfield, M.J. Botryosphaeriaceae as endophytes and latent pathogens of woody plants: Diversity, ecology and impact. *Fungal Biol. Rev.* **2007**, *21*, 90–106. [CrossRef]

132. Crous, P.W.; Groenewald, J.Z.; Slippers, B.; Wingfield, M.J.; Crous, P.W. Global food and fibre security threatened by current inefficiencies in fungal identification. *Philos. Trans. R. Soc. Lond. B Biol. Sci.* **2016**, *371*, 20160024. [CrossRef]

133. Martin, R.R.; Constable, F.; Tzanetakis, I.E. Quarantine regulations and the impact of modern detection methods. *Annu. Rev. Phytopathol.* **2016**, *54*, 189–208. [CrossRef]

134. Baker, R.; Bragard, C.; Candresse, T.; Gilioli, G.; Grégoire, J.; Holb, I.; Rafoss, T.; Rossi, V.; Schans, J.; Schrader, G.; et al. Scientific opinion on the pest categorisation of *Cryphonectria parasitica* (Murrill) Barr. 1. *EFSA J.* **2014**, *12*, 3859.

135. Jeger, M.; Bragard, C.; Caf, D.; Candresse, T.; Chatzivassiliou, E.; Dehnen-schmutz, K.; Gilioli, G.; Gregoire, J.; Anton, J.; Miret, J.; et al. Pest categorisation of *Gremmeniella abietina*. *EFSA J.* **2017**, *15*, 5030.

136. EFSA Call for Proposals GP/EFSA/ALPHA/2020/01—Entrusting Support Tasks in the Area of Plant Health Commodity Risk Assessment for High Risk Plants (Plants for Planting for Ornamental Purpose). Available online: http://www.efsa.europa.eu/en/art36grants/article36/gpefsaalpha202001-entrusting-support-tasks-area-plant-health (accessed on 12 September 2020).

137. Aschehoug, E.T.; Metlen, K.L.; Callaway, R.M.; Newcombe, G. Fungal endophytes directly increase the competitive effects of an invasive forb. *Ecology* **2012**, *93*, 3–8. [CrossRef]

138. Lin, Y.; Wang, Y.; Yang, H.; Wang, P. Detection of quiescent infection of mango stem end rot pathogen *Lasiodiplodia theobromae* in shoot and pre-plucked mango fruit by seminested PCR. *Plant Pathol. Bull.* **2009**, *18*, 225–235.

139. Chhipa, H.; Chowdhary, K.; Kaushik, N. Artificial production of agarwood oil in *Aquilaria* sp. by fungi: A review. *Phytochem. Rev.* **2017**, *16*, 835–860. [CrossRef]

140. Sajitha, K.L.; Florence, E.J.M.; Arun, S. Screening of bacterial biocontrols against sapstain fungus (*Lasiodiplodia theobromae* Pat.) of rubberwood (*Hevea brasiliensis* Muell. Arg.). *Res. Microbiol.* **2014**, *165*, 541–548. [CrossRef] [PubMed]

141. Mohali, S.; Encinas, O.; Mora, N. Manchado azul en madera de *Pinus oocarpa* y *Azadirachta indica* en Venezuela. *Fitopatol. Venez.* **2002**, *15*, 30–32.

142. Mohali, S.R.; Castro-Medina, F.; Úrbez-Torres, J.R.; Gubler, W.D. First report of *Lasiodiplodia theobromae* and *L. venezuelensis* associated with blue stain on *Ficus insipida* wood from the Natural Forest of Venezuela. *For. Pathol.* **2017**, *47*, e12355. [CrossRef]

143. Ludwig-Müller, J. Plants and endophytes: Equal partners in secondary metabolite production? *Biotechnol. Lett.* **2015**, *37*, 1325–1334. [CrossRef] [PubMed]

144. Nicoletti, R.; Fiorentino, A. Plant bioactive metabolites and drugs produced by endophytic fungi of *Spermatophyta*. *Agriculture* **2015**, *5*, 918–970. [CrossRef]

145. Caruso, G.; Abdelhamid, M.; Kalisz, A.; Sekara, A. Linking endophytic fungi to medicinal plants therapeutic activity. A case study on Asteraceae. *Agriculture* **2020**, *10*, 286. [CrossRef]

146. Salvatore, M.M.; Alves, A.; Andolfi, A. Secondary metabolites of *Lasiodiplodia theobromae*: Distribution, chemical diversity, bioactivity, and implications of their occurrence. *Toxins* **2020**, *12*, 457. [CrossRef] [PubMed]

147. Félix, C.; Salvatore, M.M.; Dellagreca, M.; Meneses, R.; Duarte, A.S.; Salvatore, F.; Naviglio, D.; Gallo, M.; Jorrín-Novo, J.V.; Alves, A.; et al. Production of toxic metabolites by two strains of *Lasiodiplodia theobromae*, isolated from a coconut tree and a human patient. *Mycologia* **2018**, *110*, 642–653. [CrossRef] [PubMed]

148. Félix, C.; Salvatore, M.M.; DellaGreca, M.; Ferreira, V.; Duarte, A.S.; Salvatore, F.; Naviglio, D.; Gallo, M.; Alves, A.; Esteves, A.C.; et al. Secondary metabolites produced by grapevine strains of *Lasiodiplodia theobromae* grown at two different temperatures. *Mycologia* **2019**, *111*, 466–476. [CrossRef] [PubMed]

149. Keller, N.P.; Turner, G.; Bennett, J.W. Fungal secondary metabolism-from biochemistry to genomics. *Nat. Rev. Microbiol.* **2005**, *3*, 937–947. [CrossRef]

150. Salvatore, M.M.; Nicoletti, R.; DellaGreca, M.; Andolfi, A. Occurrence and properties of thiosilvatins. *Mar. Drugs* **2019**, *17*, 664. [CrossRef]

151. Uranga, C.C.; Beld, J.; Mrse, A.; Córdova-Guerrero, I.; Burkart, M.D.; Hernández-Martínez, R. Fatty acid esters produced by *Lasiodiplodia theobromae* function as growth regulators in tobacco seedlings. *Biochem. Biophys. Res. Commun.* **2016**, *472*, 339–345. [CrossRef]

152. Jernerén, F.; Eng, F.; Hamberg, M.; Oliw, E.H. Linolenate 9R-dioxygenase and allene oxide synthase activities of *Lasiodiplodia theobromae*. *Lipids* **2012**, *47*, 65–73. [CrossRef]

153. Salvatore, M.M.; Félix, C.; Lima, F.; Ferreira, V.; Duarte, A.S.; Salvatore, F.; Alves, A.; Esteves, A.C.; Andolfi, A. Effect of γ-aminobutyric Acid (GABA) on the metabolome of two strains of *Lasiodiplodia theobromae* isolated from grapevine. *Molecules* **2020**, *25*, 3833. [CrossRef]

154. Zhang, Y.; Liu, H.; Li, W.; Tao, M.; Pan, Q.; Sun, Z.; Ye, W.; Li, H.; Zhang, W. 2-(2-Phenylethyl)chromones from endophytic fungal strain *Botryosphaeria rhodina* A13 from *Aquilaria sinensis*. *Chin. Herb. Med.* **2017**, *9*, 58–62. [CrossRef]

155. Valayil, M.J.; Kuriakose, G.C.; Jayabaskaran, C. Isolation, purification and characterization of a novel steroidal saponin cholestanol glucoside from *Lasiodiplodia theobromae* that induces apoptosis in A549 cells. *Anticancer Agents Med. Chem.* **2016**, *16*, 865–874. [CrossRef]

156. Vasanthakumari, M.M.; Jadhav, S.S.; Sachin, N.; Vinod, G.; Shweta, S.; Manjunatha, B.L.; Kumara, P.M.; Ravikanth, G.; Nataraja, K.N.; Uma Shaanker, R. Restoration of camptothecine production in attenuated endophytic fungus on re-inoculation into host plant and treatment with DNA methyltransferase inhibitor. *World J. Microbiol. Biotechnol.* **2015**, *31*, 1629–1639. [CrossRef]

157. Nicoletti, R. Antitumor and immunomodulatory compounds from fungi. In *Reference Module in Life Sciences (Planned for Publication in Encyclopaedia of Mycology)*; Zaragoza, O., Ed.; Elsevier: Amsterdam, The Netherlands, 2020. [CrossRef]

158. Pandi, M.; Rajapriya, P.; Suresh, G.; Ravichandran, N.; Manikandan, R.; Thiagarajan, R.; Muthumary, J. A fungal taxol from *Botryodiplodia theobromae* Pat., attenuates 7, 12 dimethyl benz(a)anthracene (DMBA)-induced biochemical changes during mammary gland carcinogenesis. *Biomed. Prev. Nutr.* **2011**, *1*, 95–102. [CrossRef]

159. Magnus, V.; Šimaga, Š.; Iskrić, S.; Kveder, S. Metabolism of Tryptophan, Indole-3-acetic acid, and related compounds in parasitic plants from the genus *Orobanche*. *Plant. Physiol.* **1982**, *69*, 853–858. [CrossRef] [PubMed]

160. Stahl, E.; Bellwon, P.; Huber, S.; Schlaeppi, K.; Bernsdorff, F.; Vallat-Michel, A.; Mauch, F.; Zeier, J. Regulatory and functional aspects of indolic metabolism in plant systemic acquired resistance. *Mol. Plant.* **2016**, *9*, 662–681. [CrossRef]

161. Bartel, B. Auxin biosynthesis. *Annu. Rev. Plant. Biol.* **1997**, *48*, 51–66. [CrossRef]

162. Félix, C.; Libório, S.; Nunes, M.; Félix, R.; Duarte, A.S.; Alves, A.; Esteves, A.C. *Lasiodiplodia theobromae* as a producer of biotechnologically relevant enzymes. *Int. J. Mol. Sci.* **2018**, *19*, 29. [CrossRef] [PubMed]

Publisher's Note: MDPI stays neutral with regard to jurisdictional claims in published maps and institutional affiliations.

 *agriculture*

Review

Biodiversity, Ecology, and Secondary Metabolites Production of Endophytic Fungi Associated with Amaryllidaceae Crops

Gianluca Caruso [1], Nadezhda Golubkina [2], Alessio Tallarita [1], Magdi T. Abdelhamid [3] and Agnieszka Sekara [4,*]

[1] Department of Agricultural Sciences, University of Naples Federico II, 80055 Portici (Naples), Italy; gcaruso@unina.it (G.C.); alessio.tallarita@yahoo.com (A.T.)
[2] Federal Scientific Center of Vegetable Production, Selectsionnaya 14 VNIISSOK, 143072 Moscow, Odintsovo, Russia; segolubkina@rambler.ru
[3] Botany Department, National Research Centre, 33 El Behouth Steet, Dokki, Cairo 12622, Egypt; mt.abdelhamid@nrc.sci.eg
[4] Department of Horticulture, Faculty of Biotechnology and Horticulture, University of Agriculture, 31-120 Krakow, Poland
* Correspondence: agnieszka.sekara@urk.edu.pl; Tel.: +48-12-6625216

Received: 28 September 2020; Accepted: 4 November 2020; Published: 6 November 2020

Abstract: Amaryllidaceae family comprises many crops of high market potential for the food and pharmaceutical industries. Nowadays, the utilization of plants as a source of bioactive compounds requires the plant/endophytic microbiome interactions, which affect all aspects of crop's quantity and quality. This review highlights the taxonomy, ecology, and bioactive chemicals synthesized by endophytic fungi isolated from plants of the Amaryllidaceae family with a focus on the detection of pharmaceutically valuable plant and fungi constituents. The fungal microbiome of Amaryllidaceae is species- and tissue-dependent, although dominating endophytes are ubiquitous and isolated worldwide from taxonomically different hosts. Root sections showed higher colonization as compared to bulbs and leaves through the adaptation of endophytic fungi to particular morphological and physiological conditions of the plant tissues. Fungal endophytes associated with Amaryllidaceae plants are a natural source of ecofriendly bioagents of unique activities, with special regard to those associated with Amarylloidae subfamily. The latter may be exploited as stimuli of alkaloids production in host tissues or can be used as a source of these compounds through in vitro synthesis. Endophytes also showed antagonistic potential against fungal, bacterial, and viral plant diseases and may find an application as alternatives to synthetic pesticides. Although Amaryllidaceae crops are cultivated worldwide and have great economic importance, the knowledge on their endophytic fungal communities and their biochemical potential has been neglected so far.

Keywords: onions; amaryllis; endosphere; endobiome; metabolome; symbiosis

---

1. Introduction

Amaryllidaceae species have been utilized as vegetables, herbs, spices, and ornamentals in all continents since ancient times. Many of them have shown widespread benefits in cuisine and healing of common diseases like atherosclerosis, diabetes, inflammation, hypertension, and cancer. These protective effects appear to be related to the presence of organosulfur compounds, predominantly allyl derivatives in Allioidae and alkaloids in the Amaryllidoidae subfamily [1]. The economic significance of Amaryllidaceae crops is substantiated by the all-year-round supply and a wide range of cultivars and landraces characterized by plant parts with different shapes and specific taste and flavor. The biochemical profile of plants determines a broad or specialized usage by food and pharmacological industry [2–4]. Considering the wide distribution, unique chemical composition, and economic importance, we selected Amaryllidaceae crops to analyze their relationship with the microbiome colonizing the root and shoot tissues.

With regard to root fungal microbiome, Amaryllidaceae symbiosis with arbuscular mycorrhizal fungi has been widely investigated at the physiological, biochemical, and genetic level, and it has been comprehensively reviewed [5–8]. However, there is selective and narrow knowledge of fungal endophytes colonizing these plants. Hardoim et al. [9] defined "endophytes" as microorganisms living in plant tissues in some periods of their life cycle or all through, so the word is referred to the habitat rather than function. Nevertheless, endophytic fungi live inside the tissues of apparently healthy and asymptomatic hosts. The case of Amaryllidaceae crops can help in understanding how fungal endophytes modulate the physiological processes in the above- and below-ground plant tissues as well as interact with a host plant and with other microbes. The effect of plant morphology on colonization and biodiversity of endophyte communities has been rarely studied. The root-associated fungi play an important role in plant nutrition by: mobilizing soil nutrients; recycling organic matter; increasing the water holding capacity and absorption; protecting against pathogens and abiotic stresses; biosynthesizing lignan, auxins, and ethylene, in order to improve the root system expansion and encourage plant fitness through thermotolerance improvement; and coping with salinity, drought, and heavy metal stress [10,11]. The underground bulb has been an additional component of plant tissue/endophyte puzzles as a perennial storage part rich in nutrient compounds, and—in some taxons containing unique bioagents—Amaryllidaceae alkaloids. Taking into account their life cycle, Amaryllidaceae have to efficiently allocate resources from bulbs to vegetative and generative tissues and vice versa during a short vegetation period. It is possible that the assimilates' flow provides a natural path for plants to partitioning the symbiotic endophytes as it was proposed for mycorrhizal fungi by Crişan et al. [12]. Foliar endophytes can stimulate host plant biomass, yield, mineral status, nutritive value, stomatal movement, and defensive mutualism against herbivores and pathogens [13,14].

The strict selection and recruitment of the endophytes by a host species was postulated and analyzed for plant-associated bacteria [15] but the mentioned mechanisms also shape fungal diversity and abundance [16]. The rhizoplane and phylloplane are considered to be a selective gate for endophytes, mediating dynamic changes in the fungal community in plant internal tissues. This is the reason for differences in microbial compositions among host species and even among their tissues [8]. Plant secondary metabolites are postulated as main regulators in colonization and development of fungal endophytes in different tissues and, vice versa, fungal metabolites can play a similar role with respect to the host and the microbiota [17,18]. Plant secondary metabolites are postulated as main regulators in colonization and development of fungal endophytes in different tissues and, vice versa, fungal metabolites can play a similar role with respect to the host and the microbiota [19]. Bioactive metabolites synthesized by fungal endophytes show multidirectional actions in plants and can be explored for the cultivation of targeted functional and medicinal crops [20]. Nowadays, endophytes are also recognized as an alternative to synthetic chemicals in crop production and protection [21]. However, to be successfully applied, endophytes need specific conditions for efficient host tissue colonization. This can be achieved by a thorough understanding of the endophyte-host relationship starting from the colonization up to well-established symbiosis, on the basis of the fungi and plant biology, covering genetic and biochemical factors as well [22–24].

This review aimed to analyze fungal endophyte communities associated with Amaryllidaceae crops, by: presenting the evidence of ubiquitous and/or species- and tissue-dependent fungal microbiome, reviewing the known chemical compounds synthesized by fungal endophytes, and emphasizing their possible vital effects on human proecological and prohealth activities.

2. Amaryllidaceae Crops—Botanical Characteristics and Biochemical Composition

According to the current taxonomy, the family Amaryllidaceae consists of three subfamilies, Agapanthoideae, Allioideae, and Amaryllidoideae, comprised of about 80 genera and approximately

2200 accepted species [25]. Amaryllidaceae are perennial or biennial geophytes or hemicryptophytes, with very diversified morphology of their underground shoots, which let distinguish three biomorphological groups: rhizomatous, bulbous, and domesticated onions [26]. In the case of a rhizomatous group (*Cryptostephanus* spp., *Clivia* spp., and some *Scadoxus* spp.), fleshy rhizomes act as storage organs, growing for several years through the successive development of the basal plate. Bulbs, composed of leaf sheaths of varying thicknesses, are formed by rhizomes. The leaves are evergreen. In the bulb group, the true bulb consists of a longitudinally compressed basal stem and fleshy, succulent, storage leaf bases. This group is well-adapted to arid and semiarid climate. Domesticated onions form storage bulbs and were grouped separately because of the diversified morphology, shaped through many centuries of breeding [26]. Most of the Amaryllidaceae plants prefer xerophytic ecosystems, with warm, dry summers and cool, wet winters, being distributed across regions, which are well-recognized biodiversity hotspots. The major center of genetic diversity is localized in Central Asia and Mediterranean basin, and the secondary one, in South Africa, western North America, and the Andes [27]. The gene pool of wild Amaryllidaceae is very rich in the centers of origin and has been explored as a source of new genes introduced into cultivated species and for the domestication of new crops useful as vegetables, herbs, medicinal plants, and ornamentals [28].

Cultivation practices were developed independently in particular regions of the Northern Hemisphere and applied to at least 20 native or introduced vegetables of Allioideae subfamily, especially of genus *Allium*. Onion (*Allium cepa* L.) and garlic are cultivated worldwide, leek (*A. ampeloprasum* L.), shallot (*A. cepa* L. Aggregatum group), and chive (*A. schoenoprasum* L.) dominate in Western and Northern Europe, kurrat (*A. ampeloprasum* L.) in Egypt and the Eastern Mediterranean, Japanese bunching onion (*A. fistulosum* L.) in Japan, rakkyo (*A. chinense* G. Don), and Chinese leek (*A. tuberosum* Rottl. ex Spr.) in China. The cultivated onion group is the result of intensive breeding and represents morphological and physiological characteristics appreciated both for cultivation and marketing. The latter include diversified shape, color, pungency, and chemical composition of bulbs, reduced bolting, long shafts in leek, fast leaf growth in chives, single heart in onion but separated in shallot [29–32]. Onions are known as a major food for preventing chronic disease [33], as a source of sulfur compounds, steroidal saponins, and flavonoids. Moreover, showing a functional food activity, they significantly contribute to the prevention of inflammatory and common lifestyle diseases [34]. Organic sulfur compounds, determining onions' pungent flavor, are the key components responsible for the therapeutic effects [35]: allicin, and ajoene, as well as volatile compounds have the ability to act as antimicrobials and antioxidants [36,37]; sulfur and phenolic compounds also show antioxidant, anticancer, anti-inflammatory activities, and can prevent chronic diseases [38]; quercetin, a bioflavonoid of onions, reveals antiproliferative and proapoptotic effects in many cancer cells, acts as a neuroprotector, and stimulates cellular defense against oxidative stress [39].

Agapanthus is the only genus in the subfamily Agapanthoideae, endemic in South Africa but naturalized around the world as ornamental. The Amaryllidoideae have a pronounced floricultural importance because this subfamily comprises popular ornamentals, including many spring-flowering bulbs (*Narcisuss* spp., *Galanthus* spp., and *Leucojum* spp.). They have been grown in European gardens since the ancient times, supplemented since 17–18 century with species of New World origin like *Hippeastrum* spp., or South African, like *Amaryllis* spp. or *Clivia* spp., widely cultivated as indoor plants. The Amaryllidoideae have been traditionally used as medicines to treat mental problems, primarily in Southern Africa [40]. Amaryllidoideae are the source of the isoquinoline alkaloids of unique structure, which were isolated from about 350 species, amongst more than 800 species belonging to this subfamily. Approximately 600 structurally diverse alkaloids were isolated to the date, chemically defined, and pharmacologically investigated, as possessing antibacterial, antifungal, antimalarial, antiviral, antitumor, analgesic, and acetylcholinesterase inhibitory activities [41–43]. The galanthamine was approved to date as the main treatment for mild to moderate Alzheimer's disease, acting as a selective, reversible competitive acetylcholinesterase inhibitor [44]. The lycorine, haemanthamine, and narciclasine series are leading anticancer bioagents in clinical research [45]. The enormous structural diversity of Amaryllidoideae alkaloids has no equivalent in the Plant Kingdom and can be explained by the chemoecological activity [46].

3. Biodiversity and Ecology of Endophytes Associated with Crops Belonging to the Amaryllidaceae Family

The endophytic symbiosis could be an implementation of the microorganisms' strategy aimed at reducing the effects of the external changeable environment through the long-term coevolution with plants providing a stable niche in their tissues [47]. The leaf surface is an attractive habitat for endophytic fungi, which are influenced by the possibility of colonization through the epidermal structures, by leaf health and nutritional status, and by competition with the other microorganisms. Several studies have been carried out to characterize the mycobiota of *A. cepa* rhizosphere and phyllosphere, but much less research has been focused on fungi colonizing internal tissues. Abdel-Gawad et al. [48] isolated and identified, based on macro- and microscopic characters, 24 genera and 38 species of fungi from the rhizoplane of onion, with dominating *Aspergillus* spp., *Cladosporium* spp. and *Penicillium* spp., and 17 genera and 35 species from the phylloplane, with dominating *Aspergillus* and *Penicillium* spp. The root and leaf surface of onion hosted a broader spectrum of species than internal tissues, confirming that plants selectively recruit endophytic microorganisms. Moreover, aboveground plant tissues are exposed to rapid fluctuations in temperature, humidity, and solar radiation, so microorganisms colonizing leaves are also affected by abiotic stress, exceeding sometimes their tolerance thresholds. Abdel-Gawad et al. [48] evidenced that the onion's fungal microbiota dependent on temperature, namely the species *Humicola grisea* (current name *Trichocladium griseum*), *Penicillium mirabile* (current name *Talaromyces verruculosus*), and *Rhizoctonia solani*, were isolated from leaves at 19 °C, whereas other species, such as *Chaetomium brasiliense* (current name *Ovatospora brasiliensis*) and *Zopfiella latipes*, at 28 °C. Moreover, the mentioned species were not specific for onion but isolated from roots of the other crops in the investigated region, namely Assiut Governorate in Egypt. Only one species, *Z. latipes*, was isolated from onion leaves for the first time in Egypt [49]. A red spider lily (*Lycoris radiata*) and golden spider lily (*L. aurea*) are ornamentals of Asian origin, introduced into many countries all over the world because of decorative flowers, but their bulbs are known as poisonous in traditional medicine systems. Zhou et al. [50] identified, using molecular (polymerase chain reaction—PCR) and morphological characteristics, 27 species of fungal endophytes belonging to 14 genera from *L. radiata*. Only *Fusarium* developed hyphae in all organs; *Stagonosporopsis* and *Glomerella* (current name *Colletotrichum*) were isolated from leaf tissues; *Phoma* from the bulb; *Galactomyces*, *Metacordyceps* (current name *Metarhizium*) and *Diaporthe* from root tissues. *Aspergillus*, *Colletotrichum*, *Diaporthe*, *Fusarium*, *Penicillium*, *Phoma*, and *Phyllosticta* were commonly isolated from a wide range of hosts but *Cylindrocarpon*, *Galactomyces*, *Sarocladium*, and *Stagonosporopsis* were described as endophytes of specific plants. In earlier studies, despite the mentioned species, *Trichoderma* sp. was isolated from a bulb of *L. radiata* [51] and *Mucor* sp. from the bulb of *L. aurea* [52]. Notably, *Metarhizium* sp., which was reported as a soil fungus [53], was isolated from *L. radiata* tissues, so this fungus seems to colonize plants occasionally [50].

The relationships between the endophyte fungi and host plant are very diversified and dynamically change from mutualism, symbiosis, and commensalism to pathogenic during plant ontogeny [54,55]. For example, *Colletotrichum*, *Diaporthe*, *Fusarium*, *Phyllosticta*, and *Phoma*, isolated from healthy tissues of *L. radiata* are commonly recognized as pathogenic, so the antifungal alkaloids can enforce symbiotic lifestyle in plants, maintaining a balance between host and its endophytes/parasites. Regarding endophytes colonization during onion's ontogeny, Mueva et al. [56] stated that the seed inoculation was more effective than seedling inoculation in terms of endophytes recovery in subsequent stages of plant development. Indeed, endophytes inoculated at the seed surface colonized seed radicle and plumule and developed internal mycelia in growing tissues. The fungal colonization and distribution in onion tissues firstly depended on inoculation technique and secondly on the endophyte selection by the host. Independently on the inoculation technique, most of the investigated endophytes, for example *Clonostachys rosea*, *Hypocrea lixii* (current name *Trichoderma lixii*), *Trichoderma asperellum*,

T. atroviride, *T. harzianum*, and *Fusarium* spp., were isolated from onion roots, followed by stems and leaves. These differences could be due to tissue morphology and physiology, microbiome interactions, and the influence of external conditions [56,57]. Onions have shallow, weakly branched root systems with sparse root hairs, inefficient in the use of soil nutrient resources. The root endophytic and mycorrhizal fungi play a significant role in supporting onions with mineral salts, that is why this species is among the most symbiosis-dependent crops [7]. Focusing on endophytic fungi colonizing shallot roots, Priyadharsini et al. [58] found that the percent of root length with fungal microsclerotia was significantly and negatively correlated with soil phosphorus level. Similarly, percents of root length with dark septate hyphae and dark septate endophyte total colonization were negatively correlated with soil zinc and copper contents. It can be concluded that colonization of shallot roots by fungal endophytes was reduced in soils rich in mineral salts. Wu et al. [59] hypothesized that the endophytic fungal community may be helpful to symbiotic plants (i.a. *Allium mongolicum*) for surviving in the extreme environments of Asian deserts. The mycobiota associated with photosynthesizing or storage leaves, for example *T. harzianum* and *T. koningii*, could act antagonistically to phytopathogens. On the other side, leaves with disease symptoms, with damaged epidermal cells and the lamellar seta shed releasing nutrients, could be secondarily colonized by opportunistic fungi such as *Botrytis cinerea*, *Penicillium aurantiogriseum*, *Alternaria alternata*, and *Cladosporium* spp. [20].

Plant storage tissues, including sugar-rich onions' bulbs, can contain specific endophytes, actively reproducing in these tissues without visible damage [60]. One of the main chemoecological roles of Amaryllidaceae alkaloids is the protection of nutrient-rich bulbs against phytopathogens and herbivores. Xiang et al. [61] isolated and sequenced six fungal endophytes from *Narcissus pseudonarcissus* bulb and only two from leaf tissues. Zhou et al. [19] found that bulbs of *L. radiata* were exclusively colonized at a higher degree than other tissues, probably because of the perennial life cycle of bulbs and annual cycle of other plant parts [62] and because of the space and carbohydrates provided by bulbs as storage sinks [63].

3.1. Biochemistry and Functions of Fungal Endophytes Associated with Allioidae Crops

Abdel-Hafez et al. [20] investigated endophytes colonizing *A. cepa* leaves, both healthy and infected by purple blotch (*Alternaria porri*). Fungi were isolated and identified according to their macroscopic and microscopic characteristics. Despite the strains detected from healthy and diseased leaves, belonging to genera *Cladosporium*, *Alternaria*, *Penicillium*, and *Stemphylium*, five species, namely *Absidia corymbifera* (current name *Lichtheimia corymbifera*), *B. cinerea*, *P. aurantiogriseum*, *P. glabrum*, and *Syncephalastrum racemosum*, were isolated only from infected leaves, while three species (*Fusarium oxysporum*, *Trichoderma harzianum*, and *T. koningii*) were isolated only from healthy ones (Table 1). *Trichoderma* spp. showed antagonistic potential against *A. porri*, through competition, lysis, antibiosis, and parasitism [64,65]. The antagonistic effect of *Epicoccum nigrum*, *Penicillium oxalicum*, and *Stachybotrys chartarum* against *A. porri* was antibiosis caused by effective lytic, as well as antimicrobial secondary metabolites produced by endophytic fungi [20]. Previously, Flori and Roberti [66] noticed the antifungal activity of the endophyte *Beauveria bassiana*, inoculated to onion roots, against *F. oxysporum* f. sp. *cepae*, causing basal rot of onion. The antifungal potential of the endophyte *Talaromyces pinophilus* (current name *Penicillium pinophilum*) against *B. cinerea* was described by Abdel-Rahim and Abo-Elyousr [67]. *T. pinophilus* was isolated from onion's inflorescences and identified with PCR amplification of the ribosomal internal transcribed spacer (ITS) region. The mycelium of *T. pinophilus* penetrated intercellularly the hyphae of *B. cinerea*, involving cell wall degrading enzymes (chitinase, lipase, and protease) in the mycoparasitic process.

Table 1. Main biological activities of endophytes isolated from Allioidae crops.

Tissues	Dominating Endophyte Species	Main Activities of the Endophyte	Main Metabolites/Enzymes Linked to Endophyte Bioactivities	Reference
Allium cepa (leaf)	*Clonostachys rosea, Hypocrea lixii, Trichoderma asperellum, T. atroviride, T. harzianum, Fusarium* sp.	Suppression of *Thrips tabacii* reproduction and viral transmitting	Volatile components	[56,57,68]
	Cladosporium cladosporioides, C. sphaerospermum	Antifungal against *Alternaria porri*	Not investigated	[20]
A. cepa (leaf healthy)	*Epicoccum nigrum*	Antifungal against *A. porri*	Flavipin 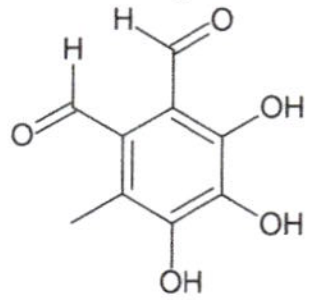	[20,69]
	Penicillium oxalicum	Antifungal against *A. porri*	Lytic extracellular enzymes (β-1,3-glucanase, chitinases, cellulases)	[20,70]
	T. harzianum	Antifungal against *A. porri*	Lytic extracellular enzymes	[20,65]
A. cepa (leaf infected with *A. porri*)	*Botrytis cinerea, Penicillium aurantiogriseum, Alternaria alternata, Cladosporium* spp.	Antifungal against *A. porri*	Not investigated	[20]
A. cepa (umbels)	*Talaromyces pinophilus*	Antifungal against *B. cinerea*	Lytic extracellular enzymes (chitinase, lipase, and protease)	[67]
A. cepa (floral stalks infected with *A. porri*)	*Trihoderma longibrachiatum, T. harzianum,*	Antifungal against *A. porri*	Lytic extracellular enzymes	[64]
A. llium sativum (leaf)	*Trichoderma brevicompactum*	Antifungal against *Rhizoctonia solani* and *B. cinerea*	4-acetoxy-12,13-epoxy-9-trichothecene (trichodermin) 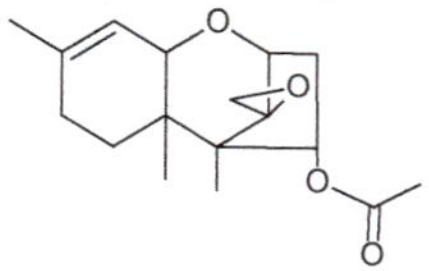	[71]
Allium schoenoprasum (leaves, roots)	*Beauveria bassiana*	Protection of the host	Increased alkaloid level	[72]

Table 1. Cont.

Tissues	Dominating Endophyte Species	Main Activities of the Endophyte	Main Metabolites/Enzymes Linked to Endophyte Bioactivities	Reference
A. schoenoprasum (bulb)	Penicillium pinophilum	Inhibition of the NCI60/ATCC panel of human cancer cell of different tissue origin	Skyrin R=H and Dicatenarin R=OH.	[73]
Allium filidens (stem, bulb)	Aspergillus terreus, Penicillium sp.	Cytotoxic against carcinoma of the cervix (HeLa), larynx (HEp-2); Inhibition of α-amylase activity	Not investigated	[74,75]
	A. terreus	Antibacterial against Pseudomonas aeruginosa	Not investigated	[74]
A. filidens (bud)	Alternaria tenuissima	Antibacterial against P. aeruginosa and Staphylococcus aureus	Not investigated	[74]
Allium longicuspis (root)	Aspergillus ochraceus	Antibacterial against P. aeruginosa and S. aureus	Not investigated	[74]
	Aspergillus versicolor	Antibacterial against E. coli, P. aeruginosa, and S. aureus	Not investigated	[74]
	Fusarium sp.	Antibacterial against E. coli	Not investigated	[74]
A. longicuspis (bulb)	Aspergillus. spectabilis	Antibacterial against P. aeruginosa and S. aureus	Not investigated	[74]
A. longicuspis (leaf)	Fusarium sambucinum	Antibacterial against E. coli	Not investigated	[74]
	Alternaria sp.	Antibacterial against E. coli, P. aeruginosa, and S. aureus	Not investigated	[74]
	A. terreus	Antibacterial against E. coli, P. aeruginosa, and S. aureus, inhibition of α-amylase activity	Not investigated	[74,75]
	Aspergillus flavus	Antibacterial against P. aeruginosa	Not investigated	[74]

Muvea et al. [56,68] showed the effect of onion inoculation with some strains of endophytic fungi on the proportion of thrips due to reduced feeding and oviposition, caused by antixenotic repellence or higher death rate of thrips. Moreover, the reduced feeding of thrips on endophyte-colonized onions could reduce the transmission of virus diseases, spread by insects.

Among endophytes of garlic that can produce bioactive compounds, Shentu et al. [71] isolated and identified, based on morphological and molecular procedures, *Trichoderma brevicompactum* with strong antifungal activities. Trichodermin, an antifungal compound of *T. brevicompactum* inhibited mycelial growth of *R. solani*, with an EC_{50} of 0.25 µg mL^{-1}, and *B. cinerea*, with an EC_{50} of 2.02 µg mL^{-1} (Table 2). A weak inhibition was noted against *Colletotrichum lindemuthianum* (EC_{50} = 25.60 µg mL^{-1}). The authors underlined that the relationship between *T. brevicompactum* and the garlic plant remained unclear. Espinoza et al. [72] investigated the chive's growth parameters and secondary metabolites as affected by inoculation with the endophyte fungus *B. bassiana*. The fungus applied to the rhizosphere, colonized plant tissues, and finally was isolated from roots and leaves, affecting total alkaloids content but not leaves yield. Koul et al. [73] isolated and morphologically and molecularly identified the fungus *Penicillium pinophilum*, from bulbs of chive's population native to snow mountain regions of India. *P. pinophilum* was a source of anticancer anthraquinones, dicatenarin, and skyrin. Both compounds inhibited human pancreatic cancer (MIA PaCa-2) cells with least IC_{50} values of 12 µg mL^{-1} and 27 µg mL^{-1} respectively, through mitochondrial-mediated apoptotic pathway. Dicatenarin cytotoxic/proapoptotic activity was more pronounced than skyrin due to the presence of an additional phenolic hydroxyl group at C-4, which increased reactive oxygen species generation [73].

Wild and endemic *Allium* species were also the object of investigation. In the latter respect, Abdulmyanova et al. [74] screened *Allium filidens* Regel and leaves of *A. longicuspis* Regel regarding endophytic fungi biodiversity and bioactivity. Among 16 isolates of endophytic fungi obtained from these plants and identified morphologically, the highest biodiversity was determined for bulbs of *A. filidens* and leaves of *A. longicuspis*. The *Penicillium* spp. were the most dominant symbionts of *A. filidens*, while *Aspergillus* spp. were commonly isolated from *A. longicuspis*. Beside cosmopolitan species, the rare endophytes *Alternaria tenuissima*, *Aspergillus spectabilis*, and *Cladosporium tenussimum* were also isolated. The endophytic fungi detected in the same host varied regarding bioactivity. For example, three strains of *Penicillium* sp. isolated from bulbs of *A. filidens* were different in cytotoxic, antibacterial, and antiamylase activity, two strains of *Alternaria* sp. from leaves of *A. longicuspis* exhibited only antibacterial activity [74,75]. Bulbs of both described *Allium* species, endemic in Afganistan, have been used in traditional Asian medicine [76].

3.2. Biochemistry and Functions of Fungal Endophytes Associated with Amaryllidoideae Crops

Amarylidoideae alkaloids can be involved in chemical crosstalk between host plant and endophytes as communication molecules that are responsible for the shaping of plant-microbe interactions. This phenomenon was more widely investigated for endophytic bacteria, which can promote the synthesis of Amaryllidaceae alkaloids [77,78], but endophytic fungi are also involved in plant-endophyte and endophyte–endophyte interspecies communication. For example, Wang et al. [79] investigated endophytic fungi and bacteria in the bulbs of the Chinese sacred lily (*Narcissus tazetta*), widely used as an ornamental and medicinal plant in Asia [79]. The authors defined nine hexacyclopeptides produced by fungus and selectively accumulated by an endophytic bacterium *Achromobacter xylosoxidans* isolated from the same tissue (Table 2). The production of targeted hexacyclopeptides by *F. solani* was possible only in planta and decreased in vitro conditions. However, the ecological basis of this chemical cross-talk needs future investigations. Yang et al. [80] isolated and identified, using morphological and molecular methods, 18 strains of endophytic fungi from *Narcissus* sp. Three species, particularly *Rhinocladiella* sp., demonstrated significant inhibitory activity against acetylcholinesterase.

Onofri et al. [81] identified, using conventional taxonomic techniques, four strains of *Cryptococcus laurentii* (current name *Papiliotrema laurentii*), C1–C3 from root tips, C4 from outer bracts of bulb of daffodil (*N. pseudonarcissus*). The authors observed that lycorine, an alkaloid of *Narcissus* bulbs, inhibited the growth of C1–C3 but not C4 strains of fungi. The inhibition was due to destroying the cellular membranes and interfering with the substrate absorption and cell metabolism, namely blocking L-galactonic acid γ-lactone conversion into ascorbate by lycorine. In contrast, *C. laurentii*, isolated from the lycorine-containing bracts of the bulb, was able to degrade lycorine and to use decomposition products as growth stimulators.

L. radiata is the main source of Amaryllidaceae alkaloids but the low yield and high costs, resulting from its complex procedures and mixed stereoisomers, limit pharmaceutical development of plant-delivered drugs [50]. *Lycoris* spp. were objects of some investigations regarding endophyte microbiota and assessment of the biological activity of their metabolites. *Penicillium* sp. isolated from *L. aurea* was able to produce galanthamine in vitro [82], and the other nonidentified fungus strain L-10 possessed antibacterial and antifungal activity against *Staphylococcus aureus* and *Candida albicans*, respectively [52]. This phenomenon confirmed the antagonism between fungal and bacterial symbionts of this plant. Both novel and known compounds, especially alkaloids, could be produced by in vitro grown endophytic fungi isolated from *Lycorsis* spp. bulbs. Moreover, the inoculation with fungal endophytes enhanced the level of various alkaloids in *L. radiata*. So, inoculation with particular fungus or consortium of fungi can be used for increasing the content of targeted alkaloids during plant cultivation [50].

Li et al. [83] investigated drimane-type sesquiterpenoids of *Aspergillus versicolor*. Among the latter compounds, Versicalin A showed moderate cytotoxic activity against HL-60 tumor cells with an IC_{50} value of 5.6 μM, while proversilin C and E showed moderate cytotoxicity against human tumor HL-60, SMMC-7721, A-549, MCF-7, and SW-480 cell lines and the normal colonic epithelial cells NCM460 with IC_{50} values ranging from 7.3 to 28.4 μM [83,84]. The synthesis of the same chemical compounds by the plant host and endophytic fungus is the phenomenon described for some other species as an example of highly specified coevolution. Moreover, this phenomenon has a great significance in the detection and production of pharmaceutically valuable plant/endophyte derived drugs [85–87].

Table 2. Main biological activities of endophytes isolated from Amaryllioidae crops.

Tissues	Dominating Endophyte Species	Main Activities of the Endophyte	Main Metabolites/Enzymes Linked to Endophyte Bioactivities	Reference
Narcissus pseudonarcissus (root tips, outer bracts)	*Papiliotrema laurentii*	Antifungal against some strains of *Cryptococcus laurentii*	Lycorine	[81]
Narcissus sp.	*Rhinocladiella* sp.	Inhibition of acetylcholinesterase	Not investigated	[80]
Narcissus tazetta (bulb)	*Fusarium solani*	Selective accumulation by an endophytic bacterium, *Achromobacter xylosoxidans*	Hexacyclopeptides	[79]
L. aurea	*Penicillium* sp.	Not investigated	Galantamine	[82]
L. radiata (bulb)	*Aspergillus versicolor*	Cytotoxic against human tumor HL-60 cells with an IC50 value of 5.6 lM.15	(±)–Versicalin A, keto-enol tautomer (+)–6a	[83]

Table 2. *Cont.*

Tissues	Dominating Endophyte Species	Main Activities of the Endophyte	Main Metabolites/Enzymes Linked to Endophyte Bioactivities	Reference
L. radiata (bulb)	*A. versicolor*	Cytotoxic against human tumor HL-60, SMMC-7721, A-549, MCF-7, and SW-480 cell lines	Proversilin C Proversilin E	[84]
Eucharis cyaneosperma	*Streptomyces recifensis*	Not investigated	Not investigated	[85]

4. Conclusions

The increasing demand for Amaryllidaceae crops is triggered not only by traditional culinary usage but also by cultural attitudes, social beliefs, modern interest in exotic and ethnic foods, and medicinal applications. However, secondary metabolites responsible for wide applications of crops in every area of human activity are synthesized by both plant and its endopytic microbiota. Indeed, endophytic fungi associated with this group of plants provide host plants with nutrients and water, alleviate biotic and abiotic stresses, increase stress tolerance, and affect metabolome profile. They are a source of metabolites of antifungal and antiparasitic activity and have a promising perspective in the application as effective biocontrol agents, replacing chemical fungicides and pesticides. Moreover, as the chemosynthesis of Amaryllidaceae alkaloids needs complicated and costly procedures, plants remain an exclusive source of these alkaloids for the pharmaceutical industry. Symbiotic endophytic fungi can be used to increase alkaloids yield in plants or as an alternative source of alkaloids and other bioactive compounds in vitro cultures. Taking into account the scant research on endophytic fungi associated with Amaryllidaceae as a prolific source of phytochemicals, the need has raised for screening investigations aimed to identify the endophytic species, as well as the molecular and genetic basis of their relationship with the host plants.

5. Review Methodology

The present review was based on the literature collected from the leading life science databases, including AGRICOLA, AGRIS, BioOne, CAB Abstracts, PubMed, SciELO, Scopus, and Web of Science. Bibliometric analysis was used for the review, evaluation, and objective representation of the structure within a presented research area, namely Amarylidaceae–fungal endophytes relationship. The most relevant aspects of the evolution, advances, and trends in the reviewed field were presented [88]. References were collected, studied, and selected considering (i) the reports of endophytes isolated from the species of Amaryllidaceae family, (ii) the reports of therapeutic utilization of the host plant or/and an endophyte, (iii) the effects of host plant phylogeny on root microbiome assembly. Plant names were verified according to the Global Biodiversity Information Facility [89] and The Plant List [90]; endophyte taxa were verified according to MycoBank database [91].

Author Contributions: Conceptualization, A.S., and G.C.; writing—original draft preparation, A.S., and G.C.; writing—review and editing, N.G., M.T.A., A.T.; visualization, A.S. and A.T.; supervision, A.S., and G.C. All authors have read and agreed to the published version of the manuscript.

Funding: This study was partially supported by the Ministry of Science and Higher Education of the Republic of Poland.

References

1. Petkova, N.T.; Ivanov, I.G.; Raeva, M.; Topuzova, M.G.; Todorova, M.M.; Denev, P.P. Fructans and antioxidants in leaves of culinary herbs from Asteraceae and Amaryllidaceae families. *Food Res.* **2019**, *3*, 407–415. [CrossRef]
2. Amalfitano, C.; Golubkina, N.A.; Del Vacchio, L.; Russo, G.; Cannoniero, M.; Somma, S.; Morano, G.; Cuciniello, A.; Caruso, G. Yield, Antioxidant Components, Oil Content, and Composition of Onion Seeds Are Influenced by Planting Time and Density. *Plants* **2019**, *8*, 293. [CrossRef]
3. Nemtinov, V.; Golubkina, N.; Koshevarov, A.; Konstanchuk, Y.; Molchanova, A.; Nadezhkin, S.; Sellitto, V.M.; Caruso, G. Health-beneficial compounds from edible and waste bulb components of sweet onion genotypes organically grown in northern Europe. *Banats J. Biotechnol.* **2019**, *10*. [CrossRef]
4. Petrovic, B.; Kopta, T.; Pokluda, R. Effect of biofertilizers on yield and morphological parameters of onion cultivars. *Folia Hortic.* **2019**, *31*, 51–59. [CrossRef]
5. Rozpądek, P.; Rąpała-Kozik, M.; Wężowicz, K.; Grandin, A.; Karlsson, S.; Ważny, R.; Anielska, T.; Turnau, K. Arbuscular mycorrhiza improves yield and nutritional properties of onion (*Allium cepa*). *Plant Physiol. Biochem.* **2016**, *107*, 264–272. [CrossRef] [PubMed]

6. Hibilik, N.; Selmaoui, K.; Msairi, S.; Touati, J.; Chliyeh, M.; Mouria, A.; Artib, M.; Touhami, A.O.; Benkirane, R.; Douira, A. Study of Arbuscular Mycorrhizal Fungi Diversity and Its Effect on Growth and Development of Leek Plants (*Allium porrum* L.). *Annu. Res. Rev.* **2018**, *24*, 1–14. [CrossRef]

7. Golubkina, N.; Krivenkov, L.; Sekara, A.; Vasileva, V.; Tallarita, A.; Caruso, G. Prospects of Arbuscular Mycorrhizal Fungi Utilization in Production of Allium Plants. *Plants* **2020**, *9*, 279. [CrossRef] [PubMed]

8. Wang, B.; Sugiyama, S. Phylogenetic signal of host plants in the bacterial and fungal root microbiomes of cultivated angiosperms. *Plant J.* **2020**. [CrossRef]

9. Hardoim, P.R.; van Overbeek, S.; Berg, G.; Pirttilä, A.M.; Compant, S.; Campisano, A.; Döring, M.; Sessitsch, A. The Hidden World within Plants: Ecological and Evolutionary Considerations for Defining Functioning of Microbial Endophytes. *Microbiol. Mol. Biol. Rev.* **2015**, *79*, 293–320. [CrossRef]

10. Aleklett, K.; Hart, M. The root microbiota—A fingerprint in the soil? *Plant Soil* **2013**, *370*, 671–686.

11. Ansari, M.W.; Trivedi, D.K.; Sahoo, R.K.; Gill, S.S.; Tuteja, N. A critical review on fungi mediated plant responses with special emphasis to *Piriformospora indica* on improved production and protection of crops. *Plant Physiol. Biochem.* **2013**, *70*, 403–410. [CrossRef] [PubMed]

12. Crişan, I.; Vidican, R.; Stoian, V.; Şandor, M.; Stoie, A. Endophytic Root Colonization Patterns in Early and Mid-Spring Geophytes from Cluj County. *Bull. Univ. Agric. Sci. Vet. Med. Cluj-Napoca* **2019**, *76*, 21–27.

13. Casas, C.; Torretta, J.P.; Exeler, N.; Omacini, M. What happens next? Legacy effects induced by grazing and grassendophyte symbiosis on thistle plants and their floral visitor. *Plant Soil* **2016**, *405*, 211–229.

14. Compant, S.; Saikkonen, K.; Mitter, B.; Campisano, A.; Mercado-Blanco, J. Editorial Special Issue: Soil, Plants and Endophytes. *Plant Soil* **2016**, *405*, 1–11. [CrossRef]

15. Johnston-Monje, D.; Lundberg, D.S.; Lazarovits, G.; Reis, V.M.; Raizada, M.N. Bacterial populations in juvenile maize rhizospheres originate from both seed and soil. *Plant Soil* **2016**, *405*, 337–355. [CrossRef]

16. Soussi, A.; Ferjani, R.; Marasco, R.; Guesmi, A.; Cherif, H.; Rolli, E.; Mapelli, F.; Ouzari, H.I.; Daffonchio, D.; Cherif, A. Plant associated microbiomes in arid lands: Diversity, ecology and biotechnological potential. *Plant Soil* **2016**, *405*, 357–370. [CrossRef]

17. Saunders, M.; Kohn, L.M. Evidence for alteration of fungal endophyte community assembly by host defense compounds. *New Phytol.* **2009**, *182*, 229–238. [CrossRef]

18. Rustamova, N.; Bozorov, K.; Efferth, T.; Yili, A. Novel secondary metabolites from endophytic fungi: Synthesis and biological properties. *Phytochem. Rev.* **2020**, *19*, 425–448. [CrossRef]

19. Zhou, J.; Liu, Z.; Wang, S.; Li, J.; Li, Y.; Chen, W.K.; Wang, R. Fungal endophytes promote the accumulation of Amaryllidaceae alkaloids in *Lycoris radiata*. *Environ. Microbiol.* **2020**, *22*, 1421–1434. [CrossRef]

20. Abdel-Hafez, S.I.; Abo-Elyousr, K.A.; Abdel-Rahim, I.R. Leaf surface and endophytic fungi associated with onion leaves and their antagonistic activity against *Alternaria porri*. *Czech Mycol.* **2015**, *67*, 1–22. [CrossRef]

21. Bamisile, B.S.; Dash, C.K.; Akutse, K.S.; Keppanan, R.; Wang, L. Fungal endophytes: Beyond herbivore management. *Front. Microbiol.* **2018**, *9*, 544. [CrossRef]

22. Vinale, F.; Nicoletti, R.; Borrelli, F.; Mangoni, A.; Parisi, O.A.; Marra, R.; Lombardi, N.; Lacatena, F.; Grauso, L.; Finizio, S.; et al. Co-culture of plant beneficial microbes as source of bioactive metabolites. *Sci. Rep.* **2017**, *7*, 1–12. [CrossRef]

23. Jaber, L.R.; Ownley, B.H. Can we use entomopathogenic fungi as endophytes for dual biological control of insect pests and plant pathogens? *Biol. Control* **2018**, *116*, 36–45. [CrossRef]

24. Caruso, G.; Abdelhamid, M.T.; Kalisz, A.; Sekara, A. Linking Endophytic Fungi to Medicinal Plants Therapeutic Activity. A Case Study on Asteraceae. *Agriculture* **2020**, *10*, 286. [CrossRef]

25. Chase, M.W.; Christenhusz, M.J.; Fay, M.F.; Byng, J.W.; Judd, W.S.; Soltis, D.E.; Mabberley, D.J.; Sennikov, A.N.; Soltis, P.S.; Stevens, P.F. An update of the Angiosperm Phylogeny Group classification for the orders and families of flowering plants: APG IV. *Bot. J. Linn. Soc.* **2016**, *181*, 1–20.

26. Kamenetsky, R.; Rabinowitch, H.D. The genus *Allium*: A developmental and horticultural analysis. *Hortic. Rev. (Am. Soc. Hortic. Sci.)* **2006**, *32*, 329.

27. Wheeler, E.J.; Mashayekhi, S.; McNeal, D.W.; Columbus, J.T.; Pires, J.C. Molecular systematics of *Allium* subgenus *Amerallium* (Amaryllidaceae) in North America. *Am. J. Bot.* **2013**, *100*, 701–711. [CrossRef] [PubMed]

28. Keusgen, M.; Schulz, H.; Glodek, J.; Krest, I.; Krüger, H.; Herchert, N.; Keller, J. Characterization of some *Allium* hybrids by aroma precursors, aroma profiles, and alliinase activity. *J. Agric. Food Chem.* **2002**, *50*, 2884–2890. [CrossRef]

29. Rabinowitch, H.D.; Currah, L. (Eds.) *Allium Crop Science: Recent Advances*; CABI Publishing: Wallingford, UK, 2002.

30. Sekara, A.; Pokluda, R.; Del Vacchio, L.; Somma, S.; Caruso, G. Interactions among genotype, environment and agronomic practices on production and quality of storage onion (*Allium cepa* L.). A review. *Hortic. Sci.* **2017**, *44*, 201–212. [CrossRef]

31. Kučová, L.; Kopta, T.; Sękara, A.; Pokluda, R. Controlling nitrate and heavy metals content in leeks (*Allium porrum* L.) using arbuscular mycorrhizal fungi inoculation. *Pol. J. Environ. Stud.* **2018**, *27*, 137–143. [CrossRef]

32. Golubkina, N.A.; Seredin, T.M.; Antoshkina, M.S.; Kosheleva, O.V.; Teliban, G.C.; Caruso, G. Yield, quality, antioxidants and elemental composition of new leek cultivars under organic or conventional systems in a greenhouse. *Horticulturae* **2018**, *4*, 39. [CrossRef]

33. Galeone, C.; Turati, F.; Zhang, Z.F.; Guercio, V.; Tavani, A.; Serraino, D.; Brennan, P.; Fabianova, E.; Lissowska, J.; Mates, D.; et al. Relation of *Allium* vegetables intake with head and neck cancers: Evidence from the inhance consortium. *Mol. Nutr. Food Res.* **2015**, *59*, 1641–1650. [CrossRef]

34. Zeng, Y.; Li, Y.; Yang, J.; Pu, X.; Du, J.; Yang, X.; Yang, T.; Yang, S. Therapeutic role of functional components in *Alliums* for preventive chronic disease in human being. *Evid. Based Complement Altern. Med.* **2017**, *3*, 9402849. [CrossRef]

35. Munday, R. Harmful and beneficial effects of organic monosulfides, disulfides, and polysulfides in animals and humans. *Chem. Res. Toxicol.* **2012**, *25*, 47–60. [CrossRef]

36. Mnayer, D.; Fabiano-Tixier, A.S.; Petitcolas, E.; Hamieh, T.; Nehme, N.; Ferrant, C.; Fernandez, X.; Chemat, F. Chemical composition, antibacterial and antioxidant activities of six essentials oils from the Alliaceae family. *Molecules* **2014**, *19*, 20034–20053. [CrossRef] [PubMed]

37. Wallock-Richards, D.; Doherty, C.J.; Doherty, L.; Clarke, D.J.; Place, M.; Govan, J.R.W.; Campopiano, D.J. Garlic revisited: Antimicrobial activity of allicin-containing garlic extracts against *Burkholderia cepacia* complex. *PLoS ONE* **2014**, *9*, e112726. [CrossRef]

38. Vlase, L.; Parvu, M.; Parvu, E.A.; Toiu, A. Chemical constituents of three *Allium* species from Romania. *Molecules* **2013**, *18*, 114–127. [CrossRef]

39. Costa, L.G.; Garrick, J.M.; Roquè, P.J.; Pellacani, C. Mechanisms of neuroprotection by quercetin: Counteracting oxidative stress and more. *Oxid. Med. Cell. Longev.* **2016**, *2016*, 2986796. [CrossRef]

40. Stafford, G.I.; Pedersen, M.E.; van Staden, J.; Jäger, A.K. Review on plants with CNS-effects used in traditional South African medicine against mental diseases. *J. Ethnopharmacol.* **2008**, *119*, 513–537. [CrossRef]

41. Nair, J.J.; Wilhelm, A.; Bonnet, S.L.; van Staden, J. Antibacterial constituents of the plant family Amaryllidaceae. *Bioorg. Med. Chem. Let.* **2017**, *27*, 4943–4951. [CrossRef] [PubMed]

42. Berkov, S.; Osorio, E.; Viladomat, F.; Bastida, J. Chemodiversity, chemotaxonomy and chemoecology of Amaryllidaceae alkaloids. *Alkaloids Chem. Biol.* **2020**, *83*, 113–185.

43. Desgagné-Penix, I. Biosynthesis of alkaloids in Amaryllidaceae plants: A review. *Phytochem. Rev.* **2020**. [CrossRef]

44. Lin, Y.T.; Chou, M.C.; Wu, S.J.; Yang, Y.H. Galantamine plasma concentration and cognitive response in Alzheimer's disease. *PeerJ* **2019**, *7*, e6887. [CrossRef]

45. Nair, J.J.; Bastida, J.; Viladomat, F.; van Staden, J. Cytotoxic agents of the crinane series of Amaryllidaceae alkaloids. *Nat. Prod. Commun.* **2012**, *7*, 12. [CrossRef]

46. Rønsted, N.; Symonds, M.R.; Birkholm, T.; Christensen, S.B.; Meerow, A.W.; Molander, M.; Stafford, G.I. Can phylogeny predict chemical diversity and potential medicinal activity of plants? A case study of Amaryllidaceae. *BMC Evol. Biol.* **2012**, *12*, 182. [CrossRef]

47. Hirano, S.S.; Upper, C.D. Bacteria in the Leaf Ecosystem with Emphasis on *Pseudomonas syringae* a Pathogen, Ice Nucleus, and Epiphyte. *Microbiol. Mol. Biol. Rev.* **2000**, *64*, 624–653. [CrossRef]

48. Abdel-Gawad, K.M.; Abdel-Mallek, A.Y.; Hussein, N.A.; Abdel-Rahim, I.R. Diversity of mycobiota associated with onion (*Allium cepa* L.) cultivated in Assiut, with a newly recorded fungal species to Egypt. *J. Microbiol. Biotechnol. Food Sci.* **2017**, *9*, 1145–1151. [CrossRef]

49. Abdel-Hafez, S.I.I.; Moharram, A.M.; Abdel-Sater, M.A. Monthly variations in the mycobiota of wheat fields in El-Kharga Oasis, Western desert, Egypt. *Bull. Fac. Sci. Assiut Univ.* **2000**, *29*, 195–211.

50. Zhou, J.; Li, P.; Meng, D.; Gu, Y.; Zheng, Z.; Yin, H.; Yin, H.; Li, J. Isolation, characterization and inoculation of Cd tolerant rice endophytes and their impacts on rice under Cd contaminated environment. *Environ. Pollut.* **2020**, *260*, 113990. [CrossRef]

51. Zhou, P.; Wu, Z.; Tan, D.; Yang, J.; Zhou, Q.; Zeng, F.; Zhang, M.; Bie, Q.; Chen, C.; Xue, Y.; et al. Atrichodermones A–C, three new secondary metabolites from the solid culture of an endophytic fungal strain, *Trichoderma atroviride*. *Fitoterapia* **2017**, *123*, 18–22. [CrossRef]

52. Yang, S.; Nan, X.H.; Lu, Y.B.; Peng, F. Study on isolation and anti-microbial activitiy of endophyte in *Lycoris aurea* Herb. *J. Tradit. Chin. Med.* **2010**, *30*, 78–80.

53. Rocha, L.F.; Inglis, P.W.; Humber, R.A.; Kipnis, A.; Luz, C. Occurrence of *Metarhizium* spp. in Central Brazilian soils. *J. Basic Microbiol.* **2013**, *53*, 251–259. [CrossRef] [PubMed]

54. Nouh, F.A.A.; Abdel-Azeem, A.M. Role of Fungi in Adaptation of Agricultural Crops to Abiotic Stresses. In *Agriculturally Important Fungi for Sustainable Agriculture*; Springer: Cham, Switzerland, 2020; pp. 55–80.

55. Pusztahelyi, T.; Holb, I.J.; Pócsi, I. Secondary metabolites in fungus-plant interactions. *Front. Plant. Sci.* **2015**, *6*, 573. [CrossRef]

56. Muvea, A.M.; Meyhöfer, R.; Subramanian, S.; Poehling, H.-M.; Ekesi, S.; Maniana, N.K. Colonization of Onions by Endophytic Fungi and Their Impacts on the Biology of Thrips tabaci. *PLoS ONE* **2014**, *9*, e108242. [CrossRef] [PubMed]

57. Muvea, A.M.; Meyhöfer, R.; Maniania, N.K.; Poehling, H.M.; Ekesi, S.; Subramanian, S. Behavioral responses of *Thrips tabaci* Lindeman to endophyte-inoculated onion plants. *J. Pest Sci.* **2015**, *88*, 555–562. [CrossRef]

58. Priyadharsini, P.; Pandey, R.; Muthukumar, T. Arbuscular mycorrhizal and dark septate fungal associations in shallot (*Allium cepa* L. var. aggregatum) under conventional agriculture. *Acta Bot. Croat.* **2012**, *71*, 159–175. [CrossRef]

59. Wu, Y.; Liu, T.; He, X. Mycorrhizal and dark septate endophytic fungi under the canopies of desert plants in Mu Us Sandy Land of China. *Front. Agric. China* **2009**, *3*, 164–170. [CrossRef]

60. Isaeva, O.V.; Glushakova, A.M.; Garbuz, S.A.; Kachalkin, A.V.; Chernov, I.Y. Endophytic yeast fungi in plant storage tissues. *Biol. Bull.* **2010**, *37*, 26–34. [CrossRef]

61. Xiang, L.; Gong, S.; Yang, L.; Hao, J.; Xue, M.; Zeng, F.; Zhang, X.; Shi, W.; Wang, H.; Yu, D. Biocontrol potential of endophytic fungi in medicinal plants from Wuhan Botanical Garden in China. *Biol. Contr.* **2015**, *94*, 47–55. [CrossRef]

62. Zheng, Y.K.; Qiao, X.G.; Miao, C.P.; Liu, K.; Chen, Y.W.; Xu, L.H.; Zhao, L.X. Diversity, distribution and biotechnological potential of endophytic fungi. *Ann. Microbiol.* **2016**, *66*, 529–542. [CrossRef]

63. Iqbal, Z.; Nasir, H.; Hiradate, S.; Fujii, Y. Plant growth inhibitory activity of *Lycoris radiata* herb. And the possible involvement of lycorine as an allelochemical. *Weed Biol. Manag.* **2006**, *6*, 221–227. [CrossRef]

64. Abo-Elyousr, K.A.M.; Abdel-Hafez, S.I.I.; Abdel-Rahim, I.R. Isolation of *Trichoderma* and evaluation of their antagonistic potential against *Alternaria porri*. *J. Phytopatol.* **2014**, *162*, 567–574. [CrossRef]

65. Siameto, E.; Okoth, S.; Amugune, N.; Chege, N. Antagonism of *Trichoderma harzianum* isolates on soil borne plant pathogenic fungi from Embu District, Kenya. *J. Yeast Fungal Res.* **2010**, *1*, 47–54.

66. Flori, P.; Roberti, R. Treatment of onion bulbs with antagonistic fungi for the control of *Fusarium oxysporum* f. sp. cepae. *Difesa Piante* **1993**, *16*, 5–12.

67. Abdel-Rahim, I.R.; Abo-Elyousr, K.A. *Talaromyces pinophilus* strain AUN-1 as a novel mycoparasite of *Botrytis cinerea*, the pathogen of onion scape and umbel blights. *Microbiol. Res.* **2018**, *212*, 1–9. [CrossRef]

68. Muvea, A.M.; Subramanian, S.; Maniania, N.K.; Poehling, H.M.; Ekesi, S.; Meyhöfer, R. Endophytic colonization of onions induces resistance against viruliferous thrips and virus replication. *Front. Plant Sci.* **2018**, *9*, 1785. [CrossRef]

69. Madrigal, C.; Tadeo, J.; Melgarejo, P. Relationship between flavipin production by *Epicoccum nigrum* and antagonism against *Monilinia laxa*. *Mycol. Res.* **1991**, *95*, 1375–1381. [CrossRef]

70. Haggag, W.M.; El Soud, M.A. Pilot-scale production and optimizing of cellulolytic *Penicillium oxalicum* for controlling of mango malformation. *Agric. Sci.* **2013**, *4*, 165–174.

71. Shentu, X.; Zhan, X.; Ma, Z.; Yu, X.; Zhang, C. Antifungal activity of metabolites of the endophytic fungus *Trichoderma brevicompactum* from garlic. *Braz. J. Microbiol.* **2014**, *45*, 248–254. [CrossRef] [PubMed]

72. Espinoza, F.; Vidal, S.; Rautenbach, F.; Lewu, F.; Nchu, F. Effects of *Beauveria bassiana* (Hypocreales) on plant growth and secondary metabolites of extracts of hydroponically cultivated chive (*Allium schoenoprasum* L. [Amaryllidaceae]). *Heliyon* **2019**, *5*, 12. [CrossRef]

73. Koul, M.; Meena, S.; Kumar, A.; Sharma, P.R.; Singamaneni, V.; Riyaz-Ul-Hassan, S.; Abid, H.; Asha, C.; Anil, P.; Prasoon, G.; et al. Secondary metabolites from endophytic fungus *Penicillium pinophilum* induce ROS-mediated apoptosis through mitochondrial pathway in pancreatic cancer cells. *Planta Med.* **2016**, *82*, 344–355. [CrossRef]

74. Abdulmyanova, L.I.; Fayzieva, F.K.; Ruzieva, D.M.; Rasulova, G.A.; Sattarova, R.S.; Gulyamova, T.G. Bioactivity of fungal endophytes associating with *Allium* plants growing in Uzbekistan. *Int. J. Curr. Microbiol. Appl. Sci.* **2016**, *5*, 769–778. [CrossRef]

75. Ruzieva, D.M.; Abdulmyanova, L.I.; Rasulova, G.A.; Sattarova, R.S.; Gulyamova, T.G. Screening of inhibitory activity against α-amylase of fungal endophytes isolated from medicinal plants in Uzbekistan. *Int. J. Curr. Microbiol. App. Sci.* **2017**, *6*, 2744–2752.

76. Keusgen, M.; Fritsch, R.M.; Hisoriev, H.; Kurbonova, P.A.; Khassanov, F.O. Wild *Allium* species (Alliaceae) used in folk medicine of Tajikistan and Uzbekistan. *J. Ethnobiol. Ethnomed.* **2006**, *2*, 18. [CrossRef]

77. Ptak, A.; Simlat, M.; Morańska, E.; Spina, R.; Dupire, F.; Kozaczka, S.; Laurain-Mattar, D. Potential of the *Leucojum aestivum* L. endophytic bacteria to promote Amaryllidaceae alkaloids biosynthesis. *Acta Biol. Cracov.* **2019**, *62*, 30.

78. Liu, Z.; Zhou, J.; Li, Y.; Wen, J.; Wang, R. Bacterial endophytes from *Lycoris radiata* promote the accumulation of Amaryllidaceae alkaloids. *Microbiol. Res.* **2020**, *239*, 126501. [CrossRef] [PubMed]

79. Wang, W.X.; Kusari, S.; Sezgin, S.; Lamshöft, M.; Kusari, P.; Kayser, O.; Spiteller, M. Hexacyclopeptides secreted by an endophytic fungus *Fusarium solani* N06 act as crosstalk molecules in *Narcissus tazetta*. *Appl. Microbiol. Biotechnol.* **2015**, *99*, 7651–7662. [CrossRef]

80. Yang, M.J.; Lu, X.L.; Wang, Y.G.; Yao, G.Y.; Yang, L.; Liu, X.F. Screening and identifying of endophytic fungi with acetylcholinesterase inhibitory activity from narcissus. *Chin. Pharmacol. Bull.* **2013**, *10*, 33.

81. Onofri, S.; Barreca, D.; Garuccio, I. Effects of lycorine on growth and effects of L-galactonic acid-γ-lactone on ascorbic acid biosynthesis in strains of *Cryptococcus laurentii* isolated from *Narcissus pseudonarcissus* roots and bulbs. *Anton. Leeuw. Int. J. G.* **2003**, *83*, 57–61. [CrossRef] [PubMed]

82. Luo, Y.; Yang, S.; Zhao, Z.; Peng, F.; Li, Y. Isolation and identification of a galanthamine-producing endophytic fungus from *Lycoris aurea* bulbs. *Northwest Pharm. J.* **2011**, *4*, 5.

83. Li, H.; Xu, Q.; Sun, W.; Zhang, R.; Wang, J.; Lai, Y.; Hu, Z.; Zhang, Y. 21-Epi-taichunamide D and (±)-versicaline A, three unusual alkaloids from the endophytic *Aspergillus versicolor* F210. *Tetrahedron Lett.* **2020**, *61*, 152219. [CrossRef]

84. Li, H.; Zhang, R.; Cao, F.; Wang, J.; Hu, Z.; Zhang, Y. Proversilins A–E, Drimane-Type Sesquiterpenoids from the Endophytic *Aspergillus versicolor*. *J. Nat. Prod.* **2020**, *83*, 2200–2206. [CrossRef]

85. Bascom-Slack, C.A.; Ma, C.; Moore, E.; Babbs, B.; Fenn, K.; Greene, J.S.; Hann, B.D.; Keehner, J.; Kelley-Swift, E.G.; Kembaiyan, V.; et al. Multiple, novel biologically active endophytic actinomycetes isolated from upper Amazonian rainforests. *Microb. Ecol.* **2009**, *58*, 374–383. [CrossRef] [PubMed]

86. Nicoletti, R.; Ferranti, P.; Caira, S.; Misso, G.; Castellano, M.; Di Lorenzo, G.; Caraglia, M. Myrtucommulone production by a strain of *Neofusicoccum australe* endophytic in myrtle (*Myrtus communis*). *World J. Microbiol. Biotechnol.* **2014**, *30*, 1047–1052. [CrossRef]

87. Nicoletti, R.; Fiorentino, A. Plant bioactive metabolites and drugs produced by endophytic fungi of Spermatophyta. *Agriculture* **2015**, *5*, 918–970. [CrossRef]

88. Zupic, I.; Cater, T. Bibliometric methods in management and organization. *Org. Res. Methods* **2015**, *18*, 429–472. [CrossRef]

89. Global Biodiversity Information Facility. Available online: https://www.gbif.org/ (accessed on 25 September 2020).

90. The Plan List. Available online: www.theplantlist.org (accessed on 25 September 2020).

91. MycoBank. Available online: http://www.mycobank.org/ (accessed on 25 September 2020).

Publisher's Note: MDPI stays neutral with regard to jurisdictional claims in published maps and institutional affiliations.

 *agriculture*

Review

Bioactive Products from Endophytic Fungi of Sages (*Salvia* spp.)

Beata Zimowska [1], Monika Bielecka [2,*], Barbara Abramczyk [3] and Rosario Nicoletti [4,5]

[1] Department of Plant Protection, University of Life Sciences, Leszczyńskiego 7, 20-069 Lublin, Poland; beata.zimowska@up.lublin.pl

[2] Department of Pharmaceutical Biotechnology, Wroclaw Medical University, Borowska 211A, 50-556 Wroclaw, Poland

[3] Department of Agricultural Microbiology, Institute of Soil Science and Plant Cultivation, Czartoryskich 8, 24-100 Puławy, Poland; babramczyk@iung.pulawy.pl

[4] Council for Agricultural Research and Economics, Research Centre for Olive, Fruit and Citrus Crops, 81100 Caserta, Italy; rosario.nicoletti@crea.gov.it

[5] Department of Agricultural Sciences, University of Naples 'Federico II', 80055 Portici, Italy

* Correspondence: monika.bielecka@umed.wroc.pl

Received: 29 October 2020; Accepted: 10 November 2020; Published: 12 November 2020

Abstract: In the aim of implementing new technologies, sustainable solutions and disruptive innovation to sustain biodiversity and reduce environmental pollution, there is a growing interest by researchers all over the world in bioprospecting endophytic microbial communities as an alternative source of bioactive compounds to be used for industrial applications. Medicinal plants represent a considerable source of endophytic fungi of outstanding importance, which highlights the opportunity of identifying and screening endophytes associated with this unique group of plants, widespread in diverse locations and biotopes, in view of assessing their biotechnological potential. As the first contribution of a series of papers dedicated to the Lamiaceae, this article reviews the occurrence and properties of endophytic fungi associated with sages (*Salvia* spp.).

Keywords: sage; endophytes; bioprospecting; bioactive compounds; medicinal plants; Lamiaceae

---

1. Introduction

Endophytic fungi are defined as fungi inhabiting tissues and organs of healthy plants during certain stages of their life cycle without causing apparent symptoms. The concept of endophytism, introduced by De Bary in 1866 and almost completely neglected for over a century, has recently become of common usage concomitantly to advances in knowledge on occurrence and functions of this component of biodiversity. Increasing attention by the scientific community is boosted by the opportunity to exploit the unique aptitudes and properties of these microbial associates of plants [1].

As a consequence of the long-term association of endophytes with medicinal plants, based on mutually beneficial relationships, the former may also participate in metabolic paths and boost their own natural biosynthetic activity, or may gain some genetic information to synthesize biologically active compounds closely related to those directly produced by the host plant [2,3]. Endophytic fungi derived from medicinal plants are becoming more and more popular, due to specific modes of action and the ability to provide multiple benefits, which make them relevant for both agricultural and pharmaceutical applications [3]. This review is devoted to an analysis of the biochemical potential of endophytic fungi reported from species of sage (*Salvia* spp.), examining the advances in this particular field made by the scientific community in recent years.

2. *Salvia*: The Largest Genus of Lamiaceae

Lamiaceae is one of the most important herbal families, including a wide variety of plants with multiple medical, culinary and industrial applications. Within the subfamily Nepetoideae, sage species are ascribed to the genus *Salvia*, a name deriving from the Latin word *"salvere"*, which refers to the curative properties of these plants. It represents the largest genus of the family counting between 700 and 1000 species [4]. Uncertainty of this number is basically due to the broad geographical range of distribution, covering all continents and climatic areas, which makes taxonomic verification problematic.

As for many plants and other organisms, the application of biomolecular techniques in taxonomy has determined several basic reassessments in classification. Until a few years ago, the genus name *Salvia* was only used for species displaying the typical morphological features of sage. Nevertheless, recent systematic work has emphasized close relationships with the genera *Dorystaechas, Meriandra, Perovskia, Rosmarinus* and *Zhumeria*, which resulted to be clearly embedded in *Salvia* in dedicated phylogenetic analyses, so that their separation is no more justified [4,5]. Although not consolidated in the common use yet, this new taxonomic sorting is basically followed in this paper. However, by reason of several peculiar aspects concerning geographical distribution and biotechnological applications, the species *Salvia rosmarinus* (=*Rosmarinus officinalis*) will be the subject of a dedicated analysis in a forthcoming paper.

Medicinal properties of sages derive from their ability to produce a multitude of bioactive secondary metabolites, many of which have been reported for antibiotic, antitumor, antiviral, antiprotozoal, insecticidal and antioxidant effects, or even to be responsible for allelopathic interactions with other plants [6]. These varied bioactivities are reflected by quite diverse chemical structures. In fact, besides flavonoids and simple phenolic compounds like caffeic, rosmarinic and salvianolic acids, which are mainly known for their radical scavenging effects, these products include monoterpenoids, sesquiterpenoids, triterpenoids and diterpenoids. Structural diversity is particularly evident within this latter grouping, including labdanes, ent-kauranes, abietanes, icetexanes, clerodanes, and pimaranes, as well as phenolic diterpenoids, such as carnosol and carnosic acid [6]. Moreover, some abietanes are rearranged, to form the important scaffold of tanshinones [7]. These latter products are particularly considered for pharmaceutical application based on their antioxidant [8], antibacterial [9], antidiabetic [10], anti-inflammatory [11], and antiproliferative [12] properties, and are currently the subject of a specific project at our laboratories, in which the species *Salvia abrotanoides* (formerly *Perovskia abrotanoides*) and *Salvia yangii* (formerly *Perovskia atriplicifolia*), regarded as an alternative source of tanshinones, are analyzed through combined metabolomics and transcriptomics approaches, also with reference to the associated endophytic fungi.

3. Ecology and Occurrence

As introduced above, endophytic fungi are polyphyletic groups of microorganisms, which asymptomatically colonize healthy tissues of different parts of living plants such as stems, leaves or roots. Their diversity is huge, and it has been estimated that every plant hosts several endophytic species, among which at least one shows host specificity [13,14]. Through the evolutionary processes, endophytic fungi have developed different symbiotic relationships with their host plants [3]. Furthermore, many species are reported to exhibit multiple ecological roles as both endophytes and pathogens. However, it is not clear whether the same genotypes can play both these roles with equal success. Understanding the mechanisms responsible for the conversion between so different outcomes of the ecological interaction represents one among many frontiers in endophyte biology [15,16]. One of the mechanisms developed by plants during the long-term co-evolution with microbial associates is the ability to produce antibiotic compounds. Simultaneously, many endophytes have developed an important transformative capacity and/or tolerance to these products which in a large part determines the colonization range of their hosts [17]. In turn, endophytes can influence growth and development of host plants, and enhance their resistance to biotic and abiotic stresses by releasing bioactive metabolites [18], to such an extent that in natural habitats some plant species require to be supported by endophytic fungi for stress tolerance and survival [19].

The diversity of endophytic fungi associated with medicinal plants is largely affected by ecological or environmental factors. Particularly, temperature, humidity and soil nutritional conditions influence the quality and quantity of secondary metabolites synthesized by the hosts, which in turn affect the population structure of endophytic fungi. The species composition of endophyte communities also differs in organ and tissue specificities, as a result of their adaptation to different physiological conditions in hosts [3,20].

The analysis of the recent literature shows that species of sage (*Salvia* spp.) host diverse communities of fungal endophytes. As many as 64 different taxa belonging to 38 genera, with a clear prevalence of Ascomycetes, have been reported so far (Table 1). Most observations concern the species *Salvia miltiorrhiza* and *S. abrotanoides*, respectively, with 28 and 24 records. There is an evident correspondence between the *Salvia* species and the geographical area. In fact, all isolations concerning *S. miltiorrhiza* come from several provinces of China, while the available findings from *S. aegyptiaca* come from Egypt, and those referring to *S. abrotanoides* derive from an Iranian study and from the activity currently in progress at our laboratories. Despite the fact that isolations have been carried out from any plant organ (Figures 1 and 2), no indications concerning a specific association with roots or the aerial parts can be inferred. The access to biomolecular methods as a taxonomic tool has generally enabled to perform reliable identification at the species level, with the exception of a Chinese study concerning seeds of *S. miltiorrhiza*, where sorting was basically limited to the class level despite the wide variation observed [21]. The repeated findings in several species/locations mostly refer to strains provisionally identified at the genus level, particularly *Alternaria*, which seems to be of quite common occurrence on sages regardless to the plant part used for isolations. At the species level, there are only two cases with more than just one record—that is, *Chaetomium globosum* and *Didymella* (=*Phoma*) *glomerata*, both from *S. miltiorrhiza* at two different locations in China. However, the recovery of these species from both roots and leaves may represent a possible indication of a more regular association with this plant, which should be taken into consideration in further studies.

Table 1. Endophytic fungi reported from *Salvia* spp.

Endophyte [1]	Plant Species/Organ	Location, Country	Reference
Acremonium sclerotigenum	*S. abrotanoides*/root	Zoshk, Iran	[22]
Alternaria alternata	*S. miltiorrhiza*/flower	Shandong, China	[23]
	S. aegyptiaca/leaf	Gebel Elba, Egypt	[24]
Alternaria chlamydosporigena	*S. abrotanoides*/root	Zoshk, Iran	[22]
Alternaria sp.	*S. miltiorrhiza*/root	Beijing, China	[25]
	S. miltiorrhiza/seed	Northwest China	[21]
	S. miltiorrhiza/root, shoot, leaf	Henan, China	[26]
	S. abrotanoides/leaf, stem	Wroclaw, Poland	this paper
	S. yangii/leaf, stem	Wroclaw, Poland	this paper
Alternaria tenuissima	*S. przewalskii*/root	Longxi, China	[27]
Aspergillus brasiliensis	*S. aegyptiaca*/leaf	Gebel Elba, Egypt	[24]
Aspergillus foeniculicola	*S. miltiorrhiza*/root	Shaanxi, China	[28]
Aspergillus nidulans	*S. aegyptiaca*/leaf	Gebel Elba, Egypt	[24]
Aspergillus niger	*S. aegyptiaca*/leaf	Gebel Elba, Egypt	[24]
Aspergillus sp.	*S. miltiorrhiza*/root	Beijing, China	[25]
	S. abrotanoides/leaf	Zoshk, Iran	[22]
Aspergillus terreus	*S. aegyptiaca*/leaf	Gebel Elba, Egypt	[24]
Aureobasidium sp.	*S. miltiorrhiza*/seed	Northwest China	[21]
Cadophora sp.	*S. miltiorrhiza*/root	Beijing, China	[25]
Canariomyces microsporus	*S. abrotanoides*/leaf	Zoshk, Iran	[22]
Cephalosporium acremonium	*S. aegyptiaca*/leaf	Gebel Elba, Egypt	[24]
Chaetomium globosum	*S. miltiorrhiza*/root	Shanluo, China	[29]
	S. miltiorrhiza/'aerial part'	Shenyang, China	[30]

Table 1. *Cont.*

Endophyte [1]	Plant Species/Organ	Location, Country	Reference
Chaetomium sp.	*S. officinalis*/stem	Beni-Mellal, Morocco	[31]
		Giza, Egypt	[32]
Cladosporium cladosporioides	*S. aegyptiaca*/leaf	Gebel Elba, Egypt	[24]
Clonostachys rosea	*S. miltiorrhiza*/root	Beijing, China	[25]
Colletotrichum gloeosporioides	*S. miltiorrhiza*/'aerial part'	Shenyang, China	[33]
Colletotrichum sp.	*S. aegyptiaca*/leaf	Gebel Elba, Egypt	[24]
	S. yangii/leaf	Wroclaw, Poland	this paper
Coniolariella hispanica	*S. abrotanoides*/root	Kalat, Iran	[22]
Curvularia papendorfii	*S. aegyptiaca*/leaf	Gebel Elba, Egypt	[24]
Diaporthe sp.	*S. miltiorrhiza*/stem	Sichuan, China	[34]
	S. abrotanoides/stem	Wroclaw, Poland	this paper
	S. yangii/stem	Wroclaw, Poland	this paper
Didymella glomerata	*S. miltiorrhiza*/root	Beijing, China	[25]
	S. miltiorrhiza/leaf	Shangluo, China	[35]
Didymella pedeiae	*S. miltiorrhiza*/root	Beijing, China	[25]
Filobasidium sp.	*S. miltiorrhiza*/seed	Northwest China	[21]
Fusarium dlaminii	*S. abrotanoides*/root	Darrud, Iran	[22]
Fusarium oxysporum	*S. aegyptiaca*/leaf	Gebel Elba, Egypt	[24]
Fusarium proliferatum	*S. miltiorrhiza*/root	Shandong, China	[23]
Fusarium redolens	*S. miltiorrhiza*/root	Beijing, China	[25]
Fusarium sp.	*S. miltiorrhiza*/root	Beijing, China	[25]
	S. abrotanoides/root, stem	Wroclaw, Poland	this paper
	S. yangii/root, stem	Wroclaw, Poland	this paper
Juxtiphoma eupyrena	*S. miltiorrhiza*/root	Beijing, China	[25]
Leptosphaeria sp.	*S. miltiorrhiza*/root	Beijing, China	[25]
Neocosmospora solani	*S. abrotanoides*/root	Kalat, Iran	[22]
Niesslia ligustica	*S. abrotanoides*/root	Darrud, Iran	[22]
Paecilomyces sp.	*S. miltiorrhiza*/root	Beijing, China	[36]
Paraphoma radicina	*S. abrotanoides*/root	Zoshk, Iran	[22]
Penicillium canescens	*S. abrotanoides*/root	Zoshk and Kalat, Iran	[22]
Penicillium charlesii	*S. abrotanoides*/root	Zoshk and Kalat, Iran	[22]
Penicillium chrysogenum	*S. abrotanoides*/root	Zoshk, Iran	[22]
Penicillium citrinum	*S. aegyptiaca*/leaf	Gebel Elba, Egypt	[24]
Penicillium commune	*S. aegyptiaca*/leaf	Gebel Elba, Egypt	[24]
Penicillium murcianum	*S. abrotanoides*/root	Kalat, Iran	[22]
Penicillium sp.	*S. abrotanoides*/root	Zoshk and Kalat, Iran	[22]
Pestalotiopsis mangiferae	*S. aegyptiaca*/leaf	Gebel Elba, Egypt	[24]
Petriella setifera	*S. miltiorrhiza*/root	Beijing, China	[25]
Phaeoacremonium rubrigenum	*S. abrotanoides*/root	Zoshk, Iran	[22]
Phoma herbarum	*S. miltiorrhiza*/seed	China	[37]
Psathyrella candolleana	*S. abrotanoides*/root	Zoshk, Iran	[22]
Purpureocillium lilacinum	*S. abrotanoides*/root	Darrud, Iran	[22]
Sarocladium kiliense	*S. miltiorrhiza*/root	Beijing, China	[25]
Schizophyllum commune	*S. miltiorrhiza*/root	Shandong, China	[23]
Simplicillium cylindrosporum	*S. abrotanoides*/root	Darrud, Iran	[22]
Talaromyces pinophilus	*S. miltiorrhiza*/'aerial part'	Shenyang, China	[38]
Talaromyces sp.	*S. abrotanoides*/root	Zoshk and Kalat, Iran	[22]
Talaromyces verruculosus	*S. abrotanoides*/root	Zoshk and Kalat, Iran	[22]
Trametes hirsuta	*S. miltiorrhiza*/root	Shandong, China	[23]
Trichocladium griseum	*S. aegyptiaca*/leaf	Gebel Elba, Egypt	[24]
Trichoderma asperellum	*S. abrotanoides*/root	Zoshk, Iran	[22]
Trichoderma atroviride	*S. miltiorrhiza*/root	Shangluo, China	[39]
Trichoderma hamatum	*S. officinalis*/root	Bonn, Germany	[40]
Trichoderma viride	*S. aegyptiaca*/leaf	Gebel Elba, Egypt	[24]
Xylomelasma sp.	*S. miltiorrhiza*/root	Beijing, China	[25]

[1] Species are reported according to the latest accepted name, which might not be the same as the one used in the corresponding reference.

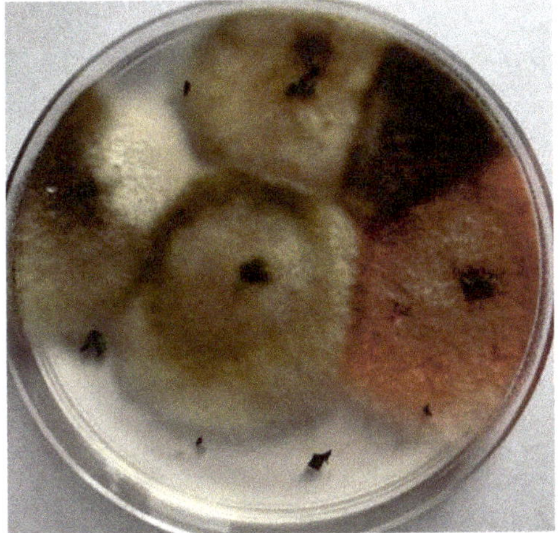

Figure 1. Isolation of endophytic fungi from leaf of *Salvia yangii* (original from work currently in progress at our laboratories).

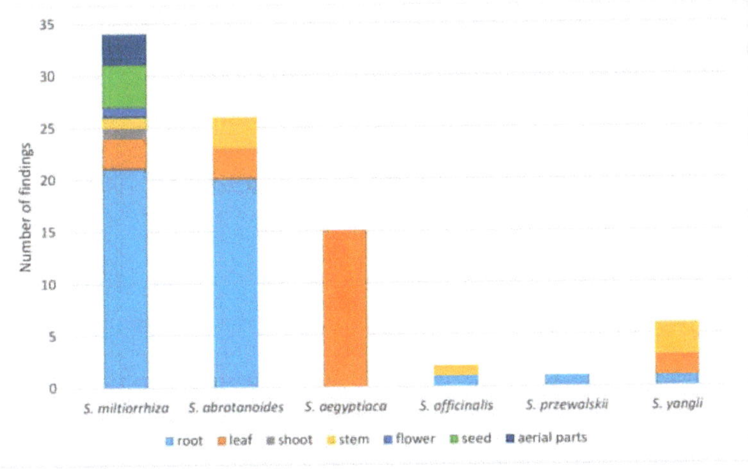

Figure 2. Graphic representation of findings concerning endophytic fungi of *Salvia* spp. based on data reported in Table 1.

4. Biochemical Properties

A large part of literature on the occurrence of endophytic fungi of *Salvia* spp. deals with their ability to produce bioactive compounds (Figure 3), focusing on structure elucidation and possible applications. Some studies have been limited to a partial characterization of culture filtrates or their extracts, highlighting general antibacterial, antioxidant or antifungal properties [23–25,41], while in other cases the basic constituents have been identified and extracted for assessments concerning their bioactivity. An annotated list of these products is reported in Table 2.

Table 2. Bioactive secondary metabolites produced by endophytic fungi from *Salvia* spp.

Secondary Metabolite	Producing Species/Strain	Bioactivity	Reference
N-Acetylanthranilic acid	*Penicillium* sp. *Talaromyces* sp.		[22]
Altenuene	*Alternaria* sp./Samif01 *Alternaria tenuissima*/SP-07		[42] [27]
2-*epi*-Altenuene	*Alternaria* sp./Samif01		[42]
2-Acetoxy-2-*epi*-altenuene	*Alternaria* sp./Samif01		[42]
3-*epi*-Dihydroaltenuene A	*Alternaria* sp./Samif01	Radical scavenging	[42]
Altenuisol	*Alternaria* sp./Samif01	Antibacterial, radical scavenging	[42]
Alternariol	*Alternaria* sp./Samif01 *Alternaria tenuissima*/SP-07	Antibacterial	[42] [27]
Alternariol-9-methyl ether	*Alternaria* sp./Samif01 *Alternaria tenuissima*/SP-07	Antibacterial, antifungal, antinematodal	[43] [27]
4-Hydroxyalternariol-9-methyl ether	*Alternaria* sp./Samif01	Antibacterial, radical scavenging	[42]
Aureonitols A–B	*Chaetomium globosum*/XL-1198		[30]
Azelaic acid	*Penicillium canescens* *Penicillium charlesii* *Penicillium* sp. *Talaromyces* sp. *Talaromyces verruculosus*		[22]
Caffeic acid	*Paraphoma radicina* *Talaromyces* sp. *Talaromyces verruculosus*		[22]
Chaetoglobosins E–F	*Chaetomium globosum*/XL-1198		[30]
Chaetomin	*Chaetomium* sp.	Cytotoxic (L5178Y mouse lymphoma)	[32]
Chaetomugilin I	*Chaetomium globosum*/XL-1198		[30]
Chaetoquadrin D	*Xylomelasma* sp./Samif07		[44]
Chaetoviridin	*Chaetomium globosum*/XL-1198		[30]
Cochliodinol, isocochliodinol, hydroperoxycochliodinol	*Chaetomium* sp.	Cytotoxic (L5178Y mouse lymphoma)	[31,32]
Colletotricholides A–B	*Colletotrichum gloeosporioides*/XL1200		[33]
Cryptotanshinone	*Coniolariella hispanica* *Paraphoma radicina* *Penicillium canescens* *Penicillium murcianum*		[22]
Daidzein	*Fusarium dlaminii* *Neocosmospora solani* *Paraphoma radicina* *Penicillium canescens*		[22]
Diaporthin	*Xylomelasma* sp./Samif07	Antibacterial, radical scavenging	[44]
2,6-Dimethyl-5-methoxyl-7-hydroxylchromone	*Xylomelasma* sp./Samif07	Antibacterial	[44]
Equisetin	*Chaetomium globosum*/XL-1198	Antibacterial, antifungal	[30]
Ferruginol	*Trichoderma atroviride* D16		[39]
Glycitein	*Talaromyces* sp.		[22]
Griseofulvin	*Talaromyces* sp.		[22]
8-Hydroxy-6-methoxy-3-methylisocoumarin	*Xylomelasma* sp./Samif07	Antibacterial	[44]
6-Hydroxymethyleugenin, 6-methoxymethyleugenin	*Xylomelasma* sp./Samif07	Antibacterial	[44]
Indole-3-acetic acid	*Penicillium canescens* *Phoma herbarum* D603 *Talaromyces* sp.		[22] [37] [22]
Indole-3-carboxylic acid, 3-formylindole	*Chaetomium* sp.		[32]
Isoeugenitol	*Xylomelasma* sp./Samif07	Antibacterial, antimycobacterial	[44]
Mandelic acid	*Paraphoma radicina* *Talaromyces* sp.		[22]
6-Methoxymellein	*Xylomelasma* sp./Samif07		[44]
Nipecotic acid	*Penicillium canescens*		[22]
Paracetamol (acetaminophen)	*Penicillium chrysogenum* *Penicillium* sp.		[22]
Pinophicin A	*Talaromyces pinophilus*	Antibacterial	[38]
Pinophol A	*Talaromyces pinophilus*	Antibacterial	[38]
Salvianolic acid C	*Didymella glomerata*/D-14		[35]
Solanapyrones A-C	*Alternaria tenuissima*/SP-07	Antibacterial	[27]
Solanapyrones P-R	*Alternaria tenuissima*/SP-07	Antibacterial	[27]
Solanidine	*Talaromyces* sp.		[22]
Stachydrine	*Fusarium dlaminii*		[22]
Tanshinone I	*Trichoderma atroviride* D16		[39]
Tanshinone IIA	*Aspergillus foeniculicola*/TR21 *Trichoderma atroviride* D16		[28] [39]
Trigonelline	*Talaromyces* sp.		[22]

Underlined compounds were first characterized from these sources.

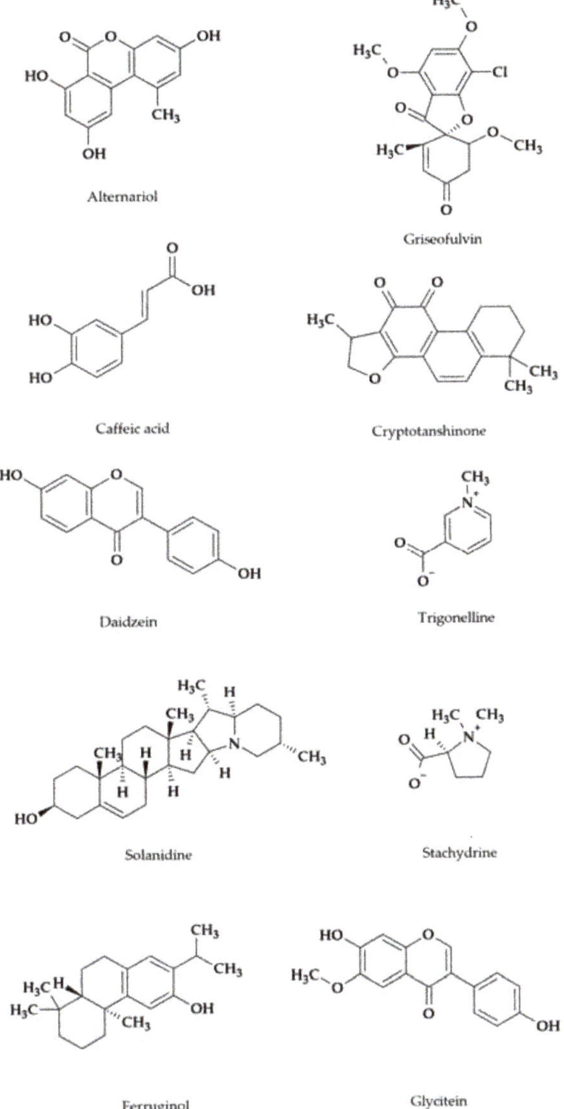

Figure 3. Chemical structure of some bioactive products from endophytic fungi of *Salvia* spp.

Confirming the assumption that endophytic fungi represent a goldmine of chemodiversity [2,45], 12 novel products were obtained from strains associated with *Salvia*. The list includes a fusicoccane diterpene pinophicin A and a polyene pinophol A from *Talaromyces pinophilus* [38]; colletotricholides A-B, two unusual eremophilane acetophenone conjugates from *Colletotrichum gloeosporioides* which are synthesized through a hybrid pathway involving polyketide and sesquiterpene synthase [33]. The novel 2,6-dimethyl-5-methoxyl-7-hydroxylchromone from *Xylomelasma* sp. displayed antibacterial activity, along with a few related eugenin derivatives and isocoumarins [44]. Moreover, there are several novel analogues of products known from *Alternaria*, such as 2-acetoxy-2-*epi*-altenuene and solanapyrones P-R [27,42], and *Chaetomium*, such as hydroperoxycochliodinol [32] and aureonitols

A-B [30]. The latter are structurally related to aureonitol, a known antiviral tetrahydrofuran [46]. As for cochliodinol derivatives, their bioactivity was found to be affected by the position of prenyl substituents in the indole ring systems, while the increased cytotoxicity of hydroperoxycochliodinol is related to the hydroperoxyl function [32].

Strains of *Alternaria* and *Chaetomium* also yielded products, such as alternariols, altenuisol, chaetoviridin and chaetoglobosins, whose antibiotic, antifungal, antiproliferative and radical scavenging properties have been previously described, and reviewed in recent papers [47,48]. More known products previously reported from other fungi are the tetramic acid derivative equisetin and the isocoumarin diaporthin, originally described as phytotoxins [49,50], and griseofulvin, a compound which has found application in dermatology and displayed interesting antitumor properties [51]. Other secondary metabolites which are used as pharmaceuticals are paracetamol, nipecotic acid, mandelic acid and azelaic acid. The latter has displayed antibiotic properties and antiproliferative effect on malignant melanocytes, and is commonly used in dermatology as an antiacne [52]. Moreover, in plants it is reported to be involved in the defense response against disease agents [53]. Other products to be mentioned are the plant hormone indole-3-acetic acid (IAA) and a couple of analogue auxins, which are considered key intermediates in the mutualistic relationship between endophytes and their host plants [2,16].

However, probably the most interesting products of endophytic fungi of *Salvia* species are a series of compounds previously identified as plant metabolites, which are treated in further detail in the next chapter.

5. Biotechnological Implications

Long-lasting evolutionary processes taking place together with their host plants have allowed endophytic fungi to work out various strategies enabling them to keep an equilibrium between virulence and plant defense in order to share common habitat. Endophytic fungi not only are able to influence plants' metabolism and physiology by producing unique and specific secondary metabolites, they were also found to produce bioactive natural products originally known exclusively from their host plants, and have elaborated strategies for detoxification by exploiting their biotransformation abilities. These properties make endophytes a perfect target for various biotechnological approaches and further commercial exploitation.

5.1. Endophytic Fungi as In Vitro Production Platforms for Plant Secondary Metabolites

The ever-increasing demand for bioactive natural compounds cannot be met at the desired levels by just relying on their extraction from plants, considering that in most instances they are produced at a specific developmental stage or under specific environmental condition, stress, or nutrient availability [54]. Medicinal plants from the *Salvia* genus are often shrubs, thus they may need several years to attain a suitable growth phase for bioactive product accumulation and extraction. Moreover, harvesting medicinally important plants from the wild makes them critically endangered and affects the environmental biodiversity [39]. As for crop plants, although cultivated in a large scale, they often produce the desired metabolites in a low yield, making the production unprofitable. Considering the limitations associated with productivity and vulnerability of plants, fungal endophytes may serve as a renewable and inexhaustible source of bioactive compounds. Many endophytes have experienced long-term symbiotic relationships with their host plants, and through long-term coexistence and direct contact, they have exchanged genetic material [17]. Horizontal gene transfer (HGT), an important evolutionary mechanism observed in prokaryotes, is also thought to be the phenomenon responsible for transmission of genetic material across phylogenetically distant species [55]. As an increasing number of reports indicate a physical clustering of genes for specialized metabolic pathways in plant genomes [56], the HGT phenomenon is believed to be responsible for rapid transfer of whole gene clusters from host plants, conferring "novel traits" to the associated fungi. As a consequence, many endophytic fungi have developed the ability to produce bioactive substances originally known from their hosts, thus raising the prospect of using such organisms as alternative and sustainable sources. HGT has been proposed to explain the production of tanshinone I, tanshinone IIA and their precursor ferruginol by *Trichoderma atroviride* D16, an endophytic fungus in *S. miltiorrhiza* [39].

Daidzein and glycitein are naturally occurring compounds found in soybeans and other legumes which are produced in plants through the phenylpropanoid pathway and structurally belonging to a class of compounds known as isoflavones. Daidzein is a phytoestrogen with possible pharmaceutical application as menopausal relief, osteoporosis, blood cholesterol lowering, and it is thought to reduce the risk of some hormone-related cancers and heart disease [57], while glycitein has a weaker estrogenic activity [58]. They both were found to be produced by endophytic fungi of *S. abrotanoides*, that is *Penicillum canescens* for daidzein and *Talaromyces* sp. for glycitein [22]. The latter is also able to synthesize trigonelline, an alkaloid originally extracted from *Trigonella foenum-graecum*, known for its antidiabetic properties [59] as well as solanidine, a potato alkaloid. Stachydrine, another alkaloid known from *Medicago sativa*, was found to be synthetized by a strain of *Fusarium dlaminii* inhabiting *S. abrotanoides* [22].

Danshen, dried roots and rhizomes of *S. miltiorrhiza*, is a well-known traditional Chinese herbal medicine [60]. It contains two kinds of bioactive compounds: tanshinones and hydrophilic phenolic acids, the latter being represented by rosmarinic acid, salvianolic acids B-C, and others. Salvianolic acids are mainly responsible for the favorable activities on cardiovascular and cerebrovascular diseases of danshen [61]. Salvianolic acid C was found in both mycelium and fermentation broth of strain D14 of *D. glomerata* in very low yields [35]. This indicates the opportunity to optimize fermentation conditions for achieving its efficient production, or alternatively to enhance its production via regulating the key enzymes involved in the biosynthetic pathway.

Caffeic acid was found in the metabolome profiles of isolates of *Talaromyces* and *Paraphoma* endophytic in *S. abrotanoides* [22]. Besides rosmarinic acid and salvianolic acid B, it is regarded as the major phenolic acid in *S. miltiorrhiza* [62]. A series of caffeic acid derivatives, obtained from *Salvia officinalis* [63,64], showed pronounced leishmanicidal activity, as well as immunomodulatory effects on macrophage functions [65]. Moreover, antibacterial, antifungal and modulatory effects of caffeic acid have been shown in recent studies [66].

Tanshinones are a group of abietane-type norditerpenoid quinones, originally found in danshen [62]. More than 40 structurally diverse tanshinones have been isolated and identified [67], among which cryptotanshinone, tanshinone IIA, and tanshinone I are the main active ingredients [68]. Although many biotechnological improvements have been implemented to increase tanshinone production from plants, at present no mature hairy root, suspension cell line, or culture system of *S. miltiorrhiza* have been developed. Thus, the extraction from roots and rhizomes of *S. miltiorrhiza* still represents the main source of tanshinones [62]. *Salvia yangii* has also been found to produce a range of tanshinones [69–72], as well as *S. abrotanoides*, although the compound assortment was found to be considerably different according to the preliminary data obtained by our working group.

Tanshinone I and tanshinone IIA display a variety of biological activities [39]. Tanshinone I is reported to induce apoptosis in leukemia cells [73], human colon cancer cells [74] and activated hepatic stellate cells [75], and displays anticancer effects in human non-small cell lung cancer [76] and human breast cancer [77]. Tanshinone IIA exerts a cardiovascular action [78], including effects against cardiomyocyte hypertrophy [79], atherosclerosis [80], hypertension [81] and ischaemic heart diseases [82]. In addition, tanshinone IIA is a potent anticarcinogenic, with possible application for the management of systemic malignancies [83].

As introduced above, tanshinone IIA is currently in short supply because of overcollection of the wild plants and environmental change [28], so that endophytic fungal strains represent an alternative source. In this respect, tanshinone I and tanshinone IIA production has been confirmed by *T. atroviride* D16 from *S. miltiorrhiza* [39]. Moreover, strain TR21 of *Aspergillus foeniculicola* was shown to produce low amount of tanshinone IIA [84]. Production of this compound by TR21 was increased in the NU152 mutant, obtained by traditional mutagenesis using ultraviolet radiation and sodium nitrate treatment [28], and in strain F-3.4 through genome shuffling [85], providing a yield of tanshinone IIA which is over 11 times higher than the original strain TR21. This study showed that the genetic basis of high-yield strains can be achieved through genome shuffling, which can shorten the breeding cycle and improve the mutagenesis efficiency in obtaining strains with good traits, to be used for industrial production.

Cryptotanshinone, another nor-abietanoid diterpenoid, which is a main bioactive compound of *S. abrotanoides* known for leishmanicidal, antiplasmodial and cytotoxic activity [86] has been found to be produced also in roots of *S. yangii* [69]. Very recently, this compound has been reported as a secondary metabolite of endophytic strains of *P. canescens*, *Penicillium murcianum*, *Paraphoma radicina*, and *Coniolariella hispanica*, independently of the host plant. Moreover, the effect of exogenous gibberellin (GA3) on *S. abrotanoides* and endophytic fungi was shown to have a positive effect on increasing the cryptotanshinone production in the plant as well as in endophytic fungi cultivated under axenic conditions [22]. Exogenous gibberellin treatment was also previously observed to promote the production of cryptotanshinone, tanshinone I and tanshinone II in *S. miltiorrhiza* [87].

The typical abietane diterpenoid, ferruginol, is mainly known from *Sequoia sempervirens* for its antibacterial and antineoplastic properties [88,89]. It has also been isolated from the roots of plants in the genus *Salvia*, for instance *Salvia viridis* [90], *S. miltiorrhiza* [91], *Salvia cilicica* [92], *Salvia deserta* [93]. As a precursor in the tanshinone pathway, ferruginol synthesis has been confirmed by the above-mentioned strain D16 of *T. atroviride* [39].

5.2. Endophytic Fungi as Biotic Elicitors

Indiscriminate collection and cutting down of medicinal plants from the wild for extraction of medicinal products have almost led to the extinction of certain plant species, making them either vulnerable or critically endangered. The biotechnological approaches involving plant cell, organ and hairy root cultures appeared to fulfill the ever-increasing demand up to a certain level [54]. Endophytes could possibly be used as alternative or more efficient elicitors, compared to other biotic and abiotic elicitation methods.

A tanshinone IIA-producing endophytic strain of *A. foeniculicola* (U104) was demonstrated to elicit production of this compound in sterile seedlings of *S. miltiorrhiza* through upregulation of several enzymes involved in its biosynthesis [94]. Likewise, mycelium extract and its polysaccharide fraction (PF) produced by *T. atroviride* D16 promoted root growth and stimulated the biosynthesis of tanshinones in hairy roots. Moreover, the transcriptional activity of genes involved in the tanshinone biosynthetic pathway increased significantly after treatment with PF, which could be effectively utilized for large-scale production of tanshinones in the *S. miltiorrhiza* hairy root culture system [95]. Later on, PF was found to more deeply regulate the metabolic profiling of roots of this plant [96]. The main component of PF resulted to be an heteropolysaccharide (PSF-W-1), whose structure has been elucidated [97]. Moreover, an enhancing role by jasmonic acid on production of tanshinone I by this fungal strain was demonstrated [26], along with Ca^{2+} triggering, peroxide reaction and protein phosphorylation, leading to an increase in leucine-rich repeat (LRR) protein synthesis [98].

Another endophytic strain from *S. miltiorrhiza* (*Phoma herbarum* D603) was found to stimulate growth and root development by producing IAA and siderophores and improving nutrition through phosphorus solubilization; moreover, it promoted the synthesis and accumulation of tanshinones by regulating the expression level of key genes in the synthetic pathway [37].

Eliciting effects on the synthesis of salvianolic acids and tanshinones, particularly dihydrotanshinone I and cryptotanshinone, have been also reported by a strain of *Chaetomium globosum* and its mycelial extract [29]. The effect of the mycelial extract was much stronger than that of the live fungus on tanshinones synthesis, which significantly increased the transcriptional activity of key genes in tanshinone biosynthetic pathway. Thus, *C. globosum* D38 was proposed to be supplemented as a biotic fertilizer in *S. miltiorrhiza* seedling culture, as it not only significantly promoted growth of the host plant, but also notably enhanced the accumulation of tanshinones and salvianolic acids.

Alternaria sp. A13 has been shown to simultaneously enhance the dry root biomass and secondary metabolite accumulation of *S. miltiorrhiza*, thus demonstrating its application potential as a bio-fertilizer in the cultivation of this plant [26]. Compared to uninoculated seedlings, *S. miltiorrhiza* seedlings colonized by *Alternaria* sp. A13 showed significant increment in the contents of total phenolic acids and lithospermic acids A and B. Examination of the related enzyme activities showed that the elicitation effect of A13 on lithospermic acid B accumulation correlated with cinnamic acid 4-hydroxylase (C4H) activity in the phenylpropanoid pathway under field conditions. A similar effect was demonstrated for a strain of *Paecilomyces* sp. which increased content of salvianolic acid B in *S. miltiorrhiza* and promoted plant growth [36].

5.3. Biotransformation/Detoxication Abilities of Endophytic Fungi

To be able to colonize host tissues, endophytes developed a strong tolerance toward host's defensive metabolites. The detoxification of plant bioactive compounds is an important transformation ability of many endophytes which, to a certain extent, decides the colonization range of their hosts [17]. Biotransformation abilities of endophytes help in detoxification of antifungal metabolites produced by the host plant, and may intervene in the production of some novel bioactive compounds [54,99,100].

Trichoderma hamatum, an endophytic fungus inhabiting roots of *Salvia officinalis* alongside other microorganisms, was found to be able to degrade caffeine [40]. Aromatic plants such as sage have been used as intercrops in coffee plantations. *Salvia officinalis* was proved to absorb caffeine from the incubation media and store it mainly in roots. The cited study demonstrated that the degradation of caffeine was initiated by the ability of the microorganisms to perform demethylations, whereas xanthine degradation may be attributed to either the plant or the microorganisms. The existence of a beneficial biochemical interaction in caffeine degradation between endophytic *T. hamatum* and sage root was proposed. Using sage with its endophyte *T. hamatum* as an intercrop may become an ecologically friendly strategy to reduce caffeine accumulation in soil.

6. Conclusions

Endophytic fungi are prospective producers of both known and novel bioactive compounds. However, to ensure feasibility of industrial application, yield and productivity enhancement strategies at several levels are required [101]. A combination of genetic, metabolic and bioprocess engineering may be used to sustain and enhance production of high value secondary metabolites from selected strains, whose biosynthetic abilities can be improved through physical and chemical mutagenesis, or various methods for genetic transformation. Improved strains can be in turn subjected to various bioprocess optimization strategies for further enhancement in yield and productivity of selected compounds.

This review of the available literature specifically concerning endophytic fungi of sages highlighted that research in the field is quickly progressing, with the aim of both refining biotechnological applications concerning tanshinone production and prospecting novel strains for further applications. The spread of reliable methods for detection and characterization of both the endophytic strains and their bioactive secondary metabolites is expected to further improve the translational perspectives.

Author Contributions: Conceptualization, B.Z.; investigation, B.Z.; resources, M.B., B.Z.; writing—original draft preparation, B.Z., M.B., B.A., R.N.; writing—review and editing, M.B., R.N.; funding acquisition, B.Z., M.B. All authors have read and agreed to the published version of the manuscript.

Funding: This research received no external funding.

Conflicts of Interest: The authors declare no conflict of interest.

References

1.   Gouda, S.; Das, G.; Sen, S.K.; Shin, H.S.; Patra, J.K. Endophytes: A treasure house of bioactive compounds of medicinal importance. *Front. Microbiol.* **2016**, *7*, 1538. [CrossRef] [PubMed]

2. Nicoletti, R.; Fiorentino, A. Plant bioactive metabolites and drugs produced by endophytic fungi of Spermatophyta. *Agriculture* **2015**, *5*, 918–970. [CrossRef]

3. Jia, M.; Chen, L.; Xin, H.L.; Zheng, C.J.; Rahman, K.; Han, T.; Qin, L.P. A friendly relationship between endophytic fungi and medicinal plants: A systematic review. *Front. Microbiol.* **2016**, *7*, 906. [CrossRef] [PubMed]

4. Drew, B.T.; González-Gallegos, J.G.; Xiang, C.L.; Kriebel, R.; Drummond, C.P.; Walker, J.B.; Sytsma, K.J. *Salvia* united: The greatest good for the greatest number. *Taxon* **2017**, *66*, 133–145. [CrossRef]

5. Will, M.; Claßen-Bockhoff, R. Time to split *Salvia* s.l. (Lamiaceae)—New insights from Old World *Salvia* phylogeny. *Mol. Phylogenet. Evol.* **2017**, *109*, 33–58. [CrossRef]

6. Jassbi, A.R.; Zare, S.; Firuzi, O.; Xiao, J. Bioactive phytochemicals from shoots and roots of *Salvia* species. *Phytochem. Rev.* **2016**, *15*, 829–867. [CrossRef]

7. Jiang, Z.; Gao, W.; Huang, L. Tanshinones, critical pharmacological components in *Salvia miltiorrhiza*. *Front. Pharmacol.* **2019**, *10*, 202. [CrossRef]

8. Cao, E.H.; Liu, X.Q.; Wang, J.J.; Xu, N.F. Effect of natural antioxidant tanshinone II-A on DNA damage by lipid peroxidation in liver cells. *Free Radic. Biol. Med.* **1996**, *20*, 801–806. [CrossRef]

9. Lee, D.S.; Lee, S.H.; Noh, J.G.; Hong, S.D. Antibacterial activities of cryptotanshinone and dihydrotanshinone i from a medicinal herb, *Salvia miltiorrhiza* Bunge. *Biosci. Biotechnol. Biochem.* **1999**, *63*, 2236–2239. [CrossRef]

10. Kim, E.J.; Jung, S.N.; Son, K.H.; Kim, S.R.; Ha, T.Y.; Park, M.G.; Jo, I.G.; Park, J.G.; Wonchae, C.; Kim, S.S.; et al. Antidiabetes and antiobesity effect of cryptotanshinone via activation of AMP-activated protein kinase. *Mol. Pharmacol.* **2007**, *72*, 62–72. [CrossRef]

11. Jang, S.I.; Jeong, S.I.; Kim, K.J.; Kim, H.J.; Yu, H.H.; Park, R.; Kim, H.M.; You, Y.O. Tanshinone IIA from *Salvia miltiorrhiza* inhibits inducible nitric oxide synthase expression and production of TNF-α, IL-1β and IL-6 in activated RAW 264.7 cells. *Planta Med.* **2003**, *69*, 1057–1059. [CrossRef] [PubMed]

12. Wang, X.; Wei, Y.; Yuan, S.; Liu, G.; Lu, Y.; Zhang, J.; Wang, W. Potential anticancer activity of tanshinone IIA against human breast cancer. *Int. J. Cancer* **2005**, *116*, 799–807. [CrossRef] [PubMed]

13. Tan, R.X.; Zou, W.X. Endophytes: A rich source of functional metabolites. *Nat. Prod. Rep.* **2001**, *18*, 448–459. [CrossRef] [PubMed]

14. Faeth, S.H.; Fagan, W.F. Fungal endophytes: Common host plant symbionts but uncommon mutualists. *Integr. Comp. Biol.* **2002**, *42*, 360–368. [CrossRef]

15. Arnold, A.E. Understanding the diversity of foliar endophytic fungi: Progress, challenges, and frontiers. *Fungal Biol. Rev.* **2007**, *21*, 51–66. [CrossRef]

16. Salvatore, M.M.; Andolfi, A.; Nicoletti, R. The thin line between pathogenicity and endophytism: The case of *Lasiodiplodia theobromae*. *Agriculture* **2020**, *10*, 488. [CrossRef]

17. Wang, Y.; Dai, C.C. Endophytes: A potential resource for biosynthesis, biotransformation, and biodegradation. *Ann. Microbiol.* **2011**, *61*, 207–215. [CrossRef]

18. Rodriguez, R.J.; White, J.F.; Arnold, A.E.; Redman, R.S. Fungal endophytes: Diversity and functional roles. *New Phytologist.* **2009**, *182*, 314–330. [CrossRef]

19. Rodriguez, R.; Redman, R. More than 400 million years of evolution and some plants still can't make it on their own: Plant stress tolerance via fungal symbiosis. *J. Exp. Bot.* **2008**, *59*, 1109–1114. [CrossRef]

20. Sun, J.; Xia, F.; Cui, L.; Liang, J.; Wang, Z.; Wei, Y. Characteristics of foliar fungal endophyte assemblages and host effective components in *Salvia miltiorrhiza* Bunge. *Appl. Microbiol. Biotechnol.* **2014**, *98*, 3143–3155. [CrossRef]

21. Chen, H.; Wu, H.; Yan, B.; Zhao, H.; Liu, F.; Zhang, H.; Sheng, Q.; Miao, F.; Liang, Z. Core microbiome of medicinal plant *Salvia miltiorrhiza* seed: A rich reservoir of beneficial microbes for secondary metabolism? *Int. J. Mol. Sci.* **2018**, *19*, 672. [CrossRef] [PubMed]

22. Teimoori-Boghsani, Y.; Ganjeali, A.; Cernava, T.; Müller, H.; Asili, J.; Berg, G. Endophytic fungi of native *Salvia abrotanoides* plants reveal high taxonomic diversity and unique profiles of secondary metabolites. *Front. Microbiol.* **2020**, *10*. [CrossRef] [PubMed]

23. Li, Y.L.; Xin, X.M.; Chang, Z.Y.; Shi, R.J.; Miao, Z.M.; Ding, J.; Hao, G.P. The endophytic fungi of *Salvia miltiorrhiza* Bunge.f. *alba* are a potential source of natural antioxidants. *Bot. Stud.* **2015**, *56*, 5. [CrossRef] [PubMed]

24. Mohamed El-Bondkly, A.A.; El-Gendy, M.M.A.A.; El-Bondkly, E.A.M.; Ahmed, A.M. Biodiversity and biological activity of the fungal microbiota derived from the medicinal plants *Salvia aegyptiaca* L. and *Balanties aegyptiaca* L. *Biocatal. Agric. Biotechnol.* **2020**, *28*, 101720. [CrossRef]

25. Jingfeng, L.; Linyun, F.; Ruiya, L.; Xiaohan, W.; Haiyu, L.; Ligang, Z. Endophytic fungi from medicinal herb *Salvia miltiorrhiza* Bunge and their antimicrobial activity. *Afr. J. Microbiol. Res.* **2013**, *7*, 5343–5349. [CrossRef]

26. Zhou, L.S.; Tang, K.; Guo, S.X. The plant growth-promoting fungus (PGPF) *Alternaria* sp. A13 markedly enhances *Salvia miltiorrhiza* root growth and active ingredient accumulation under greenhouse and field conditions. *Int. J. Mol. Sci.* **2018**, *19*, 270. [CrossRef]

27. Wang, X.Z.; Luo, X.H.; Xiao, J.; Zhai, M.M.; Yuan, Y.; Zhu, Y.; Crews, P.; Yuan, C.S.; Wu, Q.X. Pyrone derivatives from the endophytic fungus *Alternaria tenuissima* SP-07 of Chinese herbal medicine *Salvia przewalskii*. *Fitoterapia* **2014**, *99*, 184–190. [CrossRef]

28. Ma, C.; Jiang, D.; Wei, X. Mutation breeding of *Emericella foeniculicola* TR21 for improved production of tanshinone IIA. *Process Biochem.* **2011**, *46*, 2059–2063. [CrossRef]

29. Zhai, X.; Luo, D.; Li, X.; Han, T.; Jia, M.; Kong, Z.; Ji, J.; Rahman, K.; Qin, L.; Zheng, C. Endophyte *Chaetomium globosum* D38 promotes bioactive constituents accumulation and root production in *Salvia miltiorrhiza*. *Front. Microbiol.* **2018**, *8*, 2694. [CrossRef]

30. Yang, S.X.; Zhao, W.T.; Chen, H.Y.; Zhang, L.; Liu, T.K.; Chen, H.P.; Yang, J.; Yang, X.L. Aureonitols A and B, Two New C13-Polyketides from *Chaetomium globosum*, an endophytic fungus in *Salvia miltiorrhiza*. *Chem. Biodivers.* **2019**, *16*, e1900364. [CrossRef]

31. Debbab, A.; Aly, A.H.; Edrada-Ebel, R.A.; Müller, W.E.G.; Mosaddak, M.; Hakiki, A.; Ebel, R.; Proksch, P. Bioactive secondary metabolites from the endophytic fungus *Chaetomium* sp. isolated from *Salvia officinalis* growing in Morocco. *Biotechnol. Agron. Soc. Environ.* **2009**, *13*, 229–234.

32. Mallouk, S.; Mohamed, N.S.E.D.; Debbab, A. Cytotoxic hydroperoxycochliodinol derivative from endophytic *Chaetomium* sp. isolated from *Salvia officinalis*. *Chem. Nat. Compd.* **2020**, *56*, 701–705. [CrossRef]

33. Zhao, W.T.; Liu, Q.P.; Chen, H.Y.; Zhao, W.; Gao, Y.; Yang, X.L. Two novel eremophylane acetophenone conjugates from *Colletotrichum gloeosporioides*, an endophytic fungus in *Salvia miltiorrhiza*. *Fitoterapia* **2020**, *141*, 104474. [CrossRef]

34. Sang, X.; Guo, J.; Bai, L. Isolation, identification and analysis of secondary metabolites of multidrug-resistance inhibiting endophytic fungi of *Salvia miltiorrhiza* Bunge. *Chin. J. Appl. Environ. Biol.* **2014**, *20*, 621–628. [CrossRef]

35. Li, X.; Zhai, X.; Shu, Z.; Dong, R.; Ming, Q.; Qin, L.; Zheng, C. *Phoma glomerata* D14: An endophytic fungus from *Salvia miltiorrhiza* that produces salvianolic acid C. *Curr. Microbiol.* **2016**, *73*, 31–37. [CrossRef]

36. Tang, K.; Li, B.; Guo, S. An active endophytic fungus promoting growth and increasing salvianolic acid content of *Salvia miltiorrhiza*. *Mycosystema* **2014**, *33*, 594–600.

37. Chen, H.-M.; Wu, H.-X.; He, X.-Y.; Zhang, H.-H.; Miao, F.; Liang, Z.-S. Promoting tanshinone synthesis of *Salvia miltiorrhiza* root by a seed endophytic fungus, *Phoma herbarum* D603. *China J. Chin. Mater. Medica* **2020**, *45*, 65–71. [CrossRef]

38. Zhao, W.-T.; Shi, X.; Xian, P.-J.; Feng, Z.; Yang, J.; Yang, X.-L. A new fusicoccane diterpene and a new polyene from the plant endophytic fungus *Talaromyces pinophilus* and their antimicrobial activities. *Nat. Prod. Res.* **2019**, 1–7. [CrossRef]

39. Ming, Q.; Han, T.; Li, W.; Zhang, Q.; Zhang, H.; Zheng, C.; Huang, F.; Rahman, K.; Qin, L. Tanshinone IIA and tanshinone i production by *Trichoderma atroviride* D16, an endophytic fungus in *Salvia miltiorrhiza*. *Phytomedicine* **2012**, *19*, 330–333. [CrossRef]

40. Schulz, M.; Knop, M.; Muellenborn, C.; Steiner, U. Root-associated microorganisms prevent caffeine accumulation in shoots of *Salvia officinalis* L. *Int. J. Agric. For.* **2013**, *3*, 152–158. [CrossRef]

41. Huang, L.; Li, F.; Liu, R.; Guo, J.; Yang, Z.; Bai, L. Antifungal activity of an endophytic strain of *Phomopsis* sp. on *Sclerotinia sclerotiorum*, the causal agent of Sclerotinia disease. *J. Plant Pathol.* **2019**, *101*, 521–528. [CrossRef]

42. Tian, J.; Fu, L.; Zhang, Z.; Dong, X.; Xu, D.; Mao, Z.; Liu, Y.; Lai, D.; Zhou, L. Dibenzo-α-pyrones from the endophytic fungus *Alternaria* sp. Samif01: Isolation, structure elucidation, and their antibacterial and antioxidant activities. *Nat. Prod. Res.* **2017**, *31*, 387–396. [CrossRef]

43. Lou, J.; Yu, R.; Wang, X.; Mao, Z.; Fu, L.; Liu, Y.; Zhou, L. Alternariol 9-methyl ether from the endophytic fungus *Alternaria* sp. Samif01 and its bioactivities. *Brazilian J. Microbiol.* **2016**, *47*, 96–101. [CrossRef] [PubMed]

44. Lai, D.; Li, J.; Zhao, S.; Gu, G.; Gong, X.; Proksch, P.; Zhou, L. Chromone and isocoumarin derivatives from the endophytic fungus *Xylomelasma* sp. Samif07, and their antibacterial and antioxidant activities. *Nat. Prod. Res.* **2019**. [CrossRef] [PubMed]

45. Gupta, S.; Chaturvedi, P.; Kulkarni, M.G.; Van Staden, J. A critical review on exploiting the pharmaceutical potential of plant endophytic fungi. *Biotechnol. Adv.* **2020**, *39*, 107462. [CrossRef] [PubMed]

46. Sacramento, C.Q.; Marttorelli, A.; Fintelman-Rodrigues, N.; De Freitas, C.S.; De Melo, G.R.; Rocha, M.E.N.; Kaiser, C.R.; Rodrigues, K.F.; Da Costa, G.L.; Alves, C.M.; et al. Aureonitol, a fungi-derived tetrahydrofuran, inhibits influenza replication by targeting its surface glycoprotein hemagglutinin. *PLoS ONE* **2015**, *10*, e0142246. [CrossRef]

47. Fatima, N.; Muhammad, S.A.; Khan, I.; Qazi, M.A.; Shahzadi, I.; Mumtaz, A.; Hashmi, M.A.; Khan, A.K.; Ismail, T. *Chaetomium* endophytes: A repository of pharmacologically active metabolites. *Acta Physiol. Plant.* **2016**, *38*, 136. [CrossRef]

48. Eram, D.; Arthikala, M.K.; Melappa, G.; Santoyo, G. *Alternaria* species: Endophytic fungi as alternative sources of bioactive compounds. *Ital. J. Mycol.* **2018**, *47*, 40–54. [CrossRef]

49. Wheeler, M.H.; Stipanovic, R.D.; Puckhaber, L.S. Phytotoxicity of equisetin and epi-equisetin isolated from *Fusarium equiseti* and *F. pallidoroseum*. *Mycol. Res.* **1999**, *103*, 967–973. [CrossRef]

50. Arnone, A.; Assante, G.; Nasini, G.; Strada, S.; Vercesi, A. Cryphonectric acid and other minor metabolites from a hypovirulent strain of *Cryphonectria parasitica*. *J. Nat. Prod.* **2002**, *65*, 48–50. [CrossRef]

51. Nicoletti, R. Antitumor and Immunomodulatory Compounds from Fungi. In *Reference Module in Life Sciences (Planned for Publication in Encyclopaedia of Mycology)*; Zaragoza, O., Ed.; Elsevier: Amsterdam, The Netherlands, 2020.

52. Fitton, A.; Goa, K.L. Azelaic acid: A review of its pharmacological properties and therapeutic efficacy in acne and hyperpigmentary skin disorders. *Drugs* **1991**, *41*, 780–798. [CrossRef] [PubMed]

53. Kachroo, A.; Robin, G.P. Systemic signaling during plant defense. *Curr. Opin. Plant Biol.* **2013**, *16*, 527–533. [CrossRef] [PubMed]

54. Chandra, S. Endophytic fungi: Novel sources of anticancer lead molecules. *Appl. Microbiol. Biotechnol.* **2012**, *95*, 47–59. [CrossRef] [PubMed]

55. Tiwari, P.; Bae, H. Horizontal gene transfer and endophytes: An implication for the acquisition of novel traits. *Plants* **2020**, *9*, 305. [CrossRef]

56. Nützmann, H.W.; Osbourn, A. Gene clustering in plant specialized metabolism. *Curr. Opin. Biotechnol.* **2014**, *26*, 91–99. [CrossRef]

57. Holder, C.L.; Churchwell, M.I.; Doerge, D.R. Quantification of soy isoflavones, genistein and daidzein, and conjugates in rat blood using LC/ES-MS. *J. Agric. Food Chem.* **1999**, *47*, 3764–3770. [CrossRef]

58. Song, T.T.; Hendrich, S.; Murphy, P.A. Estrogenic activity of glycitein, a soy isoflavone. *J. Agric. Food Chem.* **1999**, *47*, 1607–1610. [CrossRef]

59. Zhou, J.Y.; Zhou, S.W. Trigonelline: A plant alkaloid with therapeutic potential for diabetes and central nervous system disease. *Curr. Med. Chem.* **2012**, *19*, 3523–3531. [CrossRef]

60. Tang, W.; Eisenbrand, G. Salvia miltiorrhiza Bge. In *Chinese Drugs of Plant Origin*; Springer: Berlin/Heidelberg, Germany, 1992.

61. Su, C.-Y.; Ming, Q.-L.; Rahman, K.; Han, T.; Qin, L.-P. *Salvia miltiorrhiza*: Traditional medicinal uses, chemistry, and pharmacology. *Chin. J. Nat. Med.* **2015**, *13*, 163–182. [CrossRef]

62. Lu, S. *Compendium of Plant Genomes*; Springer Nature: Cham, Switzerland, 2019; ISBN 978-3-030-24715-7.

63. Lu, Y.; Foo, L.Y. Rosmarinic acid derivatives from *Salvia officinalis*. *Phytochemistry* **1999**, *51*, 91–94. [CrossRef]

64. Lu, Y.; Foo, L.Y.; Wong, H. Sagecoumarin, a novel caffeic acid trimer from *Salvia officinalis*. *Phytochemistry* **1999**, *52*, 1149–1152. [CrossRef]

65. Radtke, O.A.; Yeap Foo, L.; Lu, Y.; Kiderlen, A.F.; Kolodziej, H. Evaluation of sage phenolics for their antileishmanial activity and modulatory effects on interleukin-6, interferon and tumour necrosis factor-α-release in RAW 264.7 Cells. *Zeitschrift Naturforsch. Sect. C J. Biosci.* **2003**, *58*, 395–400. [CrossRef] [PubMed]

66. Lima, V.N.; Oliveira-Tintino, C.D.M.; Santos, E.S.; Morais, L.P.; Tintino, S.R.; Freitas, T.S.; Geraldo, Y.S.; Pereira, R.L.S.; Cruz, R.P.; Menezes, I.R.A.; et al. Antimicrobial and enhancement of the antibiotic activity by phenolic compounds: Gallic acid, caffeic acid and pyrogallol. *Microb. Pathog.* **2016**, *99*, 56–61. [CrossRef] [PubMed]

67. Zhang, Y.; Jiang, P.; Ye, M.; Kim, S.H.; Jiang, C.; Lü, J. Tanshinones: Sources, pharmacokinetics and anti-cancer activities. *Int. J. Mol. Sci.* **2012**, *13*, 13621–13666. [CrossRef] [PubMed]

68. Zhong, G.-X.; Li, P.; Zeng, L.-J.; Guan, J.; Li, D.-Q.; Li, S.-P. Chemical characteristics of *Salvia miltiorrhiza* (Danshen) collected from different locations in China. *J. Agric. Food Chem.* **2009**, *57*, 6879–6887. [CrossRef]

69. Ślusarczyk, S.; Topolski, J.; Domaradzki, K.; Adams, M.; Hamburger, M.; Matkowski, A. Isolation and fast selective determination of nor-abietanoid diterpenoids from *Perovskia atriplicifolia* roots using LC-ESI-MS/MS with multiple reaction monitoring. *Nat. Prod. Commun.* **2015**, *10*, 1149–1152. [CrossRef]

70. Senol, F.S.; Ślusarczyk, S.; Matkowski, A.; Pérez-Garrido, A.; Girón-Rodríguez, F.; Cerón-Carrasco, J.P.; den-Haan, H.; Peña-García, J.; Pérez-Sánchez, H.; Domaradzki, K.; et al. Selective in vitro and in silico butyrylcholinesterase inhibitory activity of diterpenes and rosmarinic acid isolated from *Perovskia atriplicifolia* Benth. and *Salvia glutinosa* L. *Phytochemistry* **2017**, *133*, 33–44. [CrossRef]

71. Ślusarczyk, S.; Deniz, F.S.S.; Abel, R.; Pecio, Ł.; Pérez-Sánchez, H.; Cerón-Carrasco, J.P.; Den-Haan, H.; Banerjee, P.; Preissner, R.; Krzyżak, E.; et al. Norditerpenoids with selective anti-cholinesterase activity from the roots of *Perovskia atriplicifolia* Benth. *Int. J. Mol. Sci.* **2020**, *21*, 4475. [CrossRef]

72. Miroliaei, M.; Aminjafari, A.; Ślusarczyk, S.; Nawrot-Hadzik, I.; Rahimmalek, M.; Matkowski, A. Inhibition of glycation-induced cytotoxicity, protein glycation, and activity of proteolytic enzymes by extract from *Perovskia atriplicifolia* roots. *Pharmacogn. Mag.* **2017**, *13* (Suppl. 3), S676–S683. [CrossRef]

73. Liu, J.J.; Liu, W.-D.; Yang, H.Z.; Zhang, Y.; Fang, Z.G.; Liu, P.Q.; Lin, D.J.; Xiao, R.Z.; Hu, Y.; Wang, C.Z.; et al. Inactivation of PI3k/Akt signaling pathway and activation of caspase-3 are involved in tanshinone I-induced apoptosis in myeloid leukemia cells in vitro. *Ann. Hematol.* **2010**, *89*, 1089–1097. [CrossRef]

74. Su, C.-C.; Chen, G.-W.; Lin, J.-G. Growth inhibition and apoptosis induction by tanshinone I in human colon cancer Colo 205 cells. *Int. J. Mol. Med.* **2008**, *22*, 613–618. [CrossRef] [PubMed]

75. Kim, J.Y.; Kim, K.M.; Nan, J.X.; Zhao, Y.Z.; Park, P.H.; Lee, S.J.; Sohn, D.H. Induction of apoptosis by tanshinone I via cytochrome c release in activated hepatic stellate cells. *Pharmacol. Toxicol.* **2003**, *92*, 195–200. [CrossRef] [PubMed]

76. Lee, C.Y.; Sher, H.F.; Chen, H.W.; Liu, C.C.; Chen, C.H.; Lin, C.S.; Yang, P.C.; Tsay, H.S.; Chen, J.J.W. Anticancer effects of tanshinone I in human non-small cell lung cancer. *Mol. Cancer Ther.* **2008**, *7*, 3527–3538. [CrossRef] [PubMed]

77. Nizamutdinova, I.T.; Lee, G.W.; Lee, J.S.; Cho, M.K.; Son, K.H.; Jeon, S.J.; Kang, S.S.; Kim, Y.S.; Lee, J.H.; Seo, H.G.; et al. Tanshinone I suppresses growth and invasion of human breast cancer cells, MDA-MB-231, through regulation of adhesion molecules. *Carcinogenesis* **2008**, *29*, 1885–1892. [CrossRef]

78. Li, W.; Li, J.; Ashok, M.; Wu, R.; Chen, D.; Yang, L.; Yang, H.; Tracey, K.J.; Wang, P.; Sama, A.E.; et al. A Cardiovascular drug rescues mice from lethal sepsis by selectively attenuating a late-acting proinflammatory mediator, high mobility group box 1. *J. Immunol.* **2007**, *178*, 3856–3864. [CrossRef]

79. Yang, L.; Zou, X.; Liang, Q.; Chen, H.; Feng, J.; Yan, L.; Wang, Z.; Zhou, D.; Li, S.; Yao, S.; et al. Sodium tanshinone IIA sulfonate depresses angiotensin II-induced cardiomyocyte hypertrophy through MEK/ERK pathway. *Exp. Mol. Med.* **2007**, *39*, 65–73. [CrossRef]

80. Fang, Z.Y.; Lin, R.; Yuan, B.X.; Yang, G.D.; Liu, Y.; Zhang, H. Tanshinone IIA downregulates the CD40 expression and decreases MMP-2 activity on atherosclerosis induced by high fatty diet in rabbit. *J. Ethnopharmacol.* **2008**, *115*, 217–222. [CrossRef]

81. Chan, P.; Liu, I.M.; Li, Y.X.; Yu, W.J.; Cheng, J.T. Antihypertension induced by tanshinone IIA isolated from the roots of *Salvia miltiorrhiza*. *Evid. Based Complement. Altern. Med.* **2011**, *2011*, 392627. [CrossRef]

82. Bi, H.C.; Zuo, Z.; Chen, X.; Xu, C.S.; Wen, Y.Y.; Sun, H.Y.; Zhao, L.Z.; Pan, Y.; Deng, Y.; Liu, P.Q.; et al. Preclinical factors affecting the pharmacokinetic behaviour of tanshinone IIA, an investigational new drug isolated from *Salvia miltiorrhiza* for the treatment of ischaemic heart diseases. *Xenobiotica* **2008**, *38*, 185–222. [CrossRef]

83. Kapoor, S. Tanshinone II A: A potent, natural anti-carcinogenic agent for the management of systemic malignancies. *Chin. J. Integr. Med.* **2009**, *15*, 153. [CrossRef]

84. Wei, X.; Jing, M.; Wang, J.; Yang, X. Preliminary study on *Salvia miltiorrhiza* Bung endophytic fungus. *J. Clin. Pediatr.* **2010**, *22*, 241–246.

85. Zhang, P.; Lee, Y.; Wei, X.; Wu, J.; Liu, Q.; Wan, S. Enhanced production of tanshinone iia in endophytic fungi *Emericella foeniculicola* by genome shuffling. *Pharm. Biol.* **2018**, *56*, 357–362. [CrossRef] [PubMed]

86. Sairafianpour, M.; Christensen, J.; Steerk, D.; Budnik, B.A.; Kharazmi, A.; Bagherzadeh, K.; Jaroszewski, J.W. Leishmanicidal, antiplasmodial, and cytotoxic activity of novel diterpenoid 1,2-quinones from *Perovskia abrotanoides*: New source of tanshinones. *J. Nat. Prod.* **2001**, *64*, 1398–1403. [CrossRef] [PubMed]

87. Yuan, Y.; Lu, H. Effect of gibberellins and its synthetic inhibitor on metabolism of tanshinones. *Chin. J. Exp. Tradit. Med. Formulae* **2008**, *14*, 6–8.

88. Rungsimakan, S.; Rowan, M.G. Terpenoids, flavonoids and caffeic acid derivatives from *Salvia viridis* L. cvar. Blue Jeans. *Phytochemistry* **2014**, *108*, 177–188. [CrossRef]

89. Xiong, W.D.; Gong, J.; Xing, C. Ferruginol exhibits anticancer effects in OVCAR-3 human ovary cancer cells by inducing apoptosis, inhibition of cancer cell migration and G2/M phase cell cycle arrest. *Mol. Med. Rep.* **2017**, *16*, 7013–7017. [CrossRef]

90. Ulubelen, A.; Öksüz, S.; Kolak, U.; Bozok-Johansson, C.; Celik, C.; Voelter, W. Antibacterial diterpenes from the roots of *Salvia viridis*. *Planta Med.* **2000**, *66*, 458–462. [CrossRef]

91. Lee, S.-Y.; Choi, D.-Y.; Woo, E.-R. Inhibition of osteoclast differentiation by tanshinones from the root of *Salvia miltiorrhiza* Bunge. *Arch. Pharm. Res.* **2005**, *28*, 909–913. [CrossRef]

92. Tan, N.; Kaloga, M.; Radtke, O.A.; Kiderlen, A.F.; Öksüz, S.; Ulubelen, A.; Kolodziej, H. Abietane diterpenoids and triterpenoic acids from *Salvia cilicica* and their antileishmanial activities. *Phytochemistry* **2002**, *61*, 881–884. [CrossRef]

93. Tezuka, Y.; Kasimu, R.; Li, J.X.; Basnet, P.; Tanaka, K.; Namba, T.; Kadota, S. Constituents of roots of *Salvia deserta* Schang. *Chem. Pharm. Bull.* **1998**, *46*, 107–112. [CrossRef]

94. Jiang, Y.; Wang, L.; Lu, S.; Xue, Y.; Wei, X.; Lu, J.; Zhang, Y. Transcriptome sequencing of *Salvia miltiorrhiza* after infection by its endophytic fungi and identification of genes related to tanshinone biosynthesis. *Pharm. Biol.* **2019**, *57*, 760–769. [CrossRef] [PubMed]

95. Ming, Q.; Su, C.; Zheng, C.; Jia, M.; Zhang, Q.; Zhang, H.; Rahman, K.; Han, T.; Qin, L. Elicitors from the endophytic fungus *Trichoderma atroviride* promote *Salvia miltiorrhiza* hairy root growth and tanshinone biosynthesis. *J. Exp. Bot.* **2013**, *64*, 5687–5694. [CrossRef] [PubMed]

96. Ming, Q.; Dong, X.; Wu, S.; Zhu, B.; Jia, M.; Zheng, C.; Rahman, K.; Han, T.; Qin, L. UHPLC-HRMSn analysis reveals the dynamic metabonomic responses of *Salvia miltiorrhiza* hairy roots to polysaccharide fraction from *Trichoderma atroviride*. *Biomolecules* **2019**, *9*, 541. [CrossRef] [PubMed]

97. Wu, J.; Ming, Q.; Zhai, X.; Wang, S.; Zhu, B.; Zhang, Q.; Xu, Y.; Shi, S.; Wang, S.; Zhang, Q.; et al. Structure of a polysaccharide from *Trichoderma atroviride* and its promotion on tanshinones production in *Salvia miltiorrhiza* hairy roots. *Carbohydr. Polym.* **2019**, *223*, 115125. [CrossRef]

98. Peng, W.; Ming, Q.L.; Zhai, X.; Zhang, Q.; Rahman, K.; Wu, S.J.; Qin, L.P.; Han, T. Polysaccharide fraction extracted from endophytic fungus *Trichoderma atroviride* D16 has an influence on the proteomics profile of the *Salvia miltiorrhiza* hairy roots. *Biomolecules* **2019**, *9*, 415. [CrossRef]

99. Zikmundová, M.; Drandarov, K.; Bigler, L.; Hesse, M.; Werner, C. Biotransformation of 2-benzoxazolinone and 2-hydroxy-1,4-benzoxazin-3-one by endophytic fungi isolated from *Aphelandra tetragona*. *Appl. Environ. Microbiol.* **2002**, *68*, 4863–4870. [CrossRef]

100. Saunders, M.; Kohn, L.M. Evidence for alteration of fungal endophyte community assembly by host defense compounds. *New Phytol.* **2009**, *182*, 229–238. [CrossRef]

101. Venugopalan, A.; Srivastava, S. Endophytes as in vitro production platforms of high value plant secondary metabolites. *Biotechnol. Adv.* **2015**, *33*, 873–887. [CrossRef]

Publisher's Note: MDPI stays neutral with regard to jurisdictional claims in published maps and institutional affiliations.

 *agriculture*

Review

Endophytic Fungi of Tomato and Their Potential Applications for Crop Improvement

Martina Sinno [1,*,†], **Marta Ranesi** [1,*,†], **Laura Gioia** [1], **Giada d'Errico** [1] **and Sheridan Lois Woo** [2,3,4]

1   Department of Agricultural Sciences, University of Naples Federico II, 80055 Portici (NA), Italy; laura.gioia@unina.it (L.G.); giada.derrico@unina.it (G.d.)
2   Department of Pharmacy, University of Naples Federico II, 80131 Napoli, Italy; woo@unina.it
3   Task Force on Microbiome Studies, University of Naples Federico II, 80131 Naples, Italy
4   National Research Council, Institute for Sustainable Plant Protection, 80055 Portici, Italy
*   Correspondence: martina.sinno@unina.it (M.S.); marta.ranesi@unina.it (M.R.); Tel.: +39-340-928-4138 (M.S.); +39-329-1461-266 (M.R.)
†   These authors contributed equally to this work.

Received: 31 October 2020; Accepted: 25 November 2020; Published: 27 November 2020

Abstract: Endophytic fungi (EF) are increasingly gaining attention due to the numerous benefits many species can offer to the plant host, while reducing the application of chemicals in agriculture, thus providing advantages to human health and the environment. The growing demand for safer agrifood products and the challenge of increasing food production with a lower use of pesticides and fertilizers stimulates investigations on the use and understanding of EF. Other than direct consequences on the plant damaging agents, these microorganisms can also deliver bioactive metabolites with antimicrobial, insecticidal, or plant biostimulant activities. In tomato, EF are artificially introduced as biological control agents or naturally acquired from the surrounding environment. To date, the applications of EF to tomato has been generally limited to a restricted group of beneficial fungi. In this work, considerations are made to the effects and methods of introduction and detection of EF on tomato plants, consolidating in a review the main findings that regard pest and pathogen control, and improvement of plant performance. Moreover, a survey was undertaken of the naturally occurring constitutive endophytes present in this horticultural crop, with the aim to evaluate the potential role in the selection of new beneficial EF useful for tomato crop improvement.

Keywords: endophytes; biocontrol; biostimulants; induced systemic resistance; ISR; plant pathogens; fungal entomopathogens

1. Introduction

Different approaches can be used to discover alternatives to chemical pesticides, to prevent or control harmful plant biotic agents. In recent years, the interest in the biological control of plant pests and pathogens has surged to meet the requirements for more environmentally friendly options to synthetic chemicals. Consequently, the method for the use of microbial biological control agents (mBCAs), as natural antagonists to suppress herbivores and organisms that cause disease, has increased and improved. Fungi are among the most important mBCAs selected for this application due to their ease of isolation, selection from a vast number of known non-pathogenic strains, morphological structures for conservation and delivery, adaptation to numerous engineering fermentation technologies in industry, manageability in formulations, as well as their capability to secrete and over-express endogenous proteins or nontoxic exogenous compounds [1]. Furthermore, many beneficial fungi are known to promote plant growth and act as plant biostimulants or biofertilizers, thus their application in agriculture may also reduce the use of chemical fertilizers [2].

Among the plant favorable fungi, fungal endophytes, in particular, have been gaining increased attention because of the numerous benefits they can offer directly to the plant host with the intimate interaction established during the colonization of the plant tissues [3–7]. Since their discovery, endophytes have been isolated from different vegetative structures, many diverse plant species, both in natural uncultivated, as well as in agricultural environments [8]. These endophytic fungi (EF) represent a microbial community with an enormous reserve of biodiversity, originating from diverse ecological niches and host tissues ranging from the algae living in marine environments [9] to trees in the forest ecosystems [10].

These microorganisms have the ability to colonize plants without causing any symptoms [11], establishing a plant-fungi association inside the living plant tissue, that may occur within roots, stems, and/or leaves, and they emerge from the plant tissue only at the time of sporulation or upon senescence of the host [12–14]. The fossil record indicates that plants have had associations with endophytic [15] and mycorrhizal [16] fungi for more than 400 million years, a relationship that has likely existed since the time when plants first colonized land, thus playing a long and important role in the driving force of evolution, and life on land. In recent decades, scientific evidence has demonstrated that non-pathogenic microbes, endo- or exo-inhabitants of plants, may be associated with latent pathogens or early colonizing saprophytes that could actively grow in the living vegetative tissues only at the moment that plant defense responses waned or the plant initiated the phase of senescence [17]. The specific interaction between the host plant and its microbial partners ranges on a continuum from neutralism towards mutualism and antagonism, in which the nature of the relationship may change during the lifecycle of the plant depending upon environmental as well as intrinsic factors [18–20].

Currently, the concept of the plant microbiome considers a viewpoint on plant-microbe evolution, in which the plant and the microbiota have evolved together, with the microorganisms providing advantages and versatility to the plant in its ecosystem: an exchange of the plant as a habitat and source of nutrition, with some endophytic microorganisms producing benefits to the host plant that include the stimulation of growth and development, adaptation to the environment and abiotic stress tolerance [21,22]. More recently, it has been reported that EF can also have a protective role against attack by insects [23], pathogens [24], and nematodes [25], thus acting as multiple plant defenders or biocontrol agents.

Like most of the beneficial fungi, EF are known to secrete a vast number of bioactive secondary metabolites that are primarily responsible for the observed useful effects since they can stimulate the plant defense response and growth, as well as exert a direct antimicrobial or insecticidal effect [2,26–28]. Indeed, the remarkable advantage of these microorganisms is due to the in-depth relationship within the plant host that allows the immediate availability of the secreted active molecules within plant tissues [28,29]. In a broad sense, EF are producers of bioactive metabolites for which the plant constitutes a delivery system; in the case of insecticidal or antimicrobial molecules, the plant serves as a pipeline for the translocation of these compounds to the target pathogen or pest, thus EF act as a biopesticide [29,30].

Tomato belongs to the Solanaceae family, and it is one of the most commonly cultivated vegetable crops worldwide. *Solanum* section *Lycopersicon* includes the cultivated tomato (*Solanum lycopersicum* L.) and 12 additional wild relatives, but *S. lycopersicum* is the only domesticated species [31]. The tomato originated from South America, then spread globally through different levels of domestication, starting prior to the 15th century and continuing into Europe, arriving to its present status as one of the most highly consumed food crops of international acclaim. In 2018, the global tomato production reached more than 180 million tons, in which the cultivated area worldwide of harvested tomatoes accounted for almost 5 million ha (FAOSTAT 2018). Due to its extensive global distribution and consumption, tomato is one of the most important horticultural crops farmed, and during its cultivation, it is constantly threatened by pests and pathogens. In recent times, the use of chemical pesticides is becoming largely restricted in agriculture both due to the negative impact on human health and the environment (EU Directive 2009/128/CE), plus the risks of resistance development in pathogen/pest populations by their use.

The growing demand by consumers for safer food products plus the urgent challenge of increasing food production with lower inputs of pesticides and fertilizers stimulates both the utilization of EF as non-chemical plant beneficials, as well as the investigations that provide further insights into the interactions of these microorganisms with the crop host and the organisms that damage it. Recent studies have focused on the use of EF in protecting and improving tomato crop as an alternative to the chemical approach for crop protection (references in this review). The presence of EF in tomato can be prevalent in the plant if the fungi were introduced as mBCAs, or if the EF were naturally acquired from the surrounding environment and horizontally and/or vertically transmitted into the plant [32]. According to the increasing interest in this argument, the present paper offers a review of the principle EF that have been introduced to tomato for the biological control of various pests and pathogens, plus those applied to improve the plant performance, growth/yield, and quality. Furthermore, a section of this review is dedicated to constitutive EF isolated and identified from tomato, which can serve as a valuable source of new microbial beneficial applications with unexplored potential to improve the production of tomato and other crops.

2. Beneficial Effects of EF Introduction on Crops

The beneficial effects executed by EF include pest and pathogen control, plant growth promotion (PGP), improvement of the plant nutrient availability and uptake, and the increased tolerance to abiotic stress, hereby referred to as plant physiology improvement (PPI). EF have been confirmed to affect insect pests feeding on the plants that they colonize [23,33–35], and results in the literature indicated that these microorganisms provided protection from significant herbivory damage they cause to crops. Insect pests have been noted to be affected by EF in numerous ways such as reduction of developmental rate [36,37], deterrence of feeding on the colonized plant [38,39], retarded insect growth, higher mortality, and lower oviposition [40,41]. One of the hypothesized mechanisms underlying these effects is the bioaccumulation of secondary metabolites and mycotoxins produced by the EF within the plant tissues [34]. Moreover, these microorganisms are known to defend their plant host from pathogens attack [3–7]. This biocontrol action could be exerted through direct mechanisms including food and space competition, parasitism, and antibiosis [39,42–44]. While colonizing host plants, EF stimulate the attacked plant to create a barrier (biochemical or mechanical) that inhibits pathogenic organisms from penetrating the same tissue hence preventing the occurrence of diseases [45]. An important indirect mechanism involved in plant protection is the induction of plant resistance which is implemented by the alteration of the biochemical signaling pathways of the plant that modulate the resistance-related genes which are triggered by the endophytic colonization [46–51]. PGP implemented by EF is characterized by the improvement of the above- and/or below-ground biomass [52–54] while PPI effects include the increase of nutrients uptake, particularly nitrogen and phosphorus [55,56], and enhanced tolerance to abiotic stress including drought, salt, and heat [1]. EF can be, thus, utilized as biofertilizers as they improve the nutrient uptake firstly enhancing the plant root system and secondly, in the case of entomopathogenic endophytes, reallocating the insect-derived nitrogen to the host plant. In fact, EF, after feeding on insects in the rhizosphere, may translocate the adsorbed nitrogen to the host plant towards the association with the root system [55].

3. Introduced Endophytes of Tomato

The utilization of EF as biological control agents (BCA) represents a potential alternative that meets the growing need for more eco-sustainable agriculture. According to this new perspective, in recent years many studies have been performed, introducing EF on tomato to test their effects on plant performance. Among these introduced EF, a consistent number of species belongs to a group of fungi classified as entomopathogens, fungi that are pathogens to insects, many isolated from asymptomatic plants, including *Akanthomyces* spp., *Beauveria bassiana*, *Clonostachys rosea*, *Cordyceps farinosa* (formerly *Isaria farinosa*), *Lecanicillium* spp., and *Sarocladium* spp. (formerly *Acremonium* spp.) [39,57–61]. The natural occurrence of these fungi within the plant tissues suggests their ability to endophytically colonize a wide range of plants.

To date, in tomato as for many other horticultural crops, the artificial introduction of EF has been limited to a restricted group of beneficial microbes which include species belonging to the genus *Sarocladium*, *Beauveria*, *Metarhizium*, *Fusarium*, *Penicillium*, *Serendipita* (formerly *Piriformospora*), *Pochonia*, *Pythium*, and *Trichoderma* [36,37,62–68]. Furthermore, some introduced species belong to the Dark Septate Endophytes (DSE) such as *Neocosmospora haematococca* (formerly *Nectria haematococca*) and *Periconia* spp. DSE represent a large group of root-inhabiting endophytes not yet well defined taxonomically and/or ecologically that are distinguished as a functional group based on the presence of darkly melanized septa. DSE are ubiquitous and abundant in various ecosystems and playing an interesting role in contrasting pathogens as they can improve plant tolerance to abiotic stress [69], growth [70], and nutrient uptake [71]. In short, DSE may play an important role in the ecophysiology of plants. However, almost a century after their discovery, little is still known about the role of these mysterious and rather elusive fungal symbionts.

3.1. Biocontrol

The biocontrol of pests and pathogens has been the most documented beneficial effect explicated by the artificial introduction of endophytic fungi to tomato. In this context, 41 scientific papers report the use of mBCAs that focuses on crop protection and the consequences on the organisms that are deleterious to tomato (Table 1).

In particular, concerning the insect pests, biocontrol potential of EF was evidenced for the negative effects on *Aphis gossypii* [72], *Bemisia tabaci* [73,74], *Chortoicetes terminifera* [72], *Helicoperva armigera* [75,76], *Nesidiocoris tenuis* [77], *Spodoptera exigua* [78], *S. littoralis* [23], *Tuta absoluta* [79,80], and *Trialeurodes vaporariorum* [43]. Furthermore, *Neocosmospora solani* (formerly *Fusarium solani*) increased tomato defense against infestations of the red spider mite, *Tertranychus urticae* [46], while several endophytic species were able to induce resistance to the root-knot nematode *Meloidogyne incognita* [25,81–83]. The overall effects observed on tomato pests included: increased mortality, feeding deterrence, reduced growth rate and reproduction, reduced infestation, egg masses colonization, and increased plant defense.

Regarding disease control, EF were reported to counteract the infection of the bacterial pathogen *Clavibacter michiganensis* subsp. *michiganensis* [84], fungal pathogens *Fusarium oxysporum* f. sp. *lycopersici* [84–92], *Rhizoctonia solani* [93] and *Botrytis cinerea* [42,94]. In the above-mentioned papers, the reported effects of endophytic colonization on pathogen control were noted with reduced disease symptoms and a disease-suppressive effect.

In most papers, the authors suggested that the reduced impact of pests and diseases was due to plant resistance induced by its microbial partner. The mechanism underlying this induced resistance are subdivided into two main categories: Systemic Acquired Resistance (SAR) and Induced Systemic Resistance (ISR). SAR is induced by the plant local infection by latent pathogens and is effective against a broad range of harmful plant biotic agents, it is mediated by salicylic acid (SA) and associated with pathogenesis-related proteins [95]. This is the case of *N. solani* strain Fs-K which was reported to induce plant resistance against *Septoria lycopersici* through a SAR mechanism [87]. On the other hand, ISR is triggered by the endophytic colonization of beneficial microorganism such as plant growth-promoting rhizobacteria and EF that involves a priming process of the plant which results in more efficient activation of its defense responses against pests and pathogens [96]. It is mediated by jasmonic acid (JA) and ethylene (ET) [97]. This is the case of *Trichoderma hamatum* which is reported to induce resistance against the tomato bacterial spot caused by *Xanthomonas euvesicatoria* [98]. Nonetheless, SAR and ISR may be two distinct but overlapping mechanisms as a result of crosstalk of the two hormonal pathways [97], as noted for *Trichoderma harzianum* which induced plant resistance against *M. incognita* through priming plant defense with both SA and JA stimulation [99].

Agriculture **2020**, *10*, 587

3.2. Plant Growth Promotion and Plant Physiology Improvement

Plant growth promotion of tomato attributed to endophytic colonization has been well-documented (Table 1). Thirteen studies have indicated that there was evident PGP as demonstrated by the improvement of the root system with greater root length, biomass, and dry weight [69,83,88,94,100–102], increased plant height, shoot biomass, and fresh or dry weight [69,70,83,88,89,100,101], plus enhanced plant production with anticipated flowering and fruiting times, and increased fruit weight [102].

Moreover, a few articles reported improvement of the plant nutrient uptake and the increased tolerance to abiotic stress (PPI) (Table 1). Improved plant uptake of iron (Fe) [103], organic nitrogen (N) [70,71], and inorganic potassium (K) [70], have been demonstrated to be a consequence of the plant endophytic colonization with some fungal species namely *B. bassiana*, *Periconia macrospinosa* (DSE) plus an unidentified species also belonging to DSE. Additionally, some studies have highlighted that the presence of the endophyte confers tolerance to diverse abiotic stress such as drought [69], salinity [104], and metals [101].

Table 1. Effects of introduced fungal endophytes on tomato plants in terms of Plant Growth Promotion (PGP) and Plant Physiology Improvement (PPI), and Biocontrol (BC) of pest and pathogens.

Fungal Species	Effects	
	PGP and PPI	BC
Sarocladium strictum * *Sarocladium kiliense* *	Increased number of xylem vessels within the shoots [84]	Increased mortality of larvae of *Trialeurodes vaporariorum* [105] Reduced symptoms caused by *Fusarium oxysporum* f. sp. *lycopersici* and *Clavibacter michiganensis* subsp. *michiganensis* [84]
Beauveria bassiana	Enhanced terpene production [78] Improved iron (Fe) nutrition [103]	ISR vs. *Rhizoctonia solani* [93] ISR vs. *Botrytis cinerea* [42] ISR vs. *F. oxysporum* f. sp. *lycopersici* [85] Increased mortality of *Tuta absoluta* [79,80] Reduced incidence of *Fusarium oxysporum* f. sp. *lycopersici* and *Helicoverpa armigera* [106] Increased mortality of *Helicoperva armigera* [75,106] Increased mortality of *Bemisia tabaci* [73] Feeding deterrent for *Bemisia tabaci* [74] Increased mortality of *Spodoptera littoralis* [23] Reduced growth rate of *Spodoptera exigua* [78] Reduced reproduction of *Aphis gossypii* and reduced growth rate of *Chortoicetes terminifera* [72]
Metarhizium anisopliae	Increased plant height, root length, shoot and root dry weight [100]	Increased mortality of *Spodoptera littoralis* [23]
Fusarium oxysporum *Neocosmospora solani* *		ISR vs. *F. oxysporum* f. sp. *lycopersici* [86] ISR vs. *Meloidogyne incognita* [81,82] Fermentation broth with anti-oomycete activity vs. *Pythium ultimum*, *Phytophthora infestans* and *Phytophthora capsici* [24] Reduced infestation of *Trialeurodes vaporariorum* [42] ISR vs. *Nesidiocoris tenuis* [77] ISR vs. *F. oxysporum* f.sp. *radicis-lycopersici* [87] SAR vs. *Septoria lycopersici* [87] Increased tomato defenses against *Tertranychus urticae* [107]
Fusarium spp.	Increased roots length, shoots height and plant fresh weight [88]	ISR vs. *Fusarium oxysporum* f. sp. *radicis-lycopersici* [88]
Neocosmospora haematococca * (DSE)	Drought stress tolerance, improved plant growth, and proline accumulation [69]	
Unidentified (DSE)	Increased aboveground plant dry biomass and increased uptake of organic N and inorganic K [70] Salinity stress tolerance [104]	
Penicillium simplicissimum *	Metal stress tolerance [101] Increased shoot length and biomass under normal and Al stress conditions [101]	
Periconia macrospinosa (DSE)	Improved organic N uptake and plant biomass when organic nutrients are present [71]	
Serendipita indica *	Increased fresh weight [89] Accelerated vegetative and generative development [108]	ISR vs. Tomato yellow leaf curl virus [109] Disease-suppressive effect vs. *Verticillium dahliae* and *F. oxysporum* [89–91] Reduced infestation of *Meloidogyne incognita* [25]
Pochonia chlamydosporia	Increased root and shoot growth [83] Anticipated flowering and fruiting times, increased fruit weight and root growth [102]	Colonizes egg masses of *Meloidogyne incognita* [83]
Pythium oligandrum		ISR vs. *Ralstonia solanacearum* [110] ISR vs. *Fusarium oxysporum* f. sp. *lycopersici* [92] ISR vs. *B. cinerea* [111]
Trichoderma atroviride *Trichoderma hamatum*	Increased root and shoot growth depending on the tomato cv [94]	Reduced infestation of *Trialeurodes vaporariorum* [43] ISR vs. *Botrytis cinerea* [94] ISR vs. *Xanthomonas euvesicatoria* (tomato bacterial spot) [98] ISR and SAR vs. *Meloidogyne incognita* [99]
Trichoderma harzianum	Increased root and shoot growth depending on the tomato cv [94]	ISR vs. *Botrytis cinerea* [94] Reduced desease symptoms caused by *Alternaria solani* and *Phytophtora infestans* [112]

* scientific names are different from those present in the articles cited due to taxonomic updates to the name presently use.

3.3. Methods of Introduction and Detection

In the last decades, it has been demonstrated that several beneficial EF can be artificially introduced on tomato using different inoculation methods and numerous protocols have been developed to successfully achieve this colonization, as well as to detect the fungi within the plant tissues. The methods used for the introduction and detection of EF in tomato plants are summarized in Table 2. The inoculation of EF to tomato plants is mainly achieved with conidial suspension applied by seed treatments, root dipping, soil watering, stem injection, and leaf spraying. Alternatively, the application can be performed by mixing fungal biomass with the transplanting soil. Among the different inoculation techniques, the soil applications, mainly by watering with a conidia suspension, was the most commonly and successfully used technique applied in 18 studies. This was followed by the treatment of seed, as adopted in 15 studies, which was performed by various methods including seed soaking, seed coating, and seed dressing. Seed soaking consisted of placing the tomato seeds in a liquid conidial suspension for 2 to 24 h, before planting. Seed coating involved immerging the seeds in a conidial suspension, stirring them every 30 min for 2–3 h, to cover and adhere the spores to the seed surface, then successively air-drying under sterile conditions [93,94,113]. Seed dressing, was the less common technique, preparing and mixing the seeds in a conidia suspension with continuous shaking for several hours [74,103]. The conidia suspension usually contained a "sticker" such as Tween 80 (0.1–0.01% *v/v*) or methylcellulose (5–10% *v/v*), to ensure a more efficient adhesion of the conidia to the seed surface. Root dipping was another technique commonly used that consisted of dipping the seedling roots in a conidial or propagule suspension for 6 to 24 h prior to transplant [71,76,79,89,106,110].

It is evident that the methods of application were numerous, and the selection of the most efficient method is highly dependent on the specific EF that is employed. The majority of the studies, involving the artificial introduction of EF to tomato, were conducted in a controlled environment, usually with sterilized soil or transplanting substrate, and not in the open field, in order to facilitate the monitoring of the plant colonization. The field application of EF is a challenge that needs to take into account the enormous variability of the environment that could negatively affect the efficacy of the above-mentioned protocols for the introduction. Another critical issue is represented by the transient nature of some endophytes in plant colonization, which explains why, in most cases, the studies do not report details on the time duration of the endophytic colonization. A study by Resquin-Romero [23] indicated that the endophytic colonization of the plant was transient and that the EF–plant interaction was lost after a certain period of time after inoculation. Due to the transient nature of the endophytic colonization, as has been documented in other crops, it is recommended that parallel time-course studies should be performed to monitor the extent of endophytic development, for example, with molecular, microscopy, and/or in vivo re-isolation techniques [68,114–119]. This attests to the difficulty of establishing stable and lasting interactions between the chosen endophyte and its plant host, in the attempt to obtain the potential desired effect.

To ensure that the inoculation of the fungal species is followed by actual endophytic colonization of the plant, it is mandatory to include an experimental stage to detect the EF within the plant. Out of the 52 papers reviewed, 16 studies did not include an endophyte detection assay in their experimental workflow. It is recommended that an analysis of the EF presence should always be included in the study in order to assess the success of the endophytic colonization of the plant, plus monitor the rate of colonization. Moreover, this detection stage should follow an accurate surface-sterilization of the plant tissues to avoid the inadvertent isolation of epiphytic rather than endophytic fungi.

The methods to determine the presence of EF can be divided into three main types: the re-isolation of the EF from the plant tissue, the molecular detection by polymerase chain reaction (PCR), and morphological observation using microscopy techniques. Each method requires the sterilization of the plant material to eliminate the epiphytic microbial community, usually obtained by dipping the tissue in a diluted bleach solution for 1–3 min, that can be followed or proceeded with a brief 70% ethanol bath, completed with rinsing it at least three-times with sterile water. As a check of the efficacy of the sterilization procedure, aliquots of the rinsed water are also plated, and if bacterial or fungal growth occurs the sample is discarded.

The re-isolation of the fungal colony from the host plant tissue is the most used method to assess the endophytic colonization and is reported in 17 studies. Usually, it follows this protocol: collect, wash, and sterilize the plant material, dissect the vegetal tissues in 1 cm pieces under sterile conditions, and place the pieces on Petri dishes containing solid culture substrate. Most of the authors utilized potato dextrose agar (PDA), supplemented with antibiotics to avoid bacterial contaminations, while others used selective media for the specific EF they were interested in re-isolating [23,80,100,120]. Molecular analysis was also widely used for the identification of the EF and is reported in 12 studies. It was based on the extraction of the DNA from the pre-sterilized plant tissue, and the subsequent amplification by PCR and sequencing of amplicons for specific fungal molecular markers such as the Internal Transcribed Spacer (ITS1 and ITS2) region and the translation elongation factor (TEF). Five manuscripts included the quantitative detection of the EF within the plant tissue using a real-time PCR [77,86,87,107,121].

Eleven studies used microscopy techniques to visually examine the fungal presence within the plant tissues. These techniques included light optical microscopy using stained plant tissues, usually with trypan blue or methyl blue, scanning electron microscope (SEM), and transmission electron microscope (TEM). The microscope analysis was particularly valuable for observing and understanding the EF growth distribution patterns and translocation within the plant tissues, thus providing important information and a deeper insight of the EF colonization that was not possible in comparison to the other methods with the re-isolation or molecular detection.

Table 2. Methods of introduction and detection of fungal endophytes in tomato plant with relative cultivar.

Fungal Species	Tomato Cultivar	Method of EF Inoculation	Detection Method	Location of EF in Plant Tissues	Ref.
Sarocladium kiliense *	Haubner's Vollendung	Fungal biomass mixed with transplanting soil		Roots	[84]
S. strictum *	Haubner's Vollendung	Soil watering	Re-isolation from the plant tissue on PDA	Roots	[105]
S. strictum *	Suso RZÒ F1 hybrid	Soil watering	Re-isolation from the plant tissue on MEA	Roots	[122]
Beauveria bassiana	Platense	Seed soaking Leaf spraying Root dipping	Re-isolation from the plant tissue on PDA	Leaves	[79]
B. bassiana	Mobil	Seed coating	Re-isolation from the plant tissue on PDA		[93]
B. bassiana	Limachino—INIA	Fungal biomass mixed with transplanting substrate	Re-isolation from the plant tissue on Noble agar	Roots Stem Leaves	[42]
B. bassiana	Rio Fuego	Soil watering Leaf spraying Stem injection			[85]
B. bassiana	Ace, Early Pack, Money Maker, Peto 86, Prichard, Pusa Ruby, Strain B and LA1478	Leaf spraying Stem injection	PCR	Stem	[73]
B. bassiana	Grosse lisse	Leaf spraying	Re-isolation from the plant tissue on PDA	Leaves	[34]
B. bassiana	Harzfeuer F1	Leaf spraying	Re-isolation from the plant tissue on selective media	Leaves	[80]
B. bassiana	Regina	Conidial suspension on wounded rachis	Re-isolation from the plant tissue on selective media	Roots	[120]

Table 2. *Cont.*

Fungal Species	Tomato Cultivar	Method of EF Inoculation	Detection Method	Location of EF in Plant Tissues	Ref.
B. bassiana	Cal-J, Kilele F1, Anna F1	Seed soaking	Re-isolation from the plant tissue on SDA	Roots Stem Leaves	[123]
B. bassiana	Cal-J, Kilele, Anna	Seed soaking	Re-isolation from the plant tissue on SDA	Roots Stem Leaves	[124]
B. bassiana	Mountain Spring	Seed coating			[113]
B. bassiana	PKM1	Seed soaking Root dipping Soil watering			[76]
B. bassiana	PKM1	Seed soaking Root dipping			[106]
B. bassiana	surahi	Root dipping Stem injection Soil inoculum Leaf spray	Re-isolation from the plant tissue on PDA	Leaves	[75]
B. bassiana	Tres Cantos	Leaf spray	Re-isolation from the plant tissue on selective media	Stem Leaves	[23]
B. bassiana	Marmande-Cuarenteno	Seed soaking	Re-isolation from the plant tissue on SDCA	Stem Leaves Roots	[35]
B. bassiana	Castlemart	Seed coating	PCR	Shoot	[78]
B. bassiana	Hezuo 903	Leaf spray Root irrigation Reed dressing	PCR	Shoot	[74]
Fusarium spp.	Rio Grande	Soil watering	PCR	Root Stem	[88]
F. oxysporum	Montfavet 63-5	Root application	Real-Time qPCR	Roots Cotyledons	[86]
F. oxysporum	Furore	Soil application		Roots	[81]
F. oxysporum	Moneymaker	Soil watering		Roots	[82]
F. oxysporum	Hellfrucht/JW Frühstamm	Soil watering		Roots	[43]
Neocosmospora solani *	Pearson	Soil watering	Real-Time qPCR	Roots	[77]
N. solani *	Ace 55	Soil watering	Real-Time qPCR	Roots	[107]
N. solani *	Ace 55	Soil watering	Microscopy Real-Time qPCR	Roots	[87]
Metarhizium anisopliae	Hybrid var. 8625	Soil watering	Re-isolation from the plant tissue on selective media	Roots Shoots Leaves	[100]
M. anisopliae	Tres Cantos	Leaf spray	Re-isolation from the plant tissue on selective media	Stem Leaves	[23]
M. brunneum	Ruthje	Encapsulated mycelial biomass	Light microscopy Real-Time qPCR	Stem	[121]
Neocosmospora haematococca * (DSE)	CO-2	Soil application of mycelial biomass formulation	Light microscopy	Roots	[69]
Unidentified (DSE)	Santa Clara I-5300	Soil application of mycelial biomass	Light microscopy	Roots	[70]
Penicillium semplicissimum *	LA2710	Soil application of mycelia and culture filtrate		Roots	[101]
Periconia macrospinosa (DSE)	Hildares F1	Root dipping in propagule suspension	Light microscopy	Roots	[71]
Serendipita indica *	Hildares	Root dipping	Re-isolation from the plant tissue on PDA	Roots	[89]
S. indica *	T07-4, T07-1	Transplanting substrate application of mycelia	Light microscopy	Roots	[109]
S. indica *	Nutech	Seed coating (bioformulation)		Roots	[90]
S. indica *	Vellayani Vijay	Transplanting substrate application of mycelia	Light microscopy	Roots	[25]

Table 2. *Cont.*

Fungal Species	Tomato Cultivar	Method of EF Inoculation	Detection Method	Location of EF in Plant Tissues	Ref.
Pochonia chlamydosporia	Durinta	Plating of seedlings on fungal plate cultures	laser-scanning confocal microscopy PCR	Roots	[83]
P. chlamydosporia	Marglobe	Seed germination on fungal plate cultures	Re-isolation from the plant tissue on CMA PCR	Roots	[102]
Pythium oligandrum	Micro-Tom	Root dipping	laser scanning microscopy	Roots	[110]
P. oligandrum	Prisca	Mycelial plugs in proximity of the top root	SEM TEM	Roots	[92]
P. oligandrum	Prisca	Soil watering	TEM	Roots	[111]
Tricoderma atroviride	Hellfrucht/JW Frühstamm	Soil application		Roots	[43]
T. atroviride	Corbarino, M82, SM36, TA209	Seed coating		Roots	[94]
T. hamatum	Ohio 8245	Soil application		Roots	[98]
T. harzianum	Corbarino, M82, SM36, TA209	Seed coating		Roots	[94]
T. harzianum	Moneymaker	Soil application		Roots	[99]
T. harzianum	Arka vikas	Soil watering		Roots	[112]

* scientific names are different from those present in the articles cited due to taxonomic updates to the name presently use.

4. Constitutive Endophytes of Tomato

EF have been reported to have a crucial role in inducing plant host tolerance to stressful conditions [59], plant defense [32], and plant growth and development [125]. In all-natural or agricultural ecosystems, every plant is colonized by a diversity of soil-borne microorganisms as root endophytes, mycorrhizal fungi, and plant growth-promoting rhizobacteria. Moreover, the analysis of plant–endophyte associations in high-abiotic stress habitats revealed that at least some fungal endophytes confer habitat-specific stress tolerance to the host plants. Without the presence of the habitat-adapted fungal endophytes, these plants were unable to survive in their native habitats [126]. Thus, the naturally occurring EF constitute a poorly exploited resource, rich in terms of biodiversity, representing a pool of potentially beneficial fungi from which the selection of new strains may be obtained for useful applications in agriculture.

Seven studies focused on the naturally occurring EF of tomato and the data are summarized in Table 3. The constitutive EF were comprised of 24 different genera, among which the most represented are *Trichoderma* and *Fusarium*, which included 35 different fungal species. It is interesting to note that some of the fungi reported in Table 3 are commonly recognized as plant beneficial fungi, such as *Trichoderma* spp., *N. solani*, and *Sarocladium implicatum* (formerly *Acremonium implicatum*), while other species are known as plant pathogens, for example, *Alternaria solani*, *Stemphilyum lycopersici* and *Albifimbria verrucaria*. *A. solani* causes early blight of tomato, one of the common foliar diseases of tomato [127], *S. lycopersici* is the causal agent of leaf spot disease on pepino (*Solanum muricatum*) [128], and *A. verrucaria* produces small brown to black spots symptoms on the colonized leaves and stems [129]. Moreover, *A. verrucaria* is also known to be the responsible agent of mycotic keratitis, one of the major causes of ophthalmic morbidity and visual loss globally [130]. This highlights the importance of identifying EF to study their prospective utilization in agriculture, but also to understand the possible implications on human health.

An example of EF use for tomato improvement is provided by the work of Bogner and colleagues [32] that was conducted in five different counties of Kenya with the aim of identifying and characterizing the culturable endophytic mycobiota in the roots of tomato and screening different fungal endophytes for their biocontrol potential towards the root-knot nematode *Meloidogyne incognita*. A total of 76 fungal isolates were obtained, among which the most prevalent species associated with tomato roots were members of the *F. oxysporum* and *N. solani* species complexes. Bioassays demonstrated the ability of selected non-pathogenic fungal isolates to successfully reduce nematode penetration and subsequent galling, as well as decrease the reproduction capacity of the root-knot nematode *M. incognita*. Most isolates in the *Trichoderma asperellum* and *F. oxysporum* complex were able to reduce the root-knot nematode egg densities by 35–46% in comparison to the treatments with the nonfungal control and the other fungal isolates. Moreover, Tian and colleagues isolated an endophytic fungus from tomato root galls infected with *M. incognita* that was identified as *S. implicatum* based on morphological and molecular identification [131]. The biocontrol potential of *S. implicatum* culture filtrates was tested with the plant and nematodes in vitro, in pot and field experiments. Results from the in vitro test indicated that 96% of second-stage juveniles of *M. incognita* were killed after 48 h. The fungal compounds were also able to suppress egg hatching, the formation of root galls, and reduce the nematode population in the soil.

These findings suggest that naturally occurring EF populations in the soil represent an underestimated and valuable source of microbial diversity with positive impacts on sustainable agricultural production, due to the possibility to reduce the use of chemical products, thus benefiting the environment and human health. Moreover, this highlights the importance of promoting the constitutive endophytic populations in the soil in order to obtain the effective threshold level for biological control of organisms that compromise plant health. Many studies have demonstrated that soil type and plant genotype are the two main variables that affect the establishment of fungal species in the soil community [132–134]. The cultivation system can also influence the microbial species in the soil, whereby fungal abundance was significantly higher in organically farmed fields than the populations found in conventionally farmed that used chemicals [32,132]. In order to successfully develop applications of plant-associated EF in sustainable agricultural production, further investigations are necessary to understand the mechanisms of action and the processes employed by the fungi to produce the beneficial effects, as well as to determine how they can be efficiently utilized in actual practices.

Table 3. Naturally occurring constitutive endophytes of tomato.

Fungal Species	Tomato Cultivar	Main Results	Location of EF in Plant Tissues	Country	Ref.
Alternaria solani *Aspergillus sclerotiorum* *Cochliobolus geniculatus* *Curvularia lunata* * *Fusarium nygamai* *Fusarium sp.* *Fusarium verticillioides* *Stemphylium lycopersici* *Trichoderma asperellum* *Trichoderma lixii* *	Moneymaker	Biological control to the rootknot nematode *Meloidogyne incognita*	Root	Kenya	[32]
Fusarium spp.	Heinz 9907 Gem 611 Heinz 3402 FL 47 Mountain Fresh	No effects	Roots Crown Stem	USA	[134]
Fusarium oxysporum *Fusarium fujikuroi* *Neocosmospora solani* *	Momotaro	No effects	Stem	Japan	[135]
Ochroconis humicola *	Gohobi	Improved plant growth with organic nitrogen sources	Root	Japan	[125]
Albifimbria verrucaria * *Fusarium* spp. *Setophoma terrestris* *Trichoderma* spp.	Heinz 1706 Moneymaker	No effects	Root	Northern Italy	[133]

Table 3. *Cont.*

Fungal Species	Tomato Cultivar	Main Results	Location of EF in Plant Tissues	Country	Ref.
Sarocladium implicatum *	Lichun	Biological control suppressed *M. incognita* egg hatching and population, when inoculated to soil	Root	China	[131]
Alternaria spp. *Aspergillus fumigatus* *Aspergillus nidulans* *Chaetomium globosum* *Coniothyrium aleuritis* *Fusarium chlamydosporum* *Fusarium oxysporum* *Fusarium proliferatum* *Fusarium sp.* *Hypoxylon sp.* *Leptosphaerulina chartarum* *Meyerozyma guilliermondii* * *Neocosmospora solani* * *Nigrospora sp.* *Penicillium helicum* * *Penicillium ochrochloron* *Penicillium simplicissimum* * *Periconia macrospinosa* *Pleosporales sp.* *Rhinocladiella sp.* *Trichoderma atroviride* *Trichoderma spirale*	Big Beef	Plant growth promotion and enhanced fruit weight	Root Shoot Seed	USA	[132]

* scientific names are different from those present in the articles cited due to taxonomic updates to the name presently use.

5. Perspectives on EF Applications to Tomato

EF are ubiquitous microorganisms in the natural and agricultural environment able to colonize plants internally.

In 1994, Dreyfuss and Chapela estimated that the global fungal diversity amounts to 1.5 million species, and based on their estimates, endophytic fungi alone could account for up to 1.3 million species [136]. This perspective on EF diversity was substantiated by subsequent studies of novel plant species, in particular, a study of the fungus:plant ratio in the tropical regions, confirmed that the number of 1.3 million endophytic fungi on the planet was a good assessment [137]. Recently, Hawksworth and Lücking revised the appraisal on global fungal diversity, concluding that the above-mentioned value was too conservative, and the actual range of fungal species should be considered at 2.2–3.8 million species [138].

Although the category of EF is gaining interest in the scientific community, due to their potentially beneficial applications, the studies conducted to date on this topic are still relatively limited and require further investigations.EF can be an extraordinary source of BCAs, PGP, and bioactive molecules, that can provide multiple positive effects to crops, which make them suitable components of biostimulants and biopesticides for use in agriculture [19,26,28]. Most endophytes are considered non-pathogenic, but not all are capable of producing plant beneficial effects [139]. Moreover, even when colonization occurs and positive effects are evident, the costs to the plant in hosting the endophyte/s have to be taken into account, an aspect that has not been studied extensively and is generally underestimated [136,140–143]. It should be considered that EF constitute a rich biodiversity source requiring a greater understanding of: (1) the mechanisms of action involved, those used by the fungal colonizer and the host plant, for crosstalk and recognition that permit the establishment of the interaction; (2) the ecological, biological, and physiological functions of the EF–plant relationship over time; plus (3) the factors and conditions that determine successful colonization [26,144,145].

Understanding the mechanisms underlying the plant-endophyte association and the subsequent outcomes, or the cause and effect, is fundamental for the advancement of EF know-how for the improvement of crop production. Currently, it is well recognized that the interaction between plant and endophyte is highly influenced by three factors: the genotype of the plant and its microbiome, the fungal genome, and the environmental conditions in which the association occurs [29]. Two major challenges became apparent during the preparation of this review, that in order to develop a wider use of EF in agriculture it will be necessary: (i) to determine how to select the best endophyte–plant combination and establish a stable long-lasting interaction between this beneficial microbe and the plant host targeted for improvement; and (ii) to prove the effectiveness of this technology outside of the controlled test conditions used to date, moving from the greenhouse to the actual open field environment.

Tomato plays host to a microbial community that is vast and highly variable, depending upon the prevailing environmental conditions and the plant genotype [132,144]. It could be interesting to concentrate investigations on the constitutive fungal endophytes that are native to the tomato plant, as they, by their inherent nature, represent a massive pool of highly "tomato-adapted" fungi. This vast fungal community represents a pool of biodiversity that up to now, has been poorly exploited in the strategies to discover highly adapted beneficial microbes of specific crops of interest. In general, a greater comprehension of the mechanisms that favor, along with those that hinder, the endophytic colonization of plants, is required for wider application of EF in agriculture [144]. For example, determine the environmental conditions known to be key factors for a successful EF–plant interaction [26,132].

The artificial introduction of EF in agri-food crops also needs to be analyzed, to ascertain the possible risks that endophytic fungal colonization may present to the plant and the consumer, such as the introduction of potentially toxic metabolites (i.e., mycotoxins) to the food chain [144]. In this respect, studies should assess both the food safety of the fruits produced by EF-colonized plants, as well as evaluate the environmental effects in terms of the release or bioaccumulation of toxins in the soil or crop residues that may be a risk for the agroecosystem. Furthermore, an analysis of the outcome of the endophytic colonization on the organoleptic qualities of the agrifood products should also be taken into account.

In recent years, a growing number of studies have focused on the introduction of beneficial EF to tomato in order to exploit their biocontrol potential against pests and pathogens, as well as their growth promotion effect (references in this review). It is evident from the findings to date, that the introduction of EF represents a promising field of research and development, to which the consequences could determine a remarkable reduction of chemical use in agriculture. This outcome could be clearly observed in the field of crop protection, where it has already been well documented that the biocontrol activity of EF is able to limit the negative effects of several key tomato insect pests and pathogens, as well as nematodes. Moreover, the tomato plants harboring some EF have demonstrated enhanced tolerance to abiotic stress in the field, plus improvement in nutrient uptake, yield, and nutritional quality of the fruits.

This review reports that several EF species are good versatile BCA, controlling both pests and pathogens as demonstrated in the case of *B. bassiana*, *F. oxysporum*, *N. solani*, and *T. harzianum*, which are amenable candidates as plant beneficial microbes, also considering their additional properties as plant biostimulants. Nonetheless, a few surveyed papers considered the possibility to use EF species as a multi-use biocontrol agent, evaluating the simultaneous biocontrol of both pests and pathogens in tomato. Only recently, Jaber and Ownley underlined that some endophytic and entomopathogenic fungi conferred protection to their host plant not only against insect pests but also plant pathogens, and they proposed their use as dual biocontrol agents in agriculture [29]. Another interesting, potential application that has been poorly explored, is the possibility to use different EF species in a consortium and/or with other beneficial microbes. An example is given by the recent work of Varkey and colleagues which has proved that a consortium of rhizobacteria and fungal endophytes suppress the root-knot nematode in tomato [25]. Thus, the possibility to use EF as multiple biocontrol agents and the development of microbial consortia with synergistic beneficial effects on plant performance appears to be an interesting frontier that opens promising fields of research that deserve deeper investigations to better exploit the entire range of EF potential.

6. Conclusions

In this review, we summarized the results obtained so far with the artificial introduction of EF in tomato and the subsequent beneficial effects that were observed. The main benefits to tomato plants are attributable to the biocontrol of several insect pests and plant pathogens, as well as their ability to improve plant performance. A focus on naturally occurring, constitutive EF of tomato was also undertaken, aimed at emphasizing their possible role in the selection of new beneficial strains for future use in tomato crop improvement. Moreover, an overview was conducted on the methods of introduction and detection of EF in tomato, providing a clear synthesis of the techniques used, that could be a practical guide to other researchers approaching this interesting field of research. The potential applications of endophytic fungi in horticultural production provide many advantages to the agroecosystem in terms of reducing chemical use and establishing a biological equilibrium necessary for the establishment of sustainable agriculture.

Author Contributions: Conceptualization, M.S., M.R., and S.L.W.; literature investigation M.S., M.R., G.d., and L.G.; writing—original draft preparation, M.S. and M.R.; writing—review and editing, S.L.W., project administration, S.L.W.; funding acquisition, S.L.W. All authors have read and agreed to the published version of the manuscript.

Funding: This research was funded by the following projects: European Union Horizon 2020 Research and Innovation Program, ECOSTACK (grant agreement no. 773554); fundings and a Ph.D. bursary to M.R. in PRIN 2017 [grant number PROSPECT 2017JLN833].

Acknowledgments: Thanks are due to Nadia Lombardi of the University of Naples, Federico II, for her editorial assistance.

Conflicts of Interest: The authors declare no conflict of interest.

References

1. Khan, S.; Guo, L.; Maimaiti, Y.; Mijit, M.; Qiu, D. Entomopathogenic Fungi as Microbial Biocontrol Agent. *Mol. Plant. Breed.* **2012**, *3*, 63–79. [CrossRef]
2. Gouda, S.; Das, G.; Sen, S.K.; Shin, H.S.; Patra, J.K. Endophytes: A treasure house of bioactive compounds of medicinal importance. *Front. Microbiol.* **2016**, *7*, 1538. [CrossRef] [PubMed]
3. Bamisile, B.S.; Dash, C.K.; Akutse, K.S.; Keppanan, R.; Afolabi, O.G.; Hussain, M.; Qasim, M.; Wang, L. Prospects of endophytic fungal entomopathogens as biocontrol and plant growth promoting agents: An insight on how artificial inoculation methods affect endophytic colonization of host plants. *Microbiol. Res.* **2018**, *217*, 34–50. [CrossRef]
4. McKinnon, A.C.; Saari, S.; Moran-Diez, M.E.; Meyling, N.V.; Raad, M.; Glare, T.R. *Beauveria bassiana* as an endophyte: A critical review on associated methodology and biocontrol potential. *BioControl* **2016**, *62*, 1–17. [CrossRef]
5. Ownley, B.H.; Gwinn, K.D.; Vega, F.E. Endophytic fungal entomopathogens with activity against plant pathogens: Ecology and evolution. *BioControl* **2010**, *55*, 113–128. [CrossRef]
6. Saikkonen, K.; Saari, S.; Helander, M. Defensive mutualism between plants and endophytic fungi? *Fungal Divers.* **2010**, *41*, 101–113. [CrossRef]
7. Young, C.A.; Hume, D.E.; McCulley, R.L. Forages and Pastures Symposium: Fungal endophytes of tall fescue and perennial ryegrass: Pasture friend or foe? *J. Anim. Sci.* **2013**, *91*, 2379–2394. [CrossRef]
8. Hyde, K.D.; Soytong, K. The fungal endophyte dilemma. *Fungal Divers.* **2008**, *33*, 163–173.
9. Parthasarathy, R.; Chandrika, M.; Rao, H.C.; Kamalraj, S.; Jayabaskaran, C.; Pugazhendhi, A. Molecular profiling of marine endophytic fungi from green algae: Assessment of antibacterial and anticancer activities. *Process. Biochem.* **2020**, *96*, 11–20. [CrossRef]
10. Jia, Q.; Qu, J.; Mu, H.; Sun, H.; Wu, C. Foliar endophytic fungi: Diversity in species and functions in forest ecosystems. *Symbiosis* **2020**, *80*, 102–132. [CrossRef]
11. Stone, J.K.; Bacon, C.W.; White, J.F., Jr. An overview of endophytic microbes: Endophytism defined. In *Microbial Endophytes*; White, J.F., Jr., Bacon, C.W., Eds.; Marcel Dekker Inc.: New York, NY, USA, 2000; pp. 3–29.

12. Sherwood, M.; Carroll, G. Fungal succession on needles and young twigs of old-growth Douglas Fir. *Mycologia* **1974**, *66*, 499–506. [CrossRef]

13. Carroll, G. Fungal endophytes in stems and leaves: From latent pathogen to mutualistic symbiont. *Ecology* **1988**, *69*, 2–9. [CrossRef]

14. Stone, J.K.; Polishook, J.D.; White, J.F., Jr. Endophytic fungi. In *Biodiversity of Fungi, Inventoring and Monitoring Methods*; Mueller, G.M., Bills, G.F., Foster, M.S., Eds.; Elsevier: San Diego, CA, USA, 2004; pp. 241–270.

15. Krings, M.; Taylor, T.N.; Hass, H.; Kerp, H.; Dotzler, N.; Hermsen, E.J. Fungal endophytes in a 400-million-yr-old land plant: Infection pathways, spatial distribution, and host responses. *New Phytol.* **2007**, *174*, 648–657. [CrossRef] [PubMed]

16. Redecker, D.; Kodner, R.; Graham, L.E. Glomalean fungi from the Ordovician. *Science* **2000**, *289*, 1920–1921. [CrossRef] [PubMed]

17. Petrini, O. Fungal endophytes of tree leaves. In *Microbial Ecology of Leaves*; Andrews, J.H., Hirano, S.S., Eds.; Springer: New York, NY, USA, 1991; pp. 179–197.

18. Jia, M.; Chen, L.; Xin, H.; Zheng, C.; Rahman, K.; Han, T. A friendly relationship between endophytic fungi and medicinal Plants: A systematic review. *Front. Microbiol.* **2016**, *7*, 1–14. [CrossRef] [PubMed]

19. Kusari, S.; Hertweck, C.; Spiteller, M. Chemical ecology of endophytic fungi: Origins of secondary metabolites. *Chem. Biol.* **2012**, *19*, 792–798. [CrossRef]

20. Lo Presti, L.; Lanver, D.; Schweizer, G.; Tanaka, S.; Liang, L.; Tollot, M.; Zuccaro, A.; Reissmann, S.; Kahmann, R. Fungal effectors and plant susceptibility. *Annu. Rev. Plant. Biol.* **2015**, *66*, 513–545. [CrossRef]

21. Saikkonen, K.; Faeth, S.H.; Helander, M.; Sullivan, T.J. Fungal endophytes: A continuum of interactions with host plants. *Annu. Rev. Ecol. Syst.* **1998**, *29*, 319–343. [CrossRef]

22. Wani, Z.A.; Ashraf, N.; Mohiuddin, T.; Riyaz-Ul-Hassan, S. Plant-endophyte symbiosis, an ecological perspective. *Appl. Microbiol. Biotechnol.* **2015**, *99*, 2955–2965. [CrossRef]

23. Resquín-Romero, G.; Garrido-Jurado, I.; Delso, C.; Ríos-Moreno, A.; Quesada- Moraga, E. Transient endophytic colonizations of plants improve the outcome of foliar applications of mycoinsecticides against chewing insects. *J. Invertebr. Pathol.* **2016**, *136*, 3–31. [CrossRef]

24. Kim, H.Y.; Choi, G.J.; Lee, H.B.; Lee, S.W.; Lim, H.K.; Jang, K.S.; Son, S.W.; Lee, S.O.; Cho, K.Y.; Sung, N.D.; et al. Some fungal endophytes from vegetable crops and their anti-oomycete activities against tomato late blight. *Lett. Appl. Microbiol.* **2007**, *44*, 332–337. [CrossRef] [PubMed]

25. Varkey, S.; Anith, K.N.; Narayana, R.; Aswini, S. A consortium of rhizobacteria and fungal endophyte suppress the root-knot nematode parasite in tomato. *Rhizosphere* **2018**, *5*, 38–42. [CrossRef]

26. Segaran, G.; Sathiavelu, M. Fungal endophytes: A potent biocontrol agent and a bioactive metabolites reservoir. *Biocatal. Agric. Biotechnol.* **2019**, *21*, 101284. [CrossRef]

27. Kumar, S.; Kaushik, N. Metabolites of endophytic fungi as novel source of biofungicide: A review. *Phytochem. Rev.* **2012**, *11*, 507–522. [CrossRef]

28. Bamisile, B.S.; Dash, C.K.; Akutse, K.S.; Keppanan, R.; Wang, L. Fungal endophytes: Beyond herbivore management. *Front. Microbiol.* **2018**, *9*, 1–11. [CrossRef]

29. Jaber, L.R.; Ownley, B.H. Can we use entomopathogenic fungi as endophytes for dual biological control of insect pests and plant pathogens? *Biol. Control* **2018**, *116*, 36–45. [CrossRef]

30. Glare, T.; Caradus, J.; Gelernter, W.; Jackson, T.; Keyhani, N.; Köhl, J.; Marrone, P.; Morin, L.; Stewart, A. Have biopesticides come of age? *Trends Biotechnol.* **2012**, *30*, 250–258. [CrossRef]

31. Knapp, S.; Peralta, I.E. The Tomato (*Solanum lycopersicum* L., Solanaceae) and its botanical relatives. In *The Tomato Genome, Compendium of Plant Genomes*; Causse, M., Giovannoni, J., Bouzayen, M., Zouine, M., Eds.; Springer: Berlin/Heidelberg, Germany, 2016; pp. 7–21.

32. Bogner, C.W.; Kariuki, G.M.; Elashry, A.; Sichtermann, G.; Buch, A.K.; Mishra, B.; Thines, M.; Grundler, F.M.W.; Schouten, A. Fungal root endophytes of tomato from Kenya and their nematode biocontrol potential. *Mycol. Prog.* **2016**, *15*, 30. [CrossRef]

33. Quesada-Moraga, E.; Muñoz-Ledesma, F.; Santiago-Alvarez, C. Systemic protection of *Papaver somniferum* L. against *Iraella luteipes* (Hymenoptera: Cynipidae) by an endophytic strain of *Beauveria bassiana* (Ascomycota: Hypocreales). *Environ. Entomol* **2009**, *38*, 723–730. [CrossRef]

34. Gurulingappa, P.; McGee, P.A.; Sword, G. Endophytic *Lecanicillium lecanii* and *Beauveria bassiana* reduce the survival and fecundity of *Aphis gossypii* following contact with conidia and secondary metabolites. *Crop. Prot.* **2011**, *30*, 349–353. [CrossRef]

35. Sánchez-Rodríguez, A.R.; Raya-Díaz, S.; Zamarreño, Á.M.; García-Mina, J.M.; Del Campillo, M.C.; Quesada-Moraga, E. An endophytic *Beauveria bassiana* strain increases spike production in bread and durum wheat plants and effectively controls cotton leafworm (*Spodoptera littoralis*) larvae. *Biol. Control* **2018**, *116*, 90–102. [CrossRef]

36. Akello, J.; Sikora, R. Systemic acropedal influence of endophyte seed treatment on *Acyrthosiphon pisum* and *Aphis fabae* offspring development and reproductive fitness. *Biol. Control* **2012**, *61*, 215–221. [CrossRef]

37. Akutse, K.S.; Maniania, N.K.; Fiaboe, K.K.M.; Van den Berg, J.; Ekesi, S. Endophytic colonization of *Vicia faba* and *Phaseolus vulgaris* (Fabaceae) by fungal pathogens and their effects on the life-history parameters of *Liriomyza huidobrensis* (Diptera: Agromyzidae). *Fungal Ecol.* **2013**, *6*, 293–301. [CrossRef]

38. McGee, P. Reduced growth and deterrence from feeding of the insect pest *Helicoverpa armigera* associated with fungal endophytes from cotton. *Aust. J. Exp. Agric.* **2002**, *42*, 995–999. [CrossRef]

39. Vega, F.E. Insect pathology and fungal endophytes. *J. Invertebr. Pathol.* **2008**, *98*, 277–279. [CrossRef]

40. Lacey, L.A.; Neven, L.G. The potential of the fungus, *Muscodor albus*, as a microbial control agent of potato tuber moth (Lepidoptera: Gelechiidae) in stored potatoes. *J. Invertebr. Pathol.* **2006**, *91*, 195–198. [CrossRef] [PubMed]

41. Martinuz, A.; Schouten, A.; Sikora, R. Systemically induced resistance and microbial competitive exclusion: Implications on biological control. *Phytopathology* **2012**, *102*, 260–266. [CrossRef] [PubMed]

42. Barra-Bucarei, L.; Gerding, M. Antifungal Activity of *Beauveria bassiana* Endophyte against *Botrytis cinerea* in Two Solanaceae Crops. *Microorganisms* **2020**, *8*, 65. [CrossRef] [PubMed]

43. Menjivar, R.D.; Cabrera, J.A.; Kranz, J.; Sikora, R.A. Induction of metabolite organic compounds by mutualistic endophytic fungi to reduce the greenhouse whitefly *Trialeurodes vaporariorum* (Westwood) infection on tomato. *Plant. Soil* **2012**, *352*, 233–241. [CrossRef]

44. Gao, F.; Dai, C.; Liu, X. Mechanisms of fungal endophytes in plant protection against pathogens. *Afr. J. Microbiol. Res.* **2010**, *4*, 1346–1351.

45. Moy, M.; Belanger, F.; Duncan, R.; Freehoff, A.; Leary, C.; Meyer, W.; Sullivan, R.; White, J.F. JR. Identification of epiphyllous mycelial nets on leaves of grasses infected by Clavicipitaceous endophytes. *Symbiosis* **2000**, *28*, 291–302.

46. Conrath, U.; Beckers, G.J.M.; Flors, V.; Garcia-Agustin, P.; Jakab, G.; Mauch, F.; Newman, M.A.; Pieterse, C.M.J.; Poinssot, B.; Pozo, M.J.; et al. Priming: Getting ready for battle. *Mol. Plant Microbe Interact.* **2006**, *19*, 1062–1071. [CrossRef] [PubMed]

47. Dicke, M.; Van Loon, J.J.A.; Soler, R. Chemical complexity of volatiles from plants induced by multiple attack. *Nat. Chem. Biol.* **2009**, *5*, 317–324. [CrossRef] [PubMed]

48. Pieterse, C.M.J.; Poelman, E.H.; Van Wees, S.C.M.; Dicke, M. Induced plant responses to microbes and insects. *Front. Plant. Sci.* **2013**, *4*, 475. [CrossRef] [PubMed]

49. Poelman, E.H.; Bruinsma, M.; Zhu, F.; Weldegergis, B.T.; Boursault, A.E.; Jongema, Y.; van Loon, J.J.; Vet, L.E.; Harvey, J.A.; Dicke, M. Hyperparasitoids use herbivore-induced plant volatiles to locate their parasitoid host. *PLoS Biol.* **2012**, *10*, e1001435. [CrossRef]

50. Thakur, A.; Kaur, S.; Kaur, A.; Singh, V. Enhanced resistance to *Spodoptera litura* in endophyte infected cauliflower plants. *Environ. Entomol.* **2013**, *42*, 240–246. [CrossRef]

51. Manganiello, G.; Sacco, A.; Ercolano, M.R.; Vinale, F.; Lanzuise, S.; Pascale, A.; Napolitano, M.; Lombardi, N.; Lorito, M.; Woo, S.L. Modulation of tomato response to *Rhizoctonia solani* by *Trichoderma harzianum* and its secondary metabolite harzianic acid. *Front. Microbiol.* **2018**, *9*, 1966. [CrossRef]

52. Rodriguez, R.J.; White, J.F., Jr.; Arnold, A.E.; Redman, R.S. Fungal endophytes: Diversity and functional roles. *New Phytol.* **2009**, *182*, 314–330. [CrossRef]

53. Castillo Lopez, D.; Sword, G.A. The endophytic fungal entomopathogens *Beauveria bassiana* and *Purpureocillium lilacinum* enhance the growth of cultivated cotton (*Gossypium hirsutum*) and negatively affect survival of the cotton bollworm (*Helicoverpa zea*). *Biol. Control.* **2015**, *89*, 53–60. [CrossRef]

54. Jaber, L.R.; Enkerli, J. Effect of seed treatment duration on growth and colonization of *Vicia faba* by endophytic *Beauveria bassiana* and *Metarhizium brunneum*. *Biol. Control* **2016**, *103*, 187–195. [CrossRef]

55. Behie, S.W.; Zelisko, P.M.; Bidochka, M.J. Endophytic insect-parasitic fungi translocate nitrogen directly from insects to plants. *Science* **2012**, *336*, 1576–1577. [CrossRef]

56. Behie, S.W.; Bidochka, M.J. Nutrient transfer in plant–fungal symbioses. *Trends Plant. Sci.* **2014**, *19*, 734–740. [CrossRef]

57. Nicoletti, R.; Becchimanzi, A. Endophytism of *Lecanicillium* and *Akanthomyces*. *Agriculture* **2020**, *10*, 205. [CrossRef]

58. Bills, G.F.; Polishook, J.D. Microfungi from *Carpinus caroliniana*. *Can. J. Bot.* **1991**, *69*, 1477–1482. [CrossRef]

59. Cherry, A.J.; Lomer, C.J.; Djegui, D.; Schulthess, F. Pathogen incidence and their potential as microbial control agents in IPM of maize stem borers in West Africa. *BioControl* **1999**, *44*, 301–327. [CrossRef]

60. Pimentel, I.C.; Glienke-Blanco, C.; Gabardo, J.; Stuart, R.M.; Azevedo, J.L. Identification and colonization of endophytic fungi from soybean (*Glycine max* (L.) Merril) under different environmental conditions. *Braz. Arch. Biol. Technol* **2006**, *49*, 705–711. [CrossRef]

61. Orole, O.O.; Adejumo, T.O. Activity of fungal endophytes against four maize wilt pathogens. *Afr. J. Microbiol. Res.* **2009**, *3*, 969–973.

62. Fuller-Schaefer, C.; Jung, K.; Jaronski, S. Colonization of sugarbeet roots by entomopathogenic fungi. In Proceedings of the 38th Annual Meeting of the Society for Invertebrate Pathology, Anchorage, AK, USA, 7–11 August 2005; Volume 49.

63. Greenfield, M.; Gómez-Jiménez, M.I.; Ortiz, V.; Vega, F.E.; Kramer, M.; Parsa, S. *Beauveria bassiana* and *Metarhizium anisopliae* endophytically colonize cassava roots following soil drench inoculation. *Biol. Control* **2016**, *95*, 40–48. [CrossRef]

64. Bing, L.A.; Lewis, L.C. Suppression of *Ostrinia nubilalis* (Hübner) (Lepidoptera: Pyralidae) by endophytic *Beauveria bassiana* (Balsamo) Vuillemin. *Environ. Entomol.* **1991**, *20*, 1207–1211. [CrossRef]

65. Bing, L.A.; Lewis, L.C. Endophytic *Beauveria bassiana* (Balsamo) Vuillemin in corn: The influence of the plant growth stage and *Ostrinia nubilalis* (Hübner). *Biocontrol Sci. Technol.* **1992**, *2*, 39–47. [CrossRef]

66. Wagner, B.L.; Lewis, L.C. Colonization of corn, *Zea mays*, by the entomopathogenic fungus *Beauveria bassiana*. *Appl. Environ. Microbiol.* **2000**, *66*, 3468–3473. [CrossRef]

67. Parsa, S.; Ortiz, V.; Vega, F.E. Establishing fungal entomopathogens as endophytes: Towards endophytic biological control. *J. Vis. Exp.* **2013**, *74*, 50360. [CrossRef]

68. Russo, M.L.; Pelizza, S.A.; Cabello, M.N.; Stenglein, S.A.; Scorsetti, A.C. Endophytic colonisation of tobacco, corn, wheat and soybeans by the fungal entomopathogen *Beauveria bassiana* (Ascomycota, Hypocreales). *Biocontrol Sci. Technol.* **2015**, *25*, 475–480. [CrossRef]

69. Prema Sundara Valli, P.; Muthukumar, T. Dark Septate Root Endophytic Fungus *Nectria haematococca* Improves Tomato Growth Under Water Limiting Conditions. *Indian J. Microbiol.* **2018**, *58*, 489–495. [CrossRef]

70. Vergara, C.; Araujo, K.E.C.; Urquiaga, S.; Schultz, N.; de Balieiro, F.C.; Medeiros, P.S.; Santos, L.A.; Xavier, G.R.; Zilli, J.E. Dark Septate endophytic fungi help tomato to acquire nutrients from ground plant material. *Front. Microbiol.* **2017**, *8*, 1–12. [CrossRef]

71. Yakti, W.; Kovács, G.M.; Vági, P.; Franken, P. Impact of dark septate endophytes on tomato growth and nutrient uptake. *Plant. Ecol. Divers.* **2018**, *11*, 637–648. [CrossRef]

72. Gurulingappa, P.; Sword, G.A.; Murdoch, G.; McGee, P.A. Colonization of crop plants by fungal entomopathogens and their effects on two insect pests when in planta. *Biol. Control* **2010**, *55*, 34–41. [CrossRef]

73. El-Deeb, H.M.; Lashin, S.M.; Arab, Y.A.S. Reaction of some tomato cultivars to tomato leaf curl virus and evaluation of the endophytic colonisation with *Beauveria bassiana* on the disease incidence and its vector, *Bemisia tabaci*. *Arch. Phytopathol. Plant. Prot.* **2012**, *45*, 1538–1545. [CrossRef]

74. Wei, Q.Y.; Li, Y.Y.; Xu, C.; Wu, Y.X.; Zhang, Y.R.; Liu, H. Endophytic colonization by *Beauveria bassiana* increases the resistance of tomatoes against *Bemisia tabaci*. *Arthropod. Plant. Interact.* **2020**, *14*, 289–300. [CrossRef]

75. Qayyum, M.A.; Wakil, W.; Arif, M.J.; Sahi, S.T.; Dunlap, C.A. Infection of *Helicoverpa armigera* by endophytic *Beauveria bassiana* colonizing tomato plants. *Biol. Control* **2015**, *90*, 200–207. [CrossRef]

76. Prabhukarthikeyan, S.R.; Keerthana, U.; Archana, S.; Raguchander, T. Induced resistance in tomato plants to *Helicoverpa armigera* by mixed formulation of bacillus subtilis and *Beauveria bassiana*. *Res. J. Biotechnol.* **2017**, *12*, 53–59.

77. Garantonakis, N.; Pappas, M.L.; Varikou, K.; Skiada, V.; Broufas, G.D.; Kavroulakis, N.; Papadopoulou, K.K. Tomato inoculation with the endophytic strain *Fusarium solani* K results in reduced feeding damage by the zoophytophagous predator *Nesidiocoris tenuis*. *Front. Ecol. Evol.* **2018**, *6*, 1–7. [CrossRef]

78. Shrivastava, G.; Ownley, B.H.; Augé, R.M.; Toler, H.; Dee, M.; Vu, A.; Köllner, T.G.; Chen, F. Colonization by arbuscular mycorrhizal and endophytic fungi enhanced terpene production in tomato plants and their defense against a herbivorous insect. *Symbiosis* **2015**, *65*, 65–74. [CrossRef]

79. Allegrucci, N.; Velazquez, M.S.; Russo, M.L.; Perez, E.; Scorsetti, A.C. Endophytic colonisation of tomato by the entomopathogenic fungus *Beauveria bassiana*: The use of different inoculation techniques and their effects on the tomato leafminer *Tuta absoluta* (Lepidoptera: Gelechiidae). *J. Plant. Prot. Res.* **2017**, *57*, 205–211. [CrossRef]

80. Klieber, J.; Reineke, A. The entomopathogenic *Beauveria bassiana* has epiphytic and endophytic activity against the tomato leafminer *Tuta absoluta*. *J. Appl. Entomol.* **2016**, *140*, 580–589. [CrossRef]

81. Dababat, A.E.F.A.; Sikora, R.A. Induced resistance by the mutualistic endophyte, *Fusarium oxysporum* strain 162, toward *Meloidogyne incognita* on tomato. *Biocontrol Sci. Technol.* **2007**, *17*, 969–975. [CrossRef]

82. Martinuz, A.; Schouten, A.; Sikora, R.A. Post-infection development of *Meloidogyne incognita* on tomato treated with the endophytes *Fusarium oxysporum* strain Fo162 and *Rhizobium etli* strain G12. *BioControl* **2013**, *58*, 95–104. [CrossRef]

83. Escudero, N.; Lopez-Llorca, L.V. Effects on plant growth and root-knot nematode infection of an endophytic GFP transformant of the nematophagous fungus *Pochonia chlamydosporia*. *Symbiosis* **2012**, *57*, 33–42. [CrossRef]

84. Bargmann, C.; Schönbeck, F. *Acremonium kiliense* as inducer of resistance to wilt diseases on tomatoes. *J. Plant. Dis. Prot.* **1992**, *99*, 266–272.

85. Culebro-Ricaldi, J.M.; Ruíz-Valdiviezo, V.M.; Rodríguez-Mendiola, M.A.; Ávila-Miranda, M.E.; Gutiérrez-Miceli, F.A.; Cruz-Rodríguez, R.I.; Dendooven, L.; Montes-Molina, J.A. Antifungal properties of *Beauveria bassiana* strains against *Fusarium oxysporum* f. sp. *lycopersici* race 3 in tomato crop. *J. Environ. Biol.* **2017**, *38*, 821–827. [CrossRef]

86. Aimé, S.; Alabouvette, C.; Steinberg, C.; Olivain, C. The endophytic strain *Fusarium oxysporum* Fo47: A good candidate for priming the defense responses in tomato roots. *Mol. Plant Microbe Interact.* **2013**, *26*, 918–926. [CrossRef] [PubMed]

87. Kavroulakis, N.; Ntougias, S.; Zervakis, G.I.; Ehaliotis, C.; Haralampidis, K.; Papadopoulou, K.K. Role of ethylene in the protection of tomato plants against soil-borne fungal pathogens conferred by an endophytic *Fusarium solani* strain. *J. Exp. Bot.* **2007**, *58*, 3853–3864. [CrossRef] [PubMed]

88. Nefzi, A.; Abdallah, R.A.B.; Jabnoun-Khiareddine, H.; Ammar, N.; Daami-Remadi, M. Ability of endophytic fungi associated with *Withania somnifera* L. to control *Fusarium* Crown and Root Rot and to promote growth in tomato. *Braz. J. Microbiol.* **2019**, *50*, 481–494. [CrossRef] [PubMed]

89. Fakhro, A.; Andrade-Linares, D.R.; von Bargen, S.; Bandte, M.; Büttner, C.; Grosch, R.; Schwarz, D.; Franken, P. Impact of *Piriformospora indica* on tomato growth and on interaction with fungal and viral pathogens. *Mycorrhiza* **2010**, *20*, 191–200. [CrossRef]

90. Sarma, M.V.R.K.; Kumar, V.; Saharan, K.; Srivastava, R.; Sharma, A.K.; Prakash, A.; Sahai, V.; Bisaria, V.S. Application of inorganic carrier-based formulations of fluorescent pseudomonads and *Piriformospora indica* on tomato plants and evaluation of their efficacy. *J. Appl. Microbiol.* **2011**, *111*, 456–466. [CrossRef] [PubMed]

91. Qiang, X.; Weiss, M.; Kogel, K.H.; Schäfer, P. *Piriformospora indica* a mutualistic basidiomycete with an exceptionally large plant host range. *Mol. Plant. Pathol.* **2012**, *13*, 508–518. [CrossRef] [PubMed]

92. Benhamou, N.; Rey, P.; Chérif, M.; Hockenhull, J.; Tirilly, Y. Treatment with the mycoparasite *Pythium oligandrum* triggers induction of defense-related reactions in tomato roots when challenged with *Fusarium oxysporum* f. sp. *radicis-lycopersici*. *Phytopathology* **1997**, *87*, 108–122. [CrossRef]

93. Azadi, N.; Shirzad, A.; Mohammadi, H. A study of some biocontrol mechanisms of *Beauveria bassiana* against *Rhizoctonia* disease on tomato. *Acta Biol. Szeged.* **2016**, *60*, 119–127.

94. Tucci, M.; Ruocco, M.; De Masi, L.; De Palma, M.; Lorito, M. The beneficial effect of *Trichoderma* spp. on tomato is modulated by the plant genotype. *Mol. Plant. Pathol.* **2011**, *12*, 341–354. [CrossRef]

95. Durrant, W.E.; Dong, X. Systemic acquired resistance. *Ann. Rev. Phytopathol.* **2004**, *42*, 185–209. [CrossRef]

96. Conrath, U.; Beckers, G.J.; Langenbach, C.J.; Jaskiewicz, M.R. Priming for enhanced defense. *Ann. Rev. Phytopathol.* **2015**, *53*, 97–119. [CrossRef]

97. Vos, C.M.; Yang, Y.; De Coninck, B.; Cammue, B.P.A. Fungal (-like) biocontrol organisms in tomato disease control. *Biol. Control* **2014**, *74*, 65–81. [CrossRef]

98. Alfano, G.; Lewis Ivey, M.L.; Cakir, C.; Bos, J.I.B.; Miller, S.A.; Madden, L.V.; Kamoun, S.; Hoitink, H.A.J. Systemic modulation of gene expression in tomato by *Trichoderma hamatum* 382. *Phytopathology* **2007**, *97*, 429–437. [CrossRef]

99. Martínez-Medina, A.; Fernandez, I.; Lok, G.B.; Pozo, M.J.; Pieterse, C.M.J.; Van Wees, S.C.M. Shifting from priming of salicylic acid- to jasmonic acid-regulated defences by *Trichoderma* protects tomato against the root knot nematode *Meloidogyne incognita*. *New Phytol.* **2017**, *213*, 1363–1377. [CrossRef] [PubMed]

100. García, J.E.; Beatriz, P.J.; Alejandro, P.; Roberto, L.E. *Metarhizium anisopliae* (Metschnikoff) Sorokin promotes growth and has endophytic activity in tomato plants. *Adv. Biol. Res.* **2011**, *5*, 22–27.

101. Khan, A.L.; Waqas, M.; Hussain, J.; Al-Harrasi, A.; Hamayun, M.; Lee, I.J. Phytohormones enabled endophytic fungal symbiosis improve aluminum phytoextraction in tolerant *Solanum lycopersicum*: An examples of *Penicillium janthinellum* LK5 and comparison with exogenous GA$_3$. *J. Hazard. Mater.* **2015**, *295*, 70–78. [CrossRef] [PubMed]

102. Zavala-Gonzalez, E.A.; Escudero, N.; Lopez-Moya, F.; Aranda-Martinez, A.; Exposito, A.; Ricaño-Rodríguez, J.; Naranjo-Ortiz, M.A.; Ramírez-Lepe, M.; Lopez-Llorca, L.V. Some isolates of the nematophagous fungus *Pochonia chlamydosporia* promote root growth and reduce flowering time of tomato. *Ann. Appl. Biol.* **2015**, *166*, 472–483. [CrossRef]

103. Sánchez-Rodríguez, A.R.; Del Campillo, M.C.; Quesada-Moraga, E. *Beauveria bassiana*: An entomopathogenic fungus alleviates Fe chlorosis symptoms in plants grown on calcareous substrates. *Sci. Hortic.* **2015**, *197*, 193–202. [CrossRef]

104. Khan, A.L.; Waqas, M.; Khan, A.R.; Hussain, J.; Kang, S.M.; Gilani, S.A.; Hamayun, M.; Shin, J.H.; Kamran, M.; Al-Harrasi, A.; et al. Fungal endophyte *Penicillium janthinellum* LK5 improves growth of ABA-deficient tomato under salinity. *World J. Microbiol. Biotechnol.* **2013**, *29*, 2133–2144. [CrossRef]

105. Vidal, S. Changes in suitability of tomato for whiteflies mediated by a non-pathogenic endophytic fungus. *Entomol. Exp. Appl.* **1996**, *80*, 272–274. [CrossRef]

106. Prabhukarthikeyan, R.; Saravanakumar, D.; Raguchander, T. Combination of endophytic *Bacillus* and *Beauveria* for the management of *Fusarium* wilt and fruit borer in tomato. *Pest. Manag. Sci.* **2014**, *70*, 1742–1750. [CrossRef] [PubMed]

107. Pappas, M.L.; Liapoura, M.; Papantoniou, D.; Avramidou, M.; Kavroulakis, N.; Weinhold, A.; Broufas, G.D.; Papadopoulou, K.K. The beneficial endophytic fungus *Fusarium solani* strain K alters tomato responses against spider mites to the benefit of the plant. *Front. Plant. Sci.* **2018**, *9*, 1–17. [CrossRef] [PubMed]

108. Kost, G.; Rexer, K. Morphology and Ultrastructure of *Piriformospora indica*. *Soils Biol.* **2013**, *33*, 25–36. [CrossRef]

109. Wang, H.; Zheng, J.; Ren, X.; Yu, T.; Varma, A.; Lou, B.; Zheng, X. Effects of *Piriformospora indica* on the growth, fruit quality and interaction with Tomato yellow leaf curl virus in tomato cultivars susceptible and resistant to TYCLV. *Plant. Growth Regul.* **2015**, *76*, 303–313. [CrossRef]

110. Masunaka, A.; Nakaho, K.; Sakai, M.; Takahashi, H.; Takenaka, S. Visualization of *Ralstonia solanacearum* cells during biocontrol of bacterial wilt disease in tomato with *Pythium oligandrum*. *J. Gen. Plant Pathol.* **2009**, *75*, 281–287. [CrossRef]

111. Le Floch, G.; Vallance, J.; Benhamou, N.; Rey, P. Combining the oomycete *Pythium oligandrum* with two other antagonistic fungi: Root relationships and tomato grey mold biocontrol. *Biol. Control* **2009**, *50*, 288–298. [CrossRef]

112. Chowdappa, P.; Mohan Kumar, S.P.; Jyothi Lakshmi, M.; Upreti, K.K. Growth stimulation and induction of systemic resistance in tomato against early and late blight by *Bacillus subtilis* OTPB1 or *Trichoderma harzianum* OTPB3. *Biol. Control* **2013**, *65*, 109–117. [CrossRef]

113. Powell, W.A. Potential of *Beauveria Bassiana* 11–98 as a Biological Control Agent against Tomato Pests; and Detection of the Mycotoxic Metabolite Beauvericin in Tomato Plants Using HPLC. Master's Thesis, University of Tennessee, Knoxville, TN, USA, 2005.

114. Posada, F.; Aime, M.C.; Peterson, S.W.; Rehner, S.A.; Vega, F.E. Inoculation of coffee plants with the fungal entomopathogen *Beauveria bassiana* (Ascomycota: Hypocreales). *Mycol. Res.* **2007**, *111*, 748–757. [CrossRef]

115. Biswas, C.; Dey, P.; Satpathy, S.; Satya, P.; Mahapatra, B. Endophytic colonization of white jute (*Corchorus capsularis*) plants by different *Beauveria bassiana* strains for managing stem weevil (*Apion corchori*). *Phytoparasitica* **2013**, *41*, 17–21. [CrossRef]

116. Landa, B.B.; López-Díaz, C.; Jiménez-Fernández, D.; Montes-Borrego, M.; Muñoz-Ledesma, F.J.; Ortiz-Urquiza, A.; Quesada-Moraga, E. In-planta detection and monitorization of endophytic colonization by a *Beauveria bassiana* strain using a new-developed nested and quantitative PCR-based assay and confocal laser scanning microscopy. *J. Invertebr. Pathol.* **2013**, *114*, 128–138. [CrossRef]

117. Renuka, S.; Ramanujam, B.; Poornesha, B. Endophytic ability of different isolates of entomopathogenic fungi *Beauveria bassiana* (Balsamo) Vuillemin in stem and leaf tissues of maize (*Zea mays* L.). *Indian J. Microbiol* **2016**, *56*, 126–133. [CrossRef] [PubMed]

118. Garrido-Jurado, I.; Resquín-Romero, G.; Amarilla, S.P.; Ríos- Moreno, A.; Carrasco, L.; Quesada-Moraga, E. Transient endophytic colonization of melon plants by entomopathogenic fungi after foliar application for the control of *Bemisia tabaci* Gennadius (Hemiptera: Aleyrodidae). *J. Pest. Sci.* **2017**, *90*, 319–330. [CrossRef]

119. Rondot, Y.; Reineke, A. Endophytic *Beauveria bassiana* in grapevine *Vitis vinifera* (L.) reduces infestation with piercing-sucking insects. *Biol. Control* **2018**, *116*, 82–89. [CrossRef]

120. Nishi, O.; Sushida, H.; Higashi, Y.; Iida, Y. Epiphytic and endophytic colonisation of tomato plants by the entomopathogenic fungus *Beauveria bassiana* strain GHA. *Mycology* **2020**, *11*, 1–9. [CrossRef]

121. Krell, V.; Jakobs-Schoenwandt, D.; Vidal, S.; Patel, A.V. Encapsulation of *Metarhizium brunneum* enhances endophytism in tomato plants. *Biol. Control* **2018**, *116*, 62–73. [CrossRef]

122. Jallow, M.F.A.; Dugassa-Gobena, D.; Vidal, S. Influence of an endophytic fungus on host plant selection by a polyphagous moth via volatile spectrum changes. *Arthropod. Plant. Interact.* **2008**, *2*, 53–62. [CrossRef]

123. Omukoko, C.A. Biocontrol Mechanisms of Endophytic *Beauveria bassiana* in Three Tomato (*Lycopersum esculentum*) Varieties C. *World Dev.* **2018**, *1*, 43–52. [CrossRef]

124. Omukoko, C.A.; Turoop, L. Colonization of Tomato Varieties by *Beauveria bassiana* Isolates in the Screen House. *Int. J. Sci. Res.* **2017**, *6*, 1024–1028. [CrossRef]

125. Mahmoud, R.S.; Narisawa, K. A new fungal endophyte, *Scolecobasidium humicola*, promotes tomato growth under organic nitrogen. *PLoS ONE* **2013**, *8*, 1–8. [CrossRef]

126. Rodriguez, R.; Redman, R. More than 400 million years of evolution and some plants still can't make it on their own: Plant stress tolerance via fungal symbiosis. *J. Exp. Bot.* **2008**, *59*, 1109–1114. [CrossRef]

127. Fritz, M.; Jakobsen, I.; Lyngkjær, M.F.; Thordal-Christensen, H.; Pons-Kühnemann, J. Arbuscular mycorrhiza reduces susceptibility of tomato to *Alternaria solani*. *Mycorrhiza* **2006**, *16*, 413–419. [CrossRef] [PubMed]

128. Nasehi, A.; Kadir, J.; Nasr-Esfahani, M.; Abed-Ashtiani, F.; Golkhandan, E.; Ashkani, S. Identification of the New Pathogen (*Stemphylium lycopersici*) Causing Leaf Spot on Pepino (*Solanum muricatum*). *J. Phytopathol.* **2016**, *164*, 421–426. [CrossRef]

129. Gilardi, G.; Matic, S.; Luongo, I.; Gullino, M.L.; Garibaldi, A. First Report of Stem Necrosis and Leaf Spot of Tomato Caused by *Albifimbria verrucaria* in Italy. *Plant. Dis.* **2020**, *4*, 2026. [CrossRef]

130. Rameshkumar, G.; Sikha, M.; Ponlakshmi, M.; Selva Pandiyan, A.; Lalitha, P. A rare case of *Myrothecium* species causing mycotic keratitis: Diagnosis and management. *Med. Mycol. Case Rep.* **2019**, *25*, 53–55. [CrossRef]

131. Tian, X.; Yao, Y.; Chen, G.; Mao, Z.; Wang, X.; Xie, B. Suppression of *Meloidogyne incognita* by the endophytic fungus *Acremonium implicatum* from tomato root galls. *Int. J. Pest. Manag.* **2014**, *60*, 239–245. [CrossRef]

132. Xia, Y.; Sahib, M.R.; Amna, A.; Opiyo, S.O.; Zhao, Z.; Gao, Y.G. Culturable endophytic fungal communities associated with plants in organic and conventional farming systems and their effects on plant growth. *Sci. Rep.* **2019**, *9*, 1–10. [CrossRef]

133. Poli, A.; Lazzari, A.; Prigione, V.; Voyron, S.; Spadaro, D.; Varese, G.C. Influence of plant genotype on the cultivable fungi associated to tomato rhizosphere and roots in different soils. *Fungal Biol.* **2016**, *120*, 862–872. [CrossRef]

134. Demers, J.E.; Gugino, B.K.; del Mar Jiménez-Gasco, M. Highly diverse endophytic and soil *Fusarium oxysporum* populations associated with field-grown tomato plants. *Appl. Environ. Microbiol.* **2015**, *81*, 81–90. [CrossRef]

135. Imazaki, I.; Kadota, I. Molecular phylogeny and diversity of *Fusarium* endophytes isolated from tomato stems. *FEMS Microbiol. Ecol.* **2015**, *91*, 1–16. [CrossRef]

136. Dreyfuss, M.M.; Chapela, I.H. Potential of fungi in the discovery of novel, low-molecular weight pharmaceuticals. *Biotechnology* **1994**, *26*, 49–80. [CrossRef]

137. Hawksworth, D.L. The magnitude of fungal diversity: The 1.5 million species estimate revisited. *Mycol. Res.* **2001**, *105*, 1422–1432. [CrossRef]

138. Hawksworth, D.L.; Lücking, R. Fungal Diversity Revisited: 2.2 to 3.8 Million Species. *Microbiol Spectr.* **2017**, *5*, 79–95. [CrossRef]

139. Kuldau, G.A.; Yates, I.E. Evidence for *Fusarium* endophytes in cultivated and wild plants. In *Microbial Endophytes*; Bacon, C.W., White, J.F., Eds.; Marcel Dekker: New York, NY, USA, 2000; pp. 85–117.

140. Carroll, G.C. Beyond pest deterrence—Alternative strategies and hidden costs of endophytic mutualisms in vascular plants. In *Microbial Ecology of Leaves*; Andrews, J.H., Hirano, S.S., Eds.; Springer: New York, NY, USA, 1991; pp. 358–375.

141. Clay, K.; Schardl, C. Evolutionary origins and ecological consequences of endophyte symbiosis with grasses. *Am. Nat.* **2002**, *160*, S99–S127. [CrossRef] [PubMed]

142. Davitt, A.J.; Stansberry, M.; Rudgers, J.A. Do the costs and benefits of fungal endophyte symbiosis vary with light availability? *New Phytol.* **2010**, *188*, 824–834. [CrossRef]

143. Suryanarayanan, T.S. Endophyte research: Going beyond isolation and metabolite documentation. *Fungal Ecol.* **2013**, *6*, 561–568. [CrossRef]

144. Vega, F.E. The use of fungal entomopathogens as endophytes in biological control: A review. *Mycologia* **2018**, *110*, 4–30. [CrossRef]

145. De Silva, N.I.; Brooks, S.; Lumyong, S.; Hyde, K.D. Use of endophytes as biocontrol agents. *Fungal Biol. Rev.* **2019**, *33*, 133–148. [CrossRef]

Publisher's Note: MDPI stays neutral with regard to jurisdictional claims in published maps and institutional affiliations.

Review

A Survey of Endophytic Fungi Associated with High-Risk Plants Imported for Ornamental Purposes

Laura Gioia [1,*], Giada d'Errico [1,*], Martina Sinno [1], Marta Ranesi [1], Sheridan Lois Woo [2,3,4] and Francesco Vinale [4,5]

[1] Department of Agricultural Sciences, University of Naples Federico II, 80055 Portici, Italy; martina.sinno@unina.it (M.S.); marta.ranesi@unina.it (M.R.)
[2] Department of Pharmacy, University of Naples Federico II, 80131 Naples, Italy; woo@unina.it
[3] Task Force on Microbiome Studies, University of Naples Federico II, 80128 Naples, Italy
[4] National Research Council, Institute for Sustainable Plant Protection, 80055 Portici, Italy; frvinale@unina.it
[5] Department of Veterinary Medicine and Animal Productions, University of Naples Federico II, 80137 Naples, Italy
* Correspondence: laura.gioia@unina.it (L.G.); giada.derrico@unina.it (G.d.); Tel.: +39-2539344 (L.G. & G.d.)

Received: 31 October 2020; Accepted: 11 December 2020; Published: 17 December 2020

Abstract: An extensive literature search was performed to review current knowledge about endophytic fungi isolated from plants included in the European Food Safety Authority (EFSA) dossier. The selected genera of plants were *Acacia, Albizia, Bauhinia, Berberis, Caesalpinia, Cassia, Cornus, Hamamelis, Jasminus, Ligustrum, Lonicera, Nerium,* and *Robinia*. A total of 120 fungal genera have been found in plant tissues originating from several countries. *Bauhinia* and *Cornus* showed the highest diversity of endophytes, whereas *Hamamelis, Jasminus, Lonicera,* and *Robinia* exhibited the lowest. The most frequently detected fungi were *Aspergillus, Colletotrichum, Fusarium, Penicillium, Phyllosticta,* and *Alternaria*. Plants and plant products represent an inoculum source of several mutualistic or pathogenic fungi, including quarantine pathogens. Thus, the movement of living organisms across continents during international trade represents a serious threat to ecosystems and biosecurity measures should be taken at a global level.

Keywords: endophytic fungi; crop protection; *Acacia*; *Albizia*; *Bauhinia*; *Berberis*; *Caesalpinia*; *Cassia*; *Cornus*; *Hamamelis*; *Jasminus*; *Ligustrum*; *Lonicera*; *Nerium*; *Robinia*; EFSA; high-risk plants

---

1. Introduction

Endophytic fungi are ubiquitous to plants, and are mainly members of Ascomycota or their mitosporic stage, but they also include some taxa of Basidiomycota, Zygomycota, and Oomycota. Endophytes are organisms living within the tissues of plants [1] establishing stable relationships with their host, ranging from non-pathogenic to beneficial [2,3]. The endophytic fungi communities represent an enormous reserve of biodiversity and constitute a rich source of bioactive compounds used in agriculture [4,5]. For these reasons, they have attracted the attention of the scientific community worldwide. By definition, all or at least a significant part of the endophytic fungi life cycle occurs within the plant tissues without causing symptoms to their host [6–8]. A wide range of fungi, including pathogens and saprophytes, may be endophytes. Several pathogens live asymptomatically within plant tissues during their latency or quiescent stage, while some saprobes can also be facultative parasites [1,8,9]. Fungal endophytes are influenced by abiotic and biotic factors, occupying different habitats and locations during their life cycle phases. Even if host plants do not show any symptoms, they may represent a source of inoculum for other species [10–13]. Furthermore, changes in environmental conditions or species hosts may modify the fungal behavior, thus producing disease symptoms [8,11,14]. Large quantities of plants and plant material that are globally traded might contain asymptomatic infections of these fungi. It is generally accepted that the movement of plants and plant products by global trade and human activities is the most common way to introduce exotic pathogens and pests in non-endemic countries. Plant health is increasingly threatened by the introduction of emerging pests and/or pathogens [15,16]. Noticeable examples are represented by the invasion of alien plant pathogens into new areas [17–19]. Generally, biological invasions are the main threat to biodiversity [20], causing a decrease in species richness and diversity [20,21] or affecting local biological communities [22], as well as changing ecosystem processes [23–25].

In this scenario, the European Food Safety Authority (EFSA) Panel on Plant Health is responsible for the risk assessment, evaluations of risk reduction options, as well as guidance documents [26] in the domain of plant health for the European Union (EU) [26,27]. Commission Implementing Regulation (EU) [28] prohibits the importation of 35 so-called 'High-Risk Plants, plant products and other objects' from all third (non-EU) countries as long as no full risk assessment has been carried out. The EFSA Panel on Plant Health was requested to prepare and deliver risk assessments for these commodities [27,28], to evaluate whether the plant material will remain prohibited or removed from the list, with or without the application of additional measures [27,29]. The Commodity Risk Assessment has to be performed on the basis of technical dossiers provided by National Plant Protection Organizations of third countries. Information required for the preparation and submission of technical dossiers includes data on the pests potentially associated with the plant species or genera and on phytosanitary mitigation measures and inspections [30,31].

These plants have been identified as 'High-Risk Plants' by the EU since they 'host commonly hosted pests known to have a major impact on plant species which are of major economic, social or environmental importance to the Union' [28]. However, among these 35 plant genera, within the meaning of Art. 42 of Regulation (EU) 2016/2031, a list of only 13 taxa have been selected by the EFSA as plants mostly traded for ornamental purposes. According to this list, we have reviewed the following genera: *Acacia* Mill., *Albizia* Durazz., *Bauhinia* L., *Berberis* L., *Caesalpinia* L., *Cassia* L., *Cornus* L., *Hamamelis* L., *Jasminus* L., *Ligustrum* L., *Lonicera* L., *Nerium* L., and *Robinia* L. In this article, as much as possible, we highlight the potential risks associated with the movement of plants or materials among nations. Although other plant species may also have a significant impact, this review is limited to plants included in EU regulation [28] that do not originate within Europe. Thus, given these perspectives for future assessments, the present investigation offers an up-to-date snapshot of endophytic fungi associated with the so-called 'High-Risk Plants for ornamental purpose'. The aim is to facilitate the information required for technical dossiers, needed by the EFSA to perform the Commodity Risk Assessment of 13 plants mandated on an EU import list.

2. Endophytic Fungi Occurring in Selected Plants

Table 1 summarizes the abundance of endophytic fungi reported in association with High-Risk Plants for ornamental purposes. Herein, the number of endophytic species found in association with the examined plant genera has been taxonomically grouped by fungal genus. There are important differences in terms of fungi recovered per specific plant genus (SP) as well as in the frequency of a specific fungal genus (SF). These discrepancies could be explained by the different availability of literature data on these specific plants.

Table 1. Endophytic fungi isolated from *Acacia* (AC), *Albizia* (AL)., *Bauhinia* (BA), *Berberis* (BE), *Caesalpinia* (CP), *Cassia* (CS), *Cornus* (CO), *Hamamelis* (HA), *Jasminus* (JA), *Ligustrum* (LI), *Lonicera* (LO), *Nerium* (NE), *Robinia* (RO). Columns report the number of isolated fungal species. The total number of records calculated per fungal genus is indicated as Tot. SF. The total number of records per plant genera is indicated as Tot. SP. Fungal genera are sorted by alphabetic order.

Fungi Genera	Plant Genera													
	AC	AL	BA	BE	CP	CS	CO	HA	JA	LI	LO	NE	RO	Tot SF
Acremonium		1	3												4
Albifimbria			1												1
Alternaria	1		1	4	1		3			2		2		14
Anguillospora			1												1
Ascochyta							1								1
Ascotricha			2												2
Aspergillus	3	8	11	1	9	2	3					3		40
Aureobasidium	2						4								6
Bacillispora				1											1
Beauveria													1	1
Bipolaris		1			2							1		4
Botryosphaeria	1						2								3
Botrytis			1				1								2
Campylospora				1											1
Cercospora				1											1
Chaetomium	2		1									3		6
Chrysosporium					1										1
Cladosporium			4		1		5			1		3		14
Clonostachys			1							1			1	3
Cochliobolus	1		3									1		5
Colletotrichum	2	1	3	4	1		2	1	7	3		3		27
Coprinus						1									1
Cordyceps							1								1
Corynespora			1												1
Cryptodiaporthe							1								1
Cryptodiaporthe							1								1
Curvularia		1	5		2							2		10
Cylindrocarpon												1		1
Cyrptosporiopsis							1								1
Daldinia						1									1
Diaporthe		1	1	2			2						1	7
Didymella							2								2
Diplococcium			2												2
Diplodia	1									1				2
Discula							1								1
Dothiorella	6		2												8
Drechslera												1		1
Drepanopeziza							1								1
Elsinoe							1								1
Epicoccum					1		1								2
Eupenicillium	1														1
Eutiarosporella	1														1
Exserohilum			1												2
Fusarium	1	4	4	4	4		4			2	1	4	1	29
Fusidium			1												1
Geomyces												1		1
Geotrichum			1		1					1				3
Gibberella			2												2
Glomerella			1												1
Gloniopsis													1	1
Guignardia						1				1	1			3

Table 1. *Cont.*

Fungi Genera	AC	AL	BA	BE	CP	CS	CO	HA	JA	LI	LO	NE	RO	Tot SF
Heliscus				1										1
Helminthosporium						1	1							2
Hypoxylon						1								1
Khuskia			1											1
Kiflimonium			1											1
Lasiodiplodia	6	1	1		1					2		1		12
Lasmenia			2											2
Lecanicillium							1							1
Leptosphaerulina							1							1
Libertella										1				1
Lophiostoma							1							1
Microsphaeropsis				1										1
Moesziomyces	1												1	2
Myrmecridium			2											2
Myrothecium				1										3
Nectria					2									2
Nemania						1								1
Neocosmospora		1		1						1				3
Neofabraea							1							1
Neofusicoccum	6													6
Neonectria							2							2
Nigrospora			4			1	1	1				1		8
Nodulisporium			2			2								4
Oblongocollomyces	1													1
Paecilomyces		2												2
Papulospora						1								1
Paraboeremia			1											1
Paraphaeosphaeria	1			1										2
Penicillium	2	3	7		3	2	8			1		4		30
Periconia						1								1
Peroneutypa										1				1
Pestalotia	1		1											2
Pestalotiopsis			2				4			1				7
Peyronellaea	1													1
Pezicula								1						1
Phaeobotryosphaeria	1													1
Phoma	2		3				1					1		7
Phomopsis			3	1		2	3			3				12
Phyllosticta	1		1	1	1		1	1		1	1	1		9
Phytophthora							1							1
Pithomyces			1											1
Pleuroceras							1							1
Prathoda				1										1
Preussia	1													1
Psathyrella						1								1
Pseudopithomyces			1											1
Pseudothielavia												1		1
Puccinia				1										1
Pycnidiella										1				1
Rhizopus	1									1				2
Rosellinia		1												1
Sarocladium							1							1
Scedosporium			1											1
Sclerotinia							1							1
Scopulariopsis					1									1
Septoria							1							1

Table 1. *Cont.*

Fungi Genera	Plant Genera													
	AC	AL	BA	BE	CP	CS	CO	HA	JA	LI	LO	NE	RO	Tot SF
Simplicillium							1							1
Spegazzinia			2											2
Spencermartinsia	1													1
Sphaeria			1											1
Sporormiella			1											1
Stenella							1							1
Talaromyces		3			2		3							8
Thelioviopsis						1								1
Thelonectria							1							1
Torula												1		3
Trichoderma	1	1	2		6		1			2		1		14
Tubakia							2							2
Verticillium		1					1							2
Xylaria	1				2	1	1			2		1		8
Wickerhamomyces	1													1
Tot. SP	51	27	94	29	42	19	78	4	7	29	3	37	6	

2.1. Acacia

The *Acacia*, commonly known as wattle, belongs to the family Mimosaceae. The genus comprises more than 1350 species found throughout the world: almost 1000 are native of Australia, up to 140 species occur in Africa, 89 from Asia, and about 185 species are found in North and South America. Some Australian wattles are naturalized beyond their native range and have become invasive in many parts of Europe, South Africa, and Florida, especially in conservation areas [32]. Aboriginal communities use some *Acacia* species as sources of food and medicine. Australian acacias are widely used as wood products, ornamental plants, commercial cut flowers, and perfume crops [33].

Endophytic occurrence (Table 2) has been reported for 61 fungal isolates belonging to genera *Lasiodiplodia* (7 isolates), *Dothiorella* (8 isolates), *Neofusicoccum* (9 isolates), *Aspergillus* (3 isolates), *Chaetomium* (3 isolates), *Botryosphaeria* (1 isolate), *Colletotrichum* (2 isolates), *Aureobasidium* (2 isolates), *Spencermartinsia* (2 isolates), *Alternaria* (1 isolate), *Cochliobolus* (1 isolate), *Diplodia* (2 isolates), *Eupenicillium* (1 isolate), *Fusarium* (1 isolate), *Moesziomyces* (1 isolate), *Paraphaeosphaeria* (1 isolate), *Penicillium* (2 isolates), *Eutiarosporella* (2 isolates), *Pestalotia* (1 isolate), *Peyronellaea* (1 isolate), *Phaeobotryosphaeria* (1 isolate), *Phoma* (2 isolates), *Phyllosticta* (1 isolate), *Wickerhamomyces* (1 isolate), *Preussia* (1 isolate), *Rhizopus* (1 isolate), *Oblongocollomyces* (1 isolate), *Trichoderma* (1 isolate), and *Xylaria* (1 isolate). Plant host tissues were collected in Egypt, China, India, Australia, South Africa, La Réunion (France), France, USA, and Hawaii.

Table 2. Endophytic fungi isolated from *Acacia* species.

Species	Host Plant	Plant Part	Country	Reference
Phyllosticta sp.	*A. amara*	leaf	Masinagudi, India	[34]
Xylaria sp.	*A. amara*	leaf	Masinagudi, India	[34]
Aspergillus niger	*A. arabica*	leaf	Punjab, India	[35]
Aspergillus sp.	*A. auriculaeformis*	root	Guangdong, China	[36]
Trichoderma sp.	*A. auriculaeformis*	root	Guangdong, China	[36]
Aureobasidium pullulans	*A. baileyana*	phyllode	Melbourne, Australia	[37]
Alternaria sp.	*A. decurrens*	leaf, stem	Yunnan, China	[38]
Penicillium sp.	*A. decurrens*	leaf, stem	Yunnan, China	[38]
Peyronellaea sp.	*A. decurrens*	leaf, stem	Yunnan, China	[38]

Table 2. *Cont.*

Species	Host Plant	Plant Part	Country	Reference
Phoma sp.	*A. decurrens*	leaf, stem	Yunnan, China	[38]
Rhizopus sp.	*A. decurrens*	leaf, stem	Yunnan, China	[38]
Aureobasidium pullulans	*A. floribunda*	phyllode	Melbourne, Australia	[37]
Chaetomium globosum	*A. floribunda*	phyllode	Melbourne, Australia	[37]
Dothiorella heterophyllae	*A. heterophylla*	branch	La Réunion, France	[39]
Dothiorella reunionis	*A. heterophylla*	branch	La Réunion, France	[39]
Lasiodiplodia iranensis	*A. heterophylla*	branch	La Réunion, France	[39]
Lasiodiplodia rubropurpurea	*A. heterophylla*	branch	La Réunion, France	[39]
Neofusicoccum parvum	*A. heterophylla*	branch	La Réunion, France	[39]
Cochliobolus geniculatus	*A. hindsii*	leaf	Mexico	[40]
Colletotrichum gloeosporioides	*A. hindsii*	leaf	Mexico	[40]
Colletotrichum truncatum	*A. hindsii*	leaf	Mexico	[40]
Eupenicillium javanicum	*A. hindsii*	leaf	Mexico	[40]
Fusarium oxysporum	*A. hindsii*	leaf	Mexico	[40]
Moesziomyces bullatus	*A. hindsii*	leaf	Mexico	[40]
Paraphaeosphaeria sp.	*A. hindsii*	leaf	Mexico	[40]
Phoma sp.	*A. hindsii*	leaf	Mexico	[40]
Wickerhamomyces anomalus	*A. hindsii*	leaf	Mexico	[40]
Botryosphaeria dothidea	*A. karroo*	branch	South Africa	[41]
Diplodia allocellula	*A. karroo*	branch	South Africa	[41,42]
Dothiorella brevicollis	*A. karroo*	branch	South Africa	[41,42]
Dothiorella dulcispinae	*A. karroo*	branch	South Africa	[42]
Dothiorella pretoriensis	*A. karroo*	branch	South Africa	[41,42]
Eutiarosporella urbis-rosarum	*A. karroo*	branch	South Africa	[41,42]
Lasiodiplodia pseudotheobromae	*A. karroo*	branch	South Africa	[41]
Lasiodiplodia theobromae	*A. karroo*	branch	South Africa	[41]
Lasiodiplodia gonubiensis	*A. karroo*	branch	South Africa	[41]
Neofusicoccum kwambonambiense	*A. karroo*	branch	South Africa	[41]
Neofusicoccum protearum	*A. karroo*	branch	South Africa	[41]
Neofusicoccum vitifusiforme	*A. karroo*	branch	South Africa	[41,42]
Neofusicoccum australe	*A. karroo*	branch	South Africa	[41]
Neofusicoccum parvum	*A. karroo*	branch	South Africa	[41]
Oblongocollomyces variabilis	*A. karroo*	branch	South Africa	[41]
Phaeobotryosphaeria variabilis	*A. karroo*	branch	South Africa	[42]
Spencermartinsia viticola	*A. karroo*	branch	South Africa	[41,42]
Dothiorella koae	*A. koa*	branch	Hawaii, USA	[39]
Lasiodiplodia theobromae	*A. koa*	branch	Hawaii, USA	[39]
Lasiodiplodia exigua	*A. koa*	branch	Hawaii, USA	[39]
Neofusicoccum occulatum	*A. koa*	branch	Hawaii, USA	[39]
Neofusicoccum parvum	*A. koa*	branch	Hawaii, USA	[39]
Aspergillus ochraceus	*A. nilotica*	stem	Al-Sharqia, Egypt	[43]
Penicillium sp.	*A. nilotica*	stem	Al-Sharqia, Egypt	[43]
Pestalotia sp.	*A. nilotica*	stem	Al-Sharqia, Egypt	[43]
Chaetomium globosum	*A. podalyriifolia*	phyllode	Melbourne, Australia	[37]
Chaetomium sp.	*A. podalyriifolia*	phyllode	Melbourne, Australia	[37]
Preussia sp.	*A. victoriae*	leaf	Arizona, USA	[44]

2.2. Albizia

The genus *Albizia* (Mimosaceae) comprises almost 150 species, mostly trees and shrubs native to tropical and subtropical regions of Asia and Africa. They are common components of timber plantations, cropping, and livestock production systems [45]. *Albizia* synthesizes numerous bioactive compounds with pharmacological properties such as saponins, alkaloids, flavonoids, and phenolics [45]. The species *A. lebbeck* has been extensively introduced in seasonally dry tropical regions of Africa, Asia, the Caribbean, and South America, mainly as an ornamental plant, and has become naturalized in many areas [46].

Table 3 reports endophytes isolated from *Albizia* genera. These fungi belong to 14 different genera, most of them found in leaves and twigs of *A. lebbeck* originating from Iraq, India, Indonesia, and Egypt. Isolated fungi included different species of *Aspergillus* (9 isolates), which are dominant in comparison to other genera, followed by *Fusarium* (4 isolates), *Penicillium* (3 isolates), and *Paecilomyces* (2 isolates). One isolate for each of the following genera *Neocosmospora*, *Bipolaris*, *Colletotrichum*, *Diaporthe*, *Lasiodiplodia*, *Rosellinia*, *Acremonium*, *Trichoderma*, *Verticillium*, *Curvularia*, and *Nigrospora* has been detected.

Table 3. Endophytic fungi isolated from *Albizia* species.

Species	Host Plant	Plant Part	Country	Reference
Acremonium sp.	*A. lebbeck*	-	Indonesia	[47]
Aspergillus fumigatus	*A. lebbeck*	leaf and twig	Baghdad, Iraq	[48]
Aspergillus fumigatus	*A. lebbeck*	leaf	Al-Sharqia, Egypt	[43]
Aspergillus glaucus	*A. lebbeck*	leaf and twig	Baghdad, Iraq	[48]
Aspergillus niger	*A. lebbeck*	leaf and twig	Baghdad, Iraq	[48]
Aspergillus raperi	*A. lebbeck*	leaf and twig	Baghdad, Iraq	[48]
Aspergillus sclerotioniger	*A. lebbeck*	leaf and twig	Baghdad, Iraq	[48]
Aspergillus flavus	*A. lebbeck*	leaf and twig	Baghdad, Iraq	[48]
Aspergillus sp.	*A. lebbeck*	-	Indonesia	[47]
Aspergillus sp.	*A. lebbeck*	leaf and twig	Baghdad, Iraq	[48]
Bipolaris australiensis	*A. lebbeck*	leaf and twig	Baghdad, Iraq	[48]
Colletotrichum sp.	*A. amara*	leaf	Masinagudi, India	[34]
Curvularia cymbopogonis	*A. lebbeck*	leaf and twig	Baghdad, Iraq	[48]
Diaporthe sp.	*A. amara*	leaf	Masinagudi, India	[34]
Fusarium verticilloides	*A. lebbeck*	leaf and twig	Baghdad, Iraq	[48]
Fusarium sp.	*A. amara*	leaf	Masinagudi, India	[34]
Fusarium sp.	*A. lebbeck*	-	Indonesia	[47]
Fusarium oxysporum	*A. julibrissin*	-	-	[49]
Lasiodiplodia sp.	*A. amara*	leaf	Masinagudi, India	[34]
Neocosmospora solani	*A. lebbeck*	leaf and twig	Baghdad, Iraq	[48]
Paecilomyces variotii	*A. lebbeck*	leaf and twig	Baghdad, Iraq	[48]
Paecilomyces sp.	*A. lebbeck*	leaf and twig	Baghdad, Iraq	[48]
Penicillium sp.	*A. lebbeck*	-	Indonesia	[47]
Penicillium sp.	*A. lebbeck*	leaf	Al-Sharqia, Egypt	[43]
Penicillium sp.	*A. lebbeck*	leaf and twig	Baghdad, Iraq	[48]
Rosellinia sanctae-cruciana	*A. lebbeck*	leaf	Jammu, India	[50]
Trichoderma sp.	*A. lebbeck*	-	Indonesia	[47]
Verticillium sp.	*A. lebbeck*	-	Indonesia	[47]

2.3. Bauhinia

The genus *Bauhinia*, commonly known as the orchid tree, belongs to the family *Fabaceae*. It comprises more than 500 species of shrubs, and small trees mostly native to tropical countries (Africa, Asia, and South America). Many species are widely used as ornamental plants, forage, human food, and in folk medicine [51,52].

A total of 107 fungal endophytes have been found in *Bauhinia* plant tissues (Table 4). The most common fungi reported were: *Aspergillus* (13 isolates), *Curvularia* (8 isolates), *Penicillium* (7 isolates), *Nigrospora* (7 isolates), *Fusarium* (5 isolates), *Phoma* (3 isolates), *Cladosporium* (4 isolates), *Acremonium* (3 isolates), *Colletotrichum* (3 isolates), *Phomopsis* (3 isolates), *Cochliobolus* (3 isolates), and *Exserohilum* (3 isolates). Furthermore, other genera were found less frequently: *Myrothecium* (2 isolates), *Gibberella* (2 isolates), *Lasiodiplodia* (2 isolates), *Khuskia* (1 isolate), *Nodulisporium* (2 isolates), *Pestalotiopsis* (2 isolates), *Alternaria* (2 isolates), *Gibberella* (2 isolates), *Pithomyces* (1 isolate), *Diplococcium* (2 isolates), *Dothiorella* (2 isolates), *Ascotricha* (2 isolates), *Talaromyces* (2 isolates), *Trichoderma* (2 isolates), *Spegazzinia* (2 isolates), *Kiflimonium* (1 isolate), *Geotrichum* (1 isolate), *Corynespora* (1 isolate), *Diaporthe* (1 isolate), *Glomerella* (1 isolate), *Pestalotia* (1 isolate), *Scedosporium* (1 isolate), *Botrytis* (1 isolate), *Sporormiella* (1 isolate), *Phyllosticta* (1 isolate), *Lasmenia* (2 isolates), *Albifimbria* (1 isolate), *Myrmecridium* (2 isolates), *Sphaeria*

(1 isolate), *Paraboeremia* (1 isolate), *Pseudopithomyces* (1 isolate), *Chaetomium* (1 isolate), and *Torulomyces* (1 isolate). All host plants, namely *B. fortificata*, *B. brevipes*, *B. racemosa*, *B. guianensis*, *B. monandra*, *B. malabarica*, *B. phoenicea*, and *B. vahlii*, were from Brazil and India.

Table 4. Endophytic fungi isolated with *Bauhinia* species.

Species	Host Plant	Plant Part	Country	Reference
Acremonium sp.	*B. brevipes*	-	Brazil	[53]
	B. forficata	-	Brazil	[53]
	B. brevipes	leaf	Pirapitinga, Brazil	[54]
Albifimbriaverrucaria	*B. fortificata*	stem	Recife, Brazil	[55]
Alternaria alternata	*B. malabarica*	stem	Chennai, India	[56]
	B. racemosa	leaf	Mudumalai, India	[57]
Ascotricha sp.	*B. forficata*	-	Brazil	[53]
Ascotricha chartarum	*B. fortificata*	seed	Recife, Brazil	[55]
Aspergillus sp.	*B. forficata*	-	Brazil	[53]
	B. monandra	leaf	Recife, Brazil	[58]
	B. guianensis	-	Brazil	[53,59,60]
Aspergillus flavus	*B. malabarica*	leaf, root	Chennai, India	[56]
Aspergillus niger	*B. fortificata*	stem	Recife, Brazil	[55]
	B. malabarica	leaf, root, stem	Chennai, India	[56]
	B. racemosa	leaf	Mudumalai, India	[57]
Aspergillus ochraceus	*B. fortificata*	stem, seed	Recife, Brazil	[55]
Aspergillus tamarii	*B. malabarica*	leaf, stem	Chennai, India	[56]
Aspergillus terreus	*B. malabarica*	leaf, root	Chennai, India	[56]
Aspergillus versicolor	*B. vahlii*	leaf	Chilkigarh, India	[61]
Botrytis cinerea	*B. racemosa*	leaf	Mudumalai, India	[57]
Chaetomium globosum	*B. malabarica*	leaf	Chennai, India	[56]
Cladosporium sphaerospermum	*B. fortificata*	leaf	Recife, Brazil	[55]
Cladosporium sp.	*B. forficata*	-	Brazil	[53]
Cladosporium cladosporioides	*B. racemosa*	leaf	Mudumalai, India	[57]
Cladosporium oxysporum	*B. fortificata*	sepal	Recife, Brazil	[55]
Cochliobolus sp.	*B. forficata*	-	Brazil	[53]
Cochliobolus australiensis	*B. fortificata*	leaf	Recife, Brazil	[55]
Cochliobolus lunatus	*B. fortificata*	leaf, stem	Recife, Brazil	[55]
Colletotrichum sp.	*B. forficata*	-	Brazil	[53]
Colletotrichum coccodes	*B. guianensis*	stem	Belem, Brazil	[62]
Colletotrichum gloeosporioides	*B. racemosa*	leaf	Mudumalai, India	[57]
Corynespora cassiicola	*B. racemosa*	leaf	Mudumalai, India	[63]
Curvularia sp.	*B. monandra*	leaf	Recife, Brazil	[58]
Curvularia brachyspora	*B. malabarica*	leaf	Chennai, India	[56]
Curvularia clavata	*B. guianensis*	stem	Belem, Brazil	[62]
	B. phoenicea	leaf	Kudremukh range, India	[64]
Curvularia lunata	*B. malabarica*	leaf	Chennai, India	[56]
	B. racemosa	leaf	Mudumalai, India	[57]
	B. phoenicea	bark, leaf	Kudremukh range, India	[64]
Curvularia pallescens	*B. phoenicea*	leaf	Kudremukh range, India	[64]
Diaporthe sp.	*B. brevipes*	leaf	Pirapitinga, Brazil	[54]
Diplococcium sp.	*B. forficata*	-	Brazil	[53]
Diplococcium spicatum	*B. fortificata*	leaf	Recife, Brazil	[55]
Dothiorella sp.	*B. brevipes*	-	Brazil	[53]
		leaf	Pirapitinga, Brazil	[54]
Exserohilum rostratum	*B. racemosa*	leaf, stem	Sathyamangalam, India	[65]
	B. guianensis	stem	Belem, Brazil	[62,66]
Fusarium culmorum	*B. malabarica*	leaf, stem	Chennai, India	[56]

Table 4. *Cont.*

Species	Host Plant	Plant Part	Country	Reference
Fusarium verticillioides	*B. malabarica*	root	Chennai, India	[56]
Fusarium oxysporum	*B. malabarica*	leaf, root, stem	Chennai, India	[56]
	B. phoenicea	leaf	Kudremukh range, India	[64]
Fusarium sp.	*B. forficata*	-	Brazil	[53]
Fusidium viride	*B. vahlii*	petiole	Chilkigarh, India	[61]
Geotrichum candidum	*B. vahlii*	leaf, petiole	Chilkigarh, India	[61]
Gibberella fujikuroi	*B. fortificata*	leaf, stem	Recife, Brazil	[55]
Gibberella sp.	*B. forficata*	-	Brazil	[53]
Glomerella sp.	*B. forficata*	-	Brazil	[53]
Kiflimonium curvulum	*B. fortificata*	sepal, stem	Recife, Brazil	[55]
Khuskia sp.	*B. forficata*	-	Brazil	[53]
Lasiodiplodia theobromae	*B. racemosa*	leaf	Mudumalai, India	[57,63]
Lasmenia sp.	*B. forficata*	-	Brazil	[53]
Lasmeniabalansae	*B. fortificata*	stem	Recife, Brazil	[55]
Myrmecridium sp.	*B. forficata*	-	Brazil	[53]
Myrmecridium schulzeri	*B. fortificata*	sepal	Recife, Brazil	[55]
	B. racemosa	leaf	Mudumalai, India	[57]
Nigrospora oryzae	*B. phoenicea*	stem, leaf	Kudremukh range, India	[64]
	B. fortificata	sepal	Recife, Brazil	[55]
Nigrospora sacchari	*B. phoenicea*	leaf	Kudremukh range, India	[64]
Nigrospora sp.	*B. forficata*	-	Brazil	[53]
Nigrospora sphaerica	*B. vahlii*	stem	Chilkigarh, India	[61]
	B. racemosa	leaf, stem	Sathyamangalam, India	[65]
Nodulisporium sp.	*B. forficata*	-	Brazil	[53]
	B. fortificata	stem	Recife, Brazil	[55]
Paraboeremia putaminum	*B. fortificata*	sepal	Recife, Brazil	[55]
Penicillium commune	*B. fortificata*	sepal	Recife, Brazil	[55]
Penicillium corylophilum	*B. fortificata*	seed	Recife, Brazil	[55]
Penicillium glabrum	*B. fortificata*	stem, seed	Recife, Brazil	[55]
Penicillium implicatum	*B. fortificata*	stem	Recife, Brazil	[55]
Penicillium sp.	*B. forficata*	-	Brazil	[53]
	B. monandra	leaf	Recife, Brazil	[58]
Penicillium aurantiogriseum	*B. fortificata*	seed	Recife, Brazil	[55]
Pestalotia sp.	*B. forficata*	-	Brazil	[53]
Pestalotiopsis sp.	*B. brevipes*	leaf	Pirapitinga, Brazil	[54]
	B. brevipes	-	Brazil	[53]
	B. forficata	-	Brazil	[53]
Phoma sp.	*B. forficata*	-	Brazil	[53]
	B. brevipes	leaf	Pirapitinga, Brazil	[54]
Phomopsis sp.	*B. brevipes*	-	Brazil	[53]
	B. forficata	-	Brazil	[53]
Phomopsis diachenii	*B. fortificata*	leaf	Recife, Brazil	[55]
Phyllosticta capitalensis	*B. racemosa*	leaf	Mudumalai, India	[57]
Pithomyces sp.	*B. forficata*	-	Brazil	[53]
Pseudopithomycesatro-olivaceus	*B. fortificata*	seed	Recife, Brazil	[55]
Scedosporium apiospermum	*B. guianensis*	stem	Belem, Brazil	[62]
Spegazzinia sp.	*B. forficata*	-	Brazil	[53]
Spegazzinia tessarthra	*B. fortificata*	leaf	Recife, Brazil	[55]
Sphaeria baccata	*B. fortificata*	sepal	Recife, Brazil	[55]
Sporormiella minima	*B. racemosa*	leaf	Mudumalai, India	[57]
Talaromyces sp.	*B. forficata*	-	Brazil	[53]
Talaromyces funiculosus	*B. fortificata*	leaf	Recife, Brazil	[55]
Torulomyces lagena	*B. racemosa*	leaf	Mudumalai, India	[57]
Trichoderma piluliferum	*B. fortificata*	stem	Recife, Brazil	[55]
Trichoderma sp.	*B. forficata*	-	Brazil	[53]

2.4. Berberis

The genus *Berberis* (Berberidaceae) comprises almost 500 species of deciduous or evergreen shrubs, which occur in the temperate and subtropical regions of Europe, Asia, Africa, and America [67]. This genus has remarkable pharmacological properties [68]. Berberine and Berbamine are the main compounds produced by these plants, together with alkaloids, tannins, phenolic compounds, sterols, and triterpenes [69].

Numerous endophytic fungi belonging to 19 genera have been isolated from tissues of *Berberis* from India, China, Kenya, and the USA (Table 5). Isolated fungi included different species of *Fusarium* (4 isolates) and *Colletotrichum* (4 isolates), followed by *Alternaria* (4 isolates), *Anguillospora* (1 isolate), *Phomopsis* (1 isolate), *Campylospora* (1 isolate), *Cercospora* (1 isolate), *Clonostachys* (1 isolate), *Heliscus* (1 isolate), *Diaporthe* (2 isolates), *Microsphaeropsis* (1 isolate), *Phyllosticta* (1 isolate), *Paraphalosphaera* (1 isolate), *Prathoda* (1 isolate), *Bacillispora* (1 isolate), *Neocosmospora* (1 isolate), *Aspergillus* (1 isolate), *Myrothecium* (1 isolate), and *Puccinia* (1 isolate).

Table 5. Endophytic fungi isolated from *Berberis* species.

Species	Host Plant	Plant Part	Country	Reference
Alternaria alternata	*B. poiretii*	leaf, twig	Beijing, China	[70]
	B. aristata	leaf	Sial Sui, District Rajouri, J&K, India	[68]
Alternaria macrospora	*B. aristata*	leaf	Sial Sui, District Rajouri, J&K, India	[68]
Alternaria solani	*B. aristata*	leaf	Sial Sui, District Rajouri, J&K, India	[68]
Anguillospora crassa	*Berberis* sp.	root	Western Himalaya	[71]
Aspergillus flavus	*B. aristata*	leaf	Sial Sui, District Rajouri, J&K, India	[68]
Campylospora parvula	*Berberis* sp.	root	Western Himalaya	[71]
Cercospora citrullina	*B. aristata*	stem	Sial Sui, District Rajouri, J&K, India	[68]
Clonostachys rosea	*B. aristata*	root	Sial Sui, District Rajouri, J&K, India	[68]
Colletotrichum coccodes	*B. aristata*	root	Sial Sui, District Rajouri, J&K, India	[68]
Colletotrichum coffeanum	*B. aristata*	leaf	Sial Sui, District Rajouri, J&K, India	[68]
Colletotrichum gloeosporioides	*B. aristata*	leaf	Sial Sui, District Rajouri, J&K, India	[68]
Colletotrichum kahawae	*B. aristata*	leaf	Sial Sui, District Rajouri, J&K, India	[68]
Bacillispora aquatica	*Berberis* sp.	root	Western Himalaya	[71]
Diaporthe sp.	*B. vulgaris*	leaf, stem	Kenya	[72]
Neocosmospora falciformis	*B. aristata*	root	Sial Sui, District Rajouri, J&K, India	[68]
Fusarium lateritium	*B. aristata*	stem	Sial Sui, District Rajouri, J&K, India	[68]
Fusarium nematophilum	*B. aristata*	root	Sial Sui, District Rajouri, J&K, India	[68]
Fusarium oxysporum	*B. aristata*	stem	Sial Sui, District Rajouri, J&K, India	[68]
Fusarium solani	*B. aristata*	root	Sial Sui, District Rajouri, J&K, India	[68]
Heliscus lugdunensis	*Berberis* sp.	root	Western Himalaya	[71]
Microsphaeropsis conielloides	*B. poiretii*	twig	Beijing, China	[70]
Myrothecium inundatum	*B. aristata*	leaf	Sial Sui, District Rajouri, J&K, India	[68]
Paraphaeosphaeria sp.	*B. thunbergii*	stem	China	[73]
Phomopsis sp.	*B. poiretii*	twig	Beijing, China	[70]
Diaporthe tersa	*B. aristata*	leaf	Sial Sui, District Rajouri, J&K, India	[68]
Phyllosticta capitalensis	*B. aristata*	leaf	Sial Sui, District Rajouri, J&K, India	[68]
Prathoda longissima	*Berberis* sp.	root	Western Himalaya	[71]
Puccinia graminis f. sp. *tritici*	*B. vulgaris*	-	Pacific Northwest USA	[74]

2.5. Caesalpinia

The genus *Caesalpinia* (Fabaceae) includes approximately 200 species, mainly arboreal and shrubby species, distributed in seasonally dry tropical forests, as well as in tropical and warm temperate savannas, tropical wet forests, and tropical coastal habitats [75]. Several classes of compounds, mainly flavonoids, diterpenes, and steroids, have been isolated from *Caesalpinia* species, which have shown various medicinal properties [75]. The most common species cultivated as ornamental plants are *C. pulcherrima* and *C. echinata*.

A total of 44 fungal endophytes were isolated from leaves, stems, and bark of plants collected from India, Brazil, and Indonesia (Table 6). Fungal genera associated with different species of *Caesalpinia* were: *Aspergillus* (10 isolates), followed by *Trichoderma* (6 isolates) and *Fusarium* (4 isolates). Other isolated endophytes have been identified as *Penicillum* (3 isolates), *Curvularia* (2 isolates), *Nectria* (2 isolates), *Bipolaris* (2 isolates), *Xylaria* (2 isolates), and one isolate for the genera *Alternaria*, *Chrysosporium*, *Cladosporium*, *Colletotrichum*, *Epicoccum*, *Geotrichum*, *Helminthosporium*, *Lasiodiplodia*, *Talaromyces*, *Scopulariopsis*, and *Phyllosticta*, respectively.

Table 6. Endophytic fungi isolated from *Caesalpinia* species.

Species	Host Plant	Plant Part	Country	Reference
Alternaria alternata	*C. pulcherrima*	leaf	India	[76]
Aspergillus flavus	*C. pulcherrima*	leaf	India	[76]
Aspergillus fumigatus	*C. pulcherrima*	leaf	India	[76]
Aspergillus niger	*C. pulcherrima*	leaf	India	[76]
Aspergillus flavus var. oryzae	*C. pulcherrima*	leaf	India	[76]
Aspergillus rugulosus	*C. pulcherrima*	leaf	India	[76]
Aspergillus terreus	*C. pulcherrima*	leaf	India	[76]
	C. pyramidalis	leaf	Brazil	[53]
Aspergillus sp.	*C. echinata*	leaf	Brazil	[53]
	C. echinata	stem, bark	Brazil	[77]
Aspergillus nidulans	*C. pulcherrima*	leaf	India	[76]
Bipolaris oryzae	*C. pulcherrima*	leaf	India	[76]
Bipolaris sp.	*C. pulcherrima*	leaf	India	[76]
Chrysosporium sp.	*C. sappan*	stem	Indonesia	[78]
Cladosporium cladosporioides	*C. echinata*	leaf	Brazil	[79]
Colletotrichum gloeosporioides	*C. echinata*	leaf	Brazil	[79]
Curvularia lunata	*C. sappan*	stem	Indonesia	[78]
Curvularia pallescens	*C. echinata*	leaf	Brazil	[79]
Epicoccum sp.	*C. echinata*		Brazil	[53]
		leaf	Brazil	[53]
Fusarium sp.	*C. echinata*	stem	Brazil	[77]
		stem, bark	Brazil	[77]
	C. pulcherrima	leaf	India	[76]
Geotrichum candidum	*C. sappan*	stem	Indonesia	[78]
Helminthosporium sp.	*C. pulcherrima*	leaf	India	[76]
Lasiodiplodia theobromae	*C. echinata*	leaf	Brazil	[79]
Nectria sp.	*C. echinata*	-	Brazil	[53]
Nectria pseudotrichia	*C. echinata*	stem, bark	Brazil	[77]
		stem		[80]
Penicillium citrinum	*C. pulcherrima*	leaf	India	[76]
Penicillium chrysogenum	*C. pulcherrima*	leaf	India	[76]
Penicillium sp.	*C. sappan*	stem	Indonesia	[78]
Phyllosticta sorghina	*C. echinata*	stem, bark	Brazil	[77]
Scopulariopsis coprophila	*C. echinata*	leaf	Brazil	[79]
Talaromyces sp.	*C. echinata*	stem, bark	Brazil	[77]
		leaf	Brazil	[53]
Trichoderma atroviride	*C. pyramidalis*	stem, bark	Brazil	[81]
Trichoderma harzianum	*C. pyramidalis*	stem, bark	Brazil	[81]
Trichoderma koningiopsis	*C. pyramidalis*	stem, bark	Brazil	[81]
Trichoderma longibrachiatum	*C. pyramidalis*	stem, bark	Brazil	[81]
Trichoderma virens	*C. pyramidalis*	stem, bark	Brazil	[81]
Trichoderma sp.	*C. sappan*	stem	Indonesia	[78]
Xylaria sp.	*C. echinata*	leaf	Brazil	[53]
Xylaria berteri	*C. echinata*	stem, bark	Brazil	[77]

2.6. Cassia

The genus *Cassia* (Fabaceae) comprises about 600 species native to tropical and subtropical regions of Southeast Asia, Africa, Northern Australia, and Latin America [82,83]. In particular, *C. fistula* and *C. alata* are distributed worldwide and used as ornamental and medicinal plants for their biological and pharmacological properties [82–84]. Some investigations on phytochemicals of *Cassia* revealed that it comprises compounds like anthraquinones, alkaloids, catechols, flavonoids, phenolic compounds, saponins, steroids, tannins, and triterpenoids [83–86].

Nineteen endophytic fungi have been isolated from different tissues of *Cassia* species from Thailand, India, Malaysia, and Brazil (Table 7): *Aspergillus* (2 isolates), *Nodulisporium* (2 isolates), *Penicillium* (2 isolates), *Phomopsis* (2 isolates), *Daldinia* (1 isolate), *Coprinus* (1 isolate), *Guignardia* (1 isolate), *Hypoxylon* (1 isolate), *Nemania* (1 isolate), *Nigrospora* (1 isolate), *Papulospora* (1 isolate), *Periconia* (1 isolate), *Xylaria* (1 isolate), *Psathyrella* (1 isolate), and *Thielaoviopsis* (1 isolate).

Table 7. Endophytic fungi isolated from *Cassia* species.

Species	Host Plant	Plant Part	Country	Reference
Aspergillus flavus	*C. siamea*	leaf	Malaysia	[87]
Aspergillus sp.	*C. fistula*	leaf, stem, fruit	India	[88]
Coprinus sp.	*C. fistula*	leaf	Bangkok, Thailand	[89]
Daldinia sp.	*C. fistula*	leaf	Bangkok, Thailand	[89]
Guignardia sp.	*C. occidentalis*	leaf	Brazil	[53]
Hypoxylon sp.	*C. fistula*	leaf	Bangkok, Thailand	[89]
Nemania sp.	*C. fistula*	leaf	Bangkok, Thailand	[89]
Nigrospora sp.	*C. fistula*	leaf	Bangkok, Thailand	[89]
Nodulisporium sp.	*C. fistula*	leaf	Bangkok, Thailand	[89]
		-	-	[90]
Papulospora sp.	*C. fistula*	bark	India	[91]
Penicillium sclerotiorum	*C. fistula*	-	India	[92]
Penicillium sp.	*C. fistula*	leaf	Bangkok, Thailand	[89]
Periconia sp.	*C. fistula*	bark	India	[91]
Phomopsis cassiae	*C. spectabilis*	-	Brazil	[52]
Phomopsis sp.	*C. fistula*	leaf	Bangkok, Thailand	[89]
Psathyrella sp.	*C. fistula*	leaf	Bangkok, Thailand	[89]
Thelioviopsis sp.	*C. fistula*	leaf	India	[91]
Xylaria sp.	*C. fistula*	leaf	Bangkok, Thailand	[89]

2.7. Cornus

The genus *Cornus* (Cornaceae) consists of over 50 species of woody plants, many of which are cultivated as ornamental and medicinal trees [93]. The most widespread ornamental plants of the genus are *C. florida* and *C. stolonifera*, called the flowering dogwood, native to northern and central America [93]. The species *C. officinalis* is widely distributed in China, Korea, and Japan, and used for its several pharmacological activities. Among bioactive compounds, iridoids, tannins, and flavonoids are the major components [94].

About 90 fungal endophytes have been isolated and identified from *C. alba*, *C. alternifolia*, *C. stolonifera*, *C. controversa*, and *C. officinalis* collected in Canada, USA, Japan, China, and Korea (Table 8): *Penicillium* (8 isolates), *Fusarium* (4 isolates), *Cladosporium* (5 isolates), *Colletotrichum* (5 isolates), *Alternaria* (6 isolates), *Pestalotiopsis* (5 isolates), *Aureobasidium* (4 isolates), *Botryosphaeria* (3 isolates), *Cryptodiaporthe* (2 isolates), *Phomopsis* (3 isolates), *Talaromyces* (4 isolates), *Aspergillus* (3 isolates), *Discula* (4 isolates), *Diaporthe* (3 isolates), *Neonectria* (2 isolates), *Trichoderma* (2 isolates), *Tubakia* (2 isolates), and *Didymella* (3 isolates). Only one isolate of the following genera has been reported: *Ascochyta*, *Botrytis*, *Cyrptosporiopsis*, *Elsinoe*, *Epicoccum*, *Helminthosporium*, *Lecanicillium*, *Leptosphaerulina*, *Lophiostoma*, *Drepanopeziza*, *Nigrospora*, *Sarocladium*, *Cordyceps*, *Phyllosticta*, *Phytophthora*,

Phoma, Pleuroceras, Thelonectria, Sclerotinia, Neofabraea, Septoria, Simplicillium, Stenella, Verticillium, and *Xylaria.*

Table 8. Endophytic fungi isolated from *Cornus* species.

Species	Host Plant	Plant Part	Country	Reference
Alternaria alternata	*Cornus* spp.	leaf	Japan, USA	[95]
Alternaria sp.	*C. stolonifera*	leaf	Canada	[96]
Alternaria tenuissima	*C. officinalis*	twig, leaf	China	[97]
	Cornus spp.	leaf	Japan	[95]
Ascochyta medicaginicola	*C. officinalis*	twig	China	[97]
Aspergillus flavus var. *oryzae*	*C. alba*	leaf	-	[98]
Aspergillus sp.	*Cornus* spp.	leaf	Japan, USA	[95]
Aureobasidium pullulans	*Cornus* spp.	leaf	USA	[95]
Aureobasidium sp.	*C. stolonifera*	leaf	Canada	[96]
	Cornus spp.	leaf	Japan, USA	[95]
Botryosphaeria dothidea	*C. officinalis*	twig, leaf	China	[97]
	Cornus spp.	leaf	Japan	[95]
Botryosphaeria sp.	*Cornus* spp.	leaf	Japan	[95]
Botrytis sp.	*C. stolonifera*	leaf	Canada	[96]
Cladosporium cladosporioides	*C. stolonifera*	leaf	Canada	[96]
Cladosporium herbarum	*C. stolonifera*	leaf	Canada	[96]
Cladosporium sp.	*C. stolonifera*	leaf	Canada	[96]
	Cornus spp.	leaf	Japan	[95]
Cladosporium sphaerospermum	*C. stolonifera*	leaf	Canada	[96]
Colletotrichum acutatum	*Cornus* spp.	leaf	USA, Japan	[95]
Colletotrichum gloeosporioides	*C. officinalis*	twig, leaf	China	[97]
	C. stolonifera	leaf	Canada	[96]
Colletotrichum sp.	*Cornus* spp.	leaf	Japan	[95]
Cordycepsfarinose	*C. stolonifera*	leaf	Canada	[96]
Cryptodiaporthe corni	*C. alternifolia*	stem	USA	[99]
		bark, phloem	USA	[100]
Cyrptosporiopsis sp.	*Cornus* spp.	leaf	USA	[95]
Diaporthe amygdali	*Cornus* spp.	leaf	USA, Japan	[95]
Diaporthe lagerstroemiae	*Cornus* spp.	leaf	Japan	[95]
Didymellapinodella	*C. officinalis*	twig	China	[97]
Didymella glomerata	*Cornus* spp.	leaf	USA, Japan	[95]
	Cornus spp.	leaf	USA	[101]
		leaf	USA	[95]
Discula destructiva	*C. florida*	leaf	Germany	[102]
		leaf	Italy	[103]
Drepanopeziza populi	*C. officinalis*	twig	China	[97]
Elsinoe fawcettii	*Cornus* spp.	leaf	USA	[95]
Epicoccum nigrum	*C. stolonifera*	leaf	Canada	[96]
Fusarium lateritium	*C. controversa*	stem	Korea	[104]
Fusarium oxysporum	*C. officinalis*	root	China	[97]
Fusarium sp.	*Cornus* spp.	leaf	Japan	[95]
	C. stolonifera	leaf	Canada	[96]
Helminthosporium velutinum	*C. officinalis*	twig	China	[97]
Lecanicillium psalliotae	*C. stolonifera*	leaf	Canada	[96]
Leptosphaerulina australis	*C. officinalis*	twig	China	[97]
Lophiostoma sp.	*Cornus* spp.	leaf	USA	[95]
Neofabraea sp.	*Cornus* spp.	leaf	USA	[95]
Neonectria sp.	*Cornus* spp.	leaf	USA, Japan	[95]
Nigrospora sphaerica	*C. florida*	stem	Tennessee, USA	[105]
Penicillium brevicompactum	*C. stolonifera*	leaf	Canada	[96]
Penicillium chrysogenum	*Cornus* spp.	leaf	USA	[95]
Penicillium citrinum	*C. stolonifera*	leaf	Canada	[96]
Penicillium miczynskii	*C. stolonifera*	leaf	Canada	[96]
Penicillium simplicissimum	*Cornus* spp.	leaf	USA	[95]
Penicillium sp.	*C. stolonifera*	leaf	Canada	[96]

Table 8. *Cont.*

Species	Host Plant	Plant Part	Country	Reference
Penicillium spinulosum	*Cornus* spp.	leaf	Japan	[95]
Penicillium thomii	*C. stolonifera*	leaf	Canada	[96]
Phytophthora palmivora	*C. florida*	leaf, shoot	USA	[106]
Pestalotiopsis mangiferae	*Cornus* spp.	leaf	Japan	[95]
Pestalotiopsis microspora	*Cornus* spp.	leaf	USA, Japan	[95]
Pestalotiopsis monochaeta	*Cornus* spp.	leaf	Japan	[95]
Pestalotiopsis sp.	*Cornus* spp.	leaf	Japan	[95]
Phoma moricola	*C. officinalis*	twig	China	[97]
Phomopsis sp.	*C. stolonifera*	leaf	Canada	[96]
	Cornus spp.	leaf	USA, Japan	[95]
Phyllosticta fallopiae	*C. officinalis*	leaf	China	[97]
Phytophthora nicotianae	*C. florida*	leaf, shoot	USA	[106]
Pleuroceras tenellum	*Cornus* spp.	leaf	USA	[95]
Sarocladiumkiliense	*C. stolonifera*	leaf	Canada	[96]
Sclerotinia sclerotiorum	*C. stolonifera*	leaf	Canada	[96]
Septoria sp.	*C. stolonifera*	leaf	Canada	[96]
Simplicillium lanosoniveum	*C. officinalis*	fruit	China	[97]
Stenella sp.	*C. stolonifera*	leaf	Canada	[96]
Talaromyces assiutensis	*C. officinalis*	root	China	[97]
Talaromycescecidicola	*Cornus* spp.	leaf	USA, Japan	[95]
Talaromyces trachyspermus	*C. officinalis*	root	China	[97]
Thelonectriadiscophora	*Cornus* spp.	leaf	Japan	[95]
Trichoderma lixii	*Cornus* spp.	leaf	USA, Japan	[95]
Tubakia sp.	*Cornus* spp.	leaf	USA, Japan	[95]
Verticillium dahliae	*Cornus* spp.	leaf	USA	[95]
Xylaria sp.	*Cornus* spp.	leaf	USA	[95]

2.8. Hamamelis

Hamamelis (Hamamelidaceae), commonly known as witch hazel, comprises six species of ornamental shrubs. This genus is distributed across North America and eastern Asia. Bark extracts contain proanthocyanidins and polyphenolic fractions, with medicinal properties [107,108].

Fungal endophytes belonging to genera *Colletotrichum*, *Nigrospora*, *Pezicula*, and *Phyllosticta* have been isolated from *Hamamelis* plant tissues in the USA, China, Netherlands, Canada, and Japan (Table 9).

Table 9. Endophytic fungi isolated from *Hamamelis* species.

Species	Host Plant	Plant Part	Country	Reference
Colletotrichum acutatum	*H. virginiana*	leaf	Dutchess Co., USA	[109]
	Hamamelis sp.	leaf	Litchfield, USA	[109]
Nigrospora oryzae	*H. mollis*	leaf	China	[110]
Pezicula sporulosa	*H. mollis*	-	Netherlands	[111]
	H. virginiana	-	Canada	[112]
Phyllosticta hamamelidis	*H. japonica*	leaf	Japan	[113,114]

2.9. Jasminum

The genus *Jasminum* (Oleaceae) includes more than 200 species distributed in China, Africa, Asia, Australia, South Pacific Islands, and Europe. Jasmines are widely cultivated for ornamental, medical, and cosmetical uses. The species *J. sambac*, commonly known as Arabian Jasmine, is cultivated throughout India and tropical regions. This genus has been reported for several uses due to the following pharmaceutical activities: antimicrobial [115], antioxidant [116], antidiabetic [117], antiviral [118], and antitumor [119]. Seven species of endophytic fungi of the genus *Colletotrichum* have been reported from *J. sambac* in India and Vietnam (Table 10).

Table 10. Endophytic fungi isolated from *Jasminum* species.

Species	Host Plant	Plant Part	Country	Reference
Colletotrichum dematium	*J. sambac*	leaf	India	[120]
Colletotrichum truncatum	*J. sambac*	leaf	Vietnam	[121]
Colletotrichum jasminicola	*J. sambac*	leaf, shoot	India	[120]
Colletotrichum jasminigenum	*J. sambac*	leaf	Vietnam	[121]
Colletotrichum jasmini-sambac	*J. sambac*	leaf	Vietnam	[121]
Colletotrichum siamense	*J. sambac*	leaf	Vietnam	[121]
Colletotrichum sp.	*J. sambac*	leaf	Vietnam	[121]

2.10. Ligustrum

Ligustrum (Family *Oleaceae*) is a genus of about 50 species of shrubs and trees from warm areas of Europe to Asia [122]. Several species of the genus have been cultivated in many areas of the world as urban ornamental hedge and street trees. In particular, the most widespread species *L. lucidum* compete with and inhibit the regeneration of native flora, becoming invasive in many areas with a subtropical and temperate climate, such as North America, South America, Europe, Asia, Africa, and Oceania [123]. Due to its active constituents such as glycosides, flavonoids, phenylpropanoids, phenylethanoid, and terpenoids, *Ligustrum* spp. have been widely used as a health remedy in European, Chinese, and Japanese communities [124,125].

Collected data showed that 29 species of endophytes belonging to 20 genera have been found in plant tissues of *L. lucidum*, *L. compactum*, *L. quihoui*, *L. obsusifoilium*, and *L. vulgare* (Table 11): *Guignardia* (3 isolates), *Alternaria* (2 isolates), Colletotrichum (3 isolates), *Fusarium* (2 isolates), Xylaria (2 isolates), Pestalotiopsis (1 isolate), *Trichoderma* (2 isolates), *Lasiodiplodia* (2 isolates), Phomopsis (3 isolates), and one isolate of *Diplodia*, Geotrichum, *Libertella*, *Neocosmospora*, *Cladosporium*, *Peroneutypa*, *Penicillium*, *Phyllosticta*, *Pycnidiella*, and *Rhizopus*, respectively.

Table 11. Endophytic fungi isolated from *Ligustrum* species.

Species	Host Plant	Plant Part	Country	Reference
Alternaria alternata	*L. lucidum*	leaf, petiole	Buenos Aires, Argentina	[126]
Alternaria cheiranthi	*L. lucidum*	leaf	Buenos Aires, Argentina	[126]
Cladosporium oxysporum	*L. lucidum*	leaf	Buenos Aires, Argentina	[126]
Colletotrichum crassipes	*L. lucidum*	leaf	Buenos Aires, Argentina	[126]
Colletotrichum sp.	*L. roxburghii*	leaf	Bhavani, India	[34]
Colletotrichum gloeosporioides	*L. lucidum*	leaf	Buenos Aires, Argentina	[126]
Diplodia mutila	*L. lucidum*	stem	Buenos Aires, Argentina	[127]
Fusarium oxysporum	*L. lucidum*	-	Jiangsu, China	[128]
Fusarium lateritium	*L. lucidum*	stem	Buenos Aires, Argentina	[127]
Geotrichum candidum	*L. lucidum*	leaf	Buenos Aires, Argentina	[126]
Guignardia mangiferae	*L. compactum var. tschonski*	leaf	Kyoto, Japan	[129]
	L. quihoui	leaf	Kyoto, Japan	[129]
	L. obsusifoilium	leaf	Kyoto, Japan	[129]
Lasiodiplodia theobromae	*L. lucidum*	stem	Buenos Aires, Argentina	[127]
Lasiodiplodia sp.	*L. roxburghii*	leaf	Bhavani, India	[34]
Libertella sp.	*L. lucidum*	branches	Argentina	[130]
Neocosmospora solani	*L. lucidum*	-	Jiangsu, China	[128]
Peroneutypa scoparia	*L. lucidum*	branches	Argentina	[130]
Penicillum sp.	*L. lucidum*	leaf	China	[131]
Pestalotiopsis sp.	*L. roxburghii*	leaf	India	[132]
			Bhavani, India	[34]
Phomopsis ligustri-vulgaris	*L. lucidum*	leaf	Buenos Aires, Argentina	[126]
Phomopsis sp.	*L. vulgare*	leaf	Braunschweig, Germany	[133]
	L. roxburghii	leaf	Bhavani, India	[34]
Phyllosticta sp.	*L. roxburghii*	leaf	Bhavani, India	[34]

Table 11. *Cont.*

Species	Host Plant	Plant Part	Country	Reference
Pycnidiella resinae	*L. lucidum*	leaf	Buenos Aires, Argentina	[126]
Rhizopus microsporus	*L. lucidum*	stem	Buenos Aires, Argentina	[127]
Trichoderma harzianum	*L. lucidum*	leaf	Buenos Aires, Argentina	[126]
Trichoderma koningii	*L. lucidum*	stem	Buenos Aires, Argentina	[127]
Xylaria sp.	*L. roxburghii*	leaf	Bhavani, India	[34]
	L. lucidum	leaf	Buenos Aires, Argentina	[126]

2.11. Lonicera

Lonicera (Caprifoliaceae) is a genus that comprises more than 150 species of shrubs and twining climbers, occurring in North America, South Europe, North Africa, Philippines, and Malaysia [134]. *L. japonica* and *L. morrowii*, which are native to Asia, are ornamental species distributed in many areas of the world. In the USA, they are considered invasive plants [135]. Only 3 fungal species have been found to grow as endophytes in *Lonicera* plant tissues (Table 12): *Fusarium* sp., *Phyllosticta* sp., and *Guignarda mangiferae*.

Table 12. Endophytic fungi isolated from *Lonicera* species.

Species	Host Plant	Plant Part	Country	Reference
Fusarium sp.	*L. japonica*	leaf	Henan, China	[136]
Guignardia mangiferae	*L. morrowii*	leaf	Kyoto, Japan	[129]
Phyllosticta sp.	*L. morrowii*	leaf	Kyoto, Japan	[129]

2.12. Nerium

N. oleander, commonly called oleander, is the only species currently classified in the genus *Nerium* (Family Apocynaceae). This evergreen shrub is native or naturalized to a wide area, from the Mediterranean region to the Arabian Peninsula and Asia [137]. Several biologically active compounds have been reported in the bark (cardenolides, triterpenoidal saponins, oleanderol, rutin, dambonitol in leaves, odorosides), roots (triterpene, steroidal cardenolide, volatile oil, stearic acid, oleic acid), and flowers (gitoxigenin, uzarigenin, strospeside, odoroside H) [137–140].

Collected data showed that 38 fungi were isolated from leaves, stems, flowers, and roots of plants collected in India and China (Table 13). These isolates belong to the genera, *Fusarium* (4 isolates), *Penicillium* (4 isolates), *Cladosporium* (3 isolates), *Chaetomium* (3 isolates), *Colletotrichum* (3 isolates), *Aspergillus* (3 isolates), *Curvularia* (2 isolates), *Alternaria* (2 isolates), *Cylindrocephalum* (1 isolate), *Lasiodiplodia* (1 isolate), *Torula* (1 isolate), *Phyllosticta* (1 isolate), *Phoma* (1), *Rhizopus* (1 isolate), *Geomyces* (1 isolate), *Pseudothielavia* (1 isolate), *Trichoderma* (1 isolate), *Xylaria* (1 isolate), *Bipolaris* (1 isolate), *Cochliobolus* (1 isolate), and *Drechslera* (1 isolate).

Table 13. Endophytic fungi isolated from *Nerium* species.

Species	Host Plant	Plant Part	Country	Reference
Alternaria brassicicola	*N. oleander*	stem, flower	India	[141]
Alternaria sp.	*N. oleander*	leaf	Southern India	[142]
Aspergillus flavus	*N. oleander*	flower	Chennai, India	[143]
Aspergillus niger	*N. oleander*	flower	Chennai, India	[143]
Aspergillus sp.	*N. oleander*	stem, root	China	[144]
Bipolaris sp.	*N. oleander*	stem, flower	India	[141]
	N. oleander	stem, flower	India	[141]
Chaetomium sp.	*N. oleander*	stem	Hong Kong, China	[145]
	N. oleander	leaf	Southern India	[142]
	N. oleander	stem	Hong Kong, China	[145]
Cladosporium sp.	*N. oleander*	stem	India	[141]
	N. oleander	leaf	Southern India	[142]

Table 13. *Cont.*

Species	Host Plant	Plant Part	Country	Reference
Cochliobolus sp.	*N. oleander*	stem, flower	India	[141]
	N. oleander	stem	Hong Kong, China	[145]
Colletotrichum sp.	*N. oleander*	flower	Chennai, India	[143]
	N. oleander	leaf	Southern India	[142]
Curvularia brachyspora	*N. oleander*	stem, flower	India	[141]
Curvularia sp.	*N. oleander*	stem, flower	India	[141]
Cylindrocephalum sp.	*N. oleander*	stem, flower	India	[141]
Drechslera sp.	*N. oleander*	stem	India	[141]
Fusarium oxysporum	*N. oleander*	flower	Chennai, India	[143]
Fusarium semitectum	*N. oleander*	stem, flower	India	[141]
Fusarium sp.	*N. oleander*	stem, flower	India	[141]
	N. oleander	leaf	Southern India	[142]
Geomyces sp.	*N. oleander*	stem	China	[144]
Lasiodiplodia theobromae	*N. oleander*	flower	Chennai, India	[143]
Nigrospora sp.	*N. oleander*	root	China	[144]
	N. oleander	stem	China	[144]
Penicillium sp.	*N. oleander*	stem, flower	India	[141]
	N. oleander	root	China	[146]
	N. oleander	leaf	Southern India	[142]
Phoma sp.	*N. oleander*	stem	Hong Kong, China	[145]
Phyllosticta sp.	*N. oleander*	leaf	Southern India	[142]
Rhizopus stolonifera	*N. oleander*	flower	Chennai, India	[143]
Pseudothielavia terricola	*N. oleander*	stem	India	[141]
Torula sp.	*N. oleander*	stem	Hong Kong, China	[145]
Trichoderma sp.	*N. oleander*	stem, root	China	[144]
Xylaria sp.	*N. oleander*	leaf	Southern India	[142]

2.13. Robinia

Robinia is a genus of flowering plants of the family Fabaceae. *R. pseudoacacia*, called black locust, grows naturally on a wide range of sites. It is considered to be one of the 40 most invasive woody species all over the world [147] and it is included in the invasive alien species list of the EU [148,149]. It is used for many purposes, such as ornamental plant, for shelterbelts, land reclamation, fuelwood, and pulp production [147]. Six species of endophytic fungi were isolated from *R. pseudoacacia* in Germany, Slovakia, Hungary, and China (Table 14).

Table 14. Endophytic fungi isolated from *Robinia* species.

Species	Host Plant	Plant Part	Country	Reference
Beauveria bassiana	*R. pseudoacacia*	-	Mlyňany, Slovakia	[150]
Diaporthe oncostoma	*R. pseudoacacia*	stem	Hungary	[151]
Monodictys fluctuata	*R. pseudoacacia*	-	Germany	[152]
Fusarium sp.	*R. pseudoacacia*	-	Huaxi district, China	[153]
Gloniopsis sp.	*R. pseudoacacia*	-	Huaxi district, China	[153]
Clonostachys sp.	*R. pseudoacacia*	-	Huaxi district, China	[153]

3. An Overview of Fungal Diversity and Frequency

Investigations on the mycobiota of plants frequently reported new taxa or new species distribution, and several fungi are still undiscovered or undetected. Numerous higher plants have developed a variety of resistance mechanisms to prevent fungal infections. However, the presence of weakly pathogenic fungi in healthy plant tissues highlights the evolutionary continuum between latent pathogens and symptomless endophytes [15]. Generally, all plants have symbiotic interactions with fungal endophytes which can influence host performance in terms of disease resistance [154–156], stress tolerance [157], and biomass accumulation [158]. Fungal endophytes may also change according to plant tissues colonized [159], phenological growth stages, host genotypes [160], and geographical distribution areas [161].

In this review, a total of 428 endophytic species belonging to 122 fungal genera have been found in association with 13 plant genera (Table 1). The greatest level of fungal diversity was reported in in association with *Bauhinia* with 43 fungal genera and 94 fungal species, and *Cornus* with 44 fungal genera and 78 fungal species. The degree of fungal recovery from *Acacia* (29 genera, 51 species), *Albizia* (14 genera, 27 species), *Berberis* (17 genera, 29 species), *Caesalpinia* (19 genera, 42 species), *Cassia* (15 genera, 19 species), *Ligustrum* (20 genera, 29 species), and *Nerium* (21 genera, 37 species) was nearly half in comparison to the abundance noted in the genera *Bauhinia* and *Cornus*. Nonetheless, the lowest diversity showed for *Hamamelis* (4 species/genera), *Jasminus* (7 species, 1 genera), *Lonicera* (3 species/genera), and *Robinia* (6 species/genera) was also due to the lack of published research about fungal endophytes in these plant genera.

The literature evidenced that several fungal endophytes live in association with the investigated plants. The most representative genera in terms of abundance of isolated species were *Aspergillus* (40 spp.), *Penicillium* (30), *Fusarium* (29), *Colletotrichum* (27), *Alternaria* (14), and *Cladosporium* (14). These genera include ubiquitous and generalist fungi as well as several plant pathogens and saprobes [162–164].

It is worth noting the relative homogeneity in distribution of fungi such as *Colletotrichum*, *Fusarium*, and *Alternaria* among these plant genera. In fact, *Colletotrichum* was undetected only in *Lonicera* and *Robinia*, *Fusarium* in *Caesalpinia*, and *Hamamelis*, *Jasminus*, and *Alternaria* in *Cassia* and *Lonicera*. Although scarcely abundant, the fungal genus *Phyllosticta* was almost reported for all selected plants except for *Albizia*, *Jasminus*, *Robinia*, and *Hamamelis*. Other endophytic fungi were detected more occasionally. Future surveys may reveal the presence of additional fungal species also from less investigated plants, such as *Robinia*, *Jasminum*, and *Lonicera*.

The presence of pathogenic or saprotrophic fungi has already been discussed by several authors [165,166]. Table 1 shows that several of the listed fungi were apparently restricted to a single plant genus or at least exhibit some preference for a particular one. Some common and ubiquitous pathogens have been recovered in more than one plant host. This is the case of *F. oxysporum* (8 host plant species belonging to 7 different genera), *A. alternata*, *A. niger*, *C. gloeosporioides* (7 host plant species), *N. oryzae* (4 host plant species), *B. dothidea*, *C. globosum*, *C. acutatum* (3 host plant species), *A. ochraceus*, *A. pullulans*, and *C. truncatum* (3 host plant species).

4. The Most Common Plant Pathogens

The most frequent endophytes detected from the investigated plants are cosmopolitan and ubiquitous pathogens that may cause severe yield losses. In detail, *F. oxysporum* is responsible for the wilt of vascular tissues on numerous crops that may result in plant death, even if several strains have proved to be non-pathogenic [167]. It has been isolated from 8 different plant species belonging to 7 genera, namely *A. hindsii*, *A. julibrissin*, *B. malabarica*, *B. phoenicea*, *B. aristata*, *C. officinalis*, *L. lucidum*, and *N. oleander*. The fungus *A. alternata* may infect over 380 host plant species causing leaf spots, rots, and blights. It includes opportunistic forms in developing field crops as well as saprophytic strains that may cause harvest and post-harvest spoilage of harvested products. One of the major concerns represented by its infection is related to the production of mycotoxins that may be introduced in the food chain [168]. In this review, *A. alternata* has been found in association with 3 genera, in 7 plant species (*B. malabarica*, *B. racemosa*, *B. poiretii*, *B. aristata*, *Cornus* sp., *L. lucidum*, and *C. pulcherrima*). The saprophytic pathogen *A. niger* is responsible for the spoilage of a wide range of fruit, vegetable, and food products. It is also the causal agent of the black rot of onion bulbs, the kernel rot of maize, and the black mold rot of cherry [169,170]. It has been found within plant tissues of *A. arabica*, *A. lebbeck*, *B. fortificata*, *B. malabarica*, *B. racemosa*, *C. pulcherrima*, and *N. oleander* (7 plant species or 4 genera). Furthermore, three different species of *Colletotrichum* have been isolated from reviewed plants. *C. gloeosporioides* has been isolated from 7 plant species (3 genera), namely *A. hindsii*, *B. racemosa*,

B. aristata, C. echinata, C. officinalis, C. stolonifera, and *L. lucidum,* whereas *C. acutatum* has been found in *Cornus* spp., *Hamamelis* sp., and *H. virginiana* (3 species; 2 genera). Both *Colletotrichum* species may cause severe fruit rot mainly occurring in pre- and post-harvest [171]. Moreover, *C. truncatum,* the causal agent of anthracnose disease affecting several leguminous crops [171], has been collected from 2 plant genera, namely *A. hindsii* and *J. sambac.* Furthermore, *C. lunata,* was isolated from the tissues of 4 plant species (2 genera), including *B. malabarica, B. racemosa, B. phoenicea,* and *C. sappan,* is the causal agent of seed and seedling blight in several crops, such as rice, millet, sugarcane, and rice, and of maize leaf spot [172]. Besides, *B. dothidea* reported in association with *A. karroo, Cornus* sp., and *C. officinalis* may cause cankers, dieback, fruit rot, and blue stain in woody plants, including *Acacia, Eucalyptus, Vitis,* and *Pistachio* [12]. Concerning the species *F. lateritium,* it has been extensively investigated as the causal agent of chlorotic leaf distortion on sweet potato (*Ipomoea batatas*) in the USA [173]. This fungus has been isolated from three different plant species and genera (*B. aristata, C. controversa,* and *L. lucidum*). Moreover, the common soil-borne fungus *G. candidum,* found in association with *B. vahlii, C. sappan,* and *L. lucidum,* is the causal agent of sour-rot of tomatoes and citrus fruits, and it is also one of the most economically important post-harvest diseases of citrus [174]. Also, *C. cladosporioides,* detected in *B. racemosa, C. echinata,* and *C. stolonifera,* is the causal agent of blossom blight in strawberries [175]. Other pathogenic fungi associated with these selected plants are less widespread and some of them are subjected to containment measures in some countries. This is the case of *N. parvum, N. oryzae, L. theobromae,* and *D. destructiva.* In particular, *N. parvum,* isolated as an endophyte in three *Acacia* species (*A. heterophylla, A. karroo,* and *A. koa*), is one of the most aggressive causal agent of Botryosphaeria dieback on the grapevine and it is known as an aggressive polyphagous pathogen attacking more than 100 plant hosts [176]. Also, *N. oryzae,* reported from *H. mollis, B. phoenicea, B. racemosa,* and *B. fortificata,* may reduce plant growth and seed quality of rice plants as well as *Brassica* spp., maize, and cotton [177]. Moreover, *L. theobromae,* found in association with six different plant species (*A. karroo, A. koa, B. racemosa, C. echinata, L. lucidum,* and *N. oleander*), is the causal agent of dieback, root rot, and blights for a wide range of plant hosts, mainly located in tropical and subtropical regions [178]. Finally, *D. desctructiva,* recovered from three different species of *Cornus,* is the causal agent of the dogwood anthracnose, a devastating disease that was firstly documented in the USA and then introduced into Europe [179].

Generally, closely related organisms, including pathogenic fungi as well as those non-pathogenic, may share similar ecological niches and may potentially interact among themselves. Their co-occurrence could be due to phylogenetic evolution or some unclear biological benefits gained [180,181]. The effects of this interaction may lead to a definition of spaces for development and survival. Nevertheless, it is widely known that non-indigenous species represent one of the greatest threats to native biodiversity [11,23–25]. In fact, a fungal invasion into a new ecosystem may change the native endophytic community structure, leading to the extinction of host-specialized fungi [182]. This antagonistic phenomenon is regulated by the production of antifungal compounds, mycoparasitism, or competition for space and resources [180], as well as a synergy of these interactions [181]. Biological invasions may set in motion a long-lasting cascade of effects on the plant host and associated species in unpredictable ways. Generally, the ecological importance of native species prior to the invasion may not be quantified because of the lack of information on fungal communities, especially for non-pathogenic fungal species. As a consequence of global trade and climatic or environmental changes, studies about the impact of new organisms on the ecosystem represent innovative challenges worldwide. In view of these considerations, even if fungal pathogens found in association with investigated plants are widely distributed in the EU [182–190], the risk posed by the introduction of potentially noxious species may be very high. Thus, our results suggest the importance of monitoring imported material to avoid the introduction of such alien species.

5. Emerging and Potential Threats Due to Commercial Trade

Several species reported in this review are Quarantine Pests (QP) or Regulated as Non-Quarantine Pests (RNQP), as defined by containment measures within the importing country [191]. Among the fungal pathogens found in *Cornus* species, *Elsinoe fawcettii* is listed as a QP in the EU, Tunisia, and Israel. This fungus is the causal agent of Citrus scab and it is one of the most important pathogens in many areas of citrus production [192]. *E. fawcettii* is common in South America and its presence has been detected in other areas such as Central and South Africa, India and South-Eastern Asia, and Australia [192].

Furthermore, the following pathogens are RNQP in the EU: *F. verticilloides* (isolated from *B. malabarica* and *A. lebbeck*), *C. acutatum* (isolated from *Cornus* spp., *H. virginiana*, and *Hamamelis* sp.), *S. sclerotiorum* (isolated from *C. stolonifera*), and *V. dahliae* (isolated from *Cornus* sp.). Outside the EU, the following species are listed as QP: *L. theobromae* and *P. palmivora* (in Morocco), *A. nidulans*, *A. macrospora*, *C. kahawae*, *C. citrullina*, *C. herbarum*, *C. pallescens*, *A. brassicicola*, *F. semitectum*, *F. verticillioides*, *N. oryzae*, and *P. longissima* (in Mexico), *P. graminis* (Canada and USA), *Diaporthe tersa* (in Israel), *C. acutatum* (in Tunisia and Israel), and *C. gloeosporioides* and *P. capitalensis* (in Egypt) [192].

Organisms that move across continents may or may not become dangerous depending on several factors, and unexpected consequences may occur [193,194]. The current knowledge about the fungal community associated with ornamental plants and their interaction with the environment is fragmentary. Fungi species generally well known as pathogens, are not necessarily pathogenic when isolated as endophytes [6–8]. Genetic mutation can occur in virulent pathogens, transforming the original pathogen into a nonpathogenic strain [9]. Likewise, even though some endophytes are mutualistic, this does not imply that they will not have negative impacts if introduced in a new ecosystem [6,9]. Alien pathogens can often encounter more susceptible host plants and different microbial and abiotic environments without their own 'natural enemies'. The so-called 'risk pathway' defined by international protocols tend to assume that the pathogen will attack a plant host taxonomically similar to that of the susceptibile plant species in its native countries. However, an invasive pathogen may spread to new target hosts, when introduced in a new ecosystem, and novel pathogen combinations can occur [11]. The disease outcomes of these combinations may be extremely complex and the invasive pathogen populations can reach explosive distribution levels that are usually difficult to eradicate once established [23–25]. Beyond the damage which may occur on the host plant species and local microbial communities, biological invasions may affect entire ecosystems and the connected ecosystem processes and services, such as soil fertility, fire control, hydrology, and recreation and tourism amenities [23–25]. In response to expanding global trade, several EU regulations [27–29] and international protocols [195,196] are aimed at regulating over-dissemination and accidental introduction of plant diseases. However, despite existing laws and efforts to prevent the introduction of potential pathogens at ports of entry, many of them will evade detection and establish alien populations [197,198]. Many pathogenic fungi may be undetected, transported in the form of inocula as endophytes, propagules, mycelium, or spores of vegetative material. In addition, large import volumes often permit the inspection of only a small proportion of the introduced plants. According to the precautionary principle, all imported plant species should be considered as a potential threat (vectors of fungi), therefore the presence and establishment may not depend on the number of arrivals. As a consequence, even a reduced amount of infected plants, which can easily escape phytosanitary inspections, may cause the introduction and the spread of diseases with devastating outcomes [199]. The development of tools, such as new molecular diagnosistics [200] and volatile compounds detection devices [201], that allow the rapid and on-site identification of potentially invasive species and the screening of large volumes of plants, clearly appears to be essential [202]. Despite increasing trade, targeted investment in biosecurity may be effective to reduce pathogen introduction and limit the establishment of alien microorganisms. Thus, we highlight the importance of surveillance due to the potential risk of accidental introductions in the absence of effective biosecurity measures.

6. Conclusions

Globalization has led to intensified movement of people, plants, and plant products, and an increase in the unintentional introduction of non-native fungal species into new ecosystems. Many plant pathogens are biological opportunistic invaders causing several billion dollars in losses to crops, pastures, and forests annually, worldwide. Consideration needs to be given to building resilience in the new environments, from the perspective of pathogen introductions. In particular, the monitoring of plants and plant products, plus early identification-detection of pathogen risks are key steps towards ensuring successful regulation to exclude potential disorders caused by pathogens. This review demonstrated the broad fungal diversity recovered from a small group of ornamental plants that have been relatively unexplored as fungal hosts. Even if the reviewed plant genera are not recognized as sources of significant forest diseases, that have had an ecosystemic impact on a continental scale in the past, we highlight the risk represented by plants as inoculum sources of potentially harmful organisms. Overall, many other species not listed by the EU have represented or may cause important impact in many ecosystemic, environmental, and ecological issues. Our literature search revealed that fungal species may also be introduced through a few hundred plants and invade new ecosystems. In this context, it is important to underline that the amount of imported plant material may not be related to a specific risk, but needs to be considered and evaluated to estimate the negative impacts on agriculture, forestry, and public health, associated with non-indigenous species in European ecosystems. For example, little is known about the effects of invasive species on ecosystem services, although some historic pest invasions (e.g., chestnut blight from North America to Europe) have destroyed host tree species in their locations. The true challenge lies in preventing further damage to natural and managed ecosystems. For this reason, preventative policies need to take into account the means through which pathogens gain access to the EU. The accidental introduction of potentially harmful pathogens also links to other issues of major policy concern (i.e., biotechnology, human health, climate change, etc.) that should be addressed through improved international cooperation and a holistic approach. We should expect that some strategies should be continued or further established to prevent or monitor future introductions, especially at airports, seaports, and other ports of entry, to reduce risks to an acceptable level and preserve natural and agricultural ecosystems.

Author Contributions: Conceptualization, L.G., G.d., F.V. and S.L.W.; literature investigation L.G., G.d., M.S. and M.R.; writing—original draft preparation, L.G. and G.d.; writing—review and editing, project administration, F.V. and S.L.W.; funding acquisition, F.V. and S.L.W. All authors have read and agreed to the published version of the manuscript.

Funding: This research was funded by the following projects: MIURPON (Grant number Linfa 03PE_00026_1; Grant number Marea 03PE_00106); POR FESR CAMPANIA 2014/2020-O.S. 1.1 (Grant number Bioagro 559); MISE CRESO (Grant number Protection n. F/050421/01-03/X32); PSR Veneto 16.1.1 (Grant number Divine n. 3589659); PSR Campania 2014/2020 Misura 16-Tipologia di intervento 16.1–Azione 2 'Sostegno ai Progetti Operativi di Innovazione (POI)'—Progetto 'DI.O.N.IS.O.', C.U.P. B98H19005010009; European Union Horizon 2020 Research and Innovation Program, ECOSTACK (Grant agreement no. 773554); PRIN 2017 (Grant number PROSPECT 2017JLN833).

Conflicts of Interest: The authors declare no conflict of interest.

References

1. Andrews, J.H.; Hirano, S.S. *Microbial Ecology of Leaves*; Springer: Berlin/Heidelberg, Germany; New York, NY, USA, 1991; pp. 467–479.
2. Ragozzino, A.; d'Errico, G. Interactions between nematodes and fungi: A concise review. *Redia* **2011**, *94*, 123–125.
3. d'Errico, G.; Mormile, P.; Malinconico, M.; Bolletti Censi, S.; Lanzuise, S.; Crasto, A.; Woo, S.L.; Marra, R.; Lorito, M.; Vinale, F. *Trichoderma* spp. and a carob (*Ceratonia siliqua*) galactomannan to control the root-knot nematode *Meloidogyne incognita* on tomato plants. *Can. J. Plant. Pathol.* **2020**, 1–8. [CrossRef]

4. Vinale, F.; Nicoletti, R.; Lacatena, F.; Marra, R.; Sacco, A.; Lombardi, N.; d'Errico, G.; Digilio, M.C.; Lorito, M.; Woo, S.L. Secondary metabolites from the endophytic fungus *Talaromyces pinophilus*. *Nat. Prod. Res.* **2017**, *31*, 1778–1785. [CrossRef] [PubMed]

5. d'Errico, G.; Aloj, V.; Flematti, G.R.; Sivasithamparam, K.; Worth, C.M.; Lombardi, N.; Ritini, A.; Marra, R.; Lorito, M.; Vinale, F. Metabolites of a *Drechslera* sp. endophyte with potential as biocontrol and bioremediation agent. *Nat. Prod. Res.* **2020**, 1–9. [CrossRef] [PubMed]

6. Stone, J.K.; Bacon, C.W.; White, J.F. An overview of endophytic microbes: Endophytism defined. *Microb. Endophytes.* **2000**, *3*, 29–33.

7. Wilson, D. Fungal endophytes which invade insect galls: Insect pathogens, benign saprophytes, or fungal inquilines? *Oecologia* **1995**, *103*, 255–260. [CrossRef]

8. Petrini, O. Fungal endophytes of tree leaves. In *Microbial Ecology of Leaves*; Andrews, J.H., Hirano, S.S., Eds.; Springer: New York, NY, USA, 1991; pp. 179–197.

9. Osono, T. Role of phyllosphere fungi of forest trees in the development of decomposer fungal communities and decomposition processes of leaf litter. *Can. J. Microbiol.* **2006**, *52*, 701–716. [CrossRef]

10. Redman, R.S.; Dunigan, D.D.; Rodriguez, R.J. Fungal symbiosis from mutualism to parasitism: Who controls the outcome, host or invader? *New Phytol.* **2001**, *151*, 705–716. [CrossRef]

11. Dunn, A.M.; Hatcher, M.J. Parasites and biological invasions: Parallels, interactions, and control. *Trends Parasit.* **2015**, *31*, 189–199. [CrossRef]

12. Marsberg, A.; Kemler, M.; Jami, F.; Nagel, J.H.; Postma-Smidt, A.; Naidoo, S.; Wingfield, M.J.; Crous, P.W.; Spatafora, J.W.; Hesse, C.N.; et al. *Botryosphaeria dothidea*: A latent pathogen of global importance to woody plant health. *Mol. Plant. Pathol.* **2017**, *18*, 477–488. [CrossRef]

13. Slippers, B.; Smit, W.A.; Crous, P.W.; Coutinho, T.A.; Wingfield, B.D.; Wingfield, M.J. Taxonomy, phylogeny and identification of Botryosphaeriaceae associated with pome and stone fruit trees in South Africa and other regions of the world. *Plant. Pathol.* **2007**, *56*, 128–139. [CrossRef]

14. Facon, B.; Genton, B.J.; Shykoff, J.; Jarne, P.; Estoup, A.; David, P. A general eco-evolutionary framework for understanding bioinvasions. *Trends Ecol. Evol.* **2006**, *21*, 130–135. [CrossRef]

15. Saul, W.C.; Jeschke, J.; Heger, T. The role of eco-evolutionary experience in invasion success. *NeoBiota* **2013**, *17*, 57. [CrossRef]

16. Pautasso, M.; Schlegel, M.; Holdenrieder, O. Forest health in a changing world. *Microb. Ecol.* **2015**, *69*, 826–842. [CrossRef] [PubMed]

17. Grünwald, N.J. Genome sequences of *Phytophthora* enable translational plant disease management and accelerate research. *Can. J. Plant Pathol.* **2012**, *34*, 13–19. [CrossRef]

18. Cleary, M.; Nguyen, D.; Marčiulynienė, D.; Berlin, A.; Vasaitis, R.; Stenlid, J. Friend or foe? Biological and ecological traits of the European ash dieback pathogen *Hymenoscyphus fraxineus* in its native environment. *Sci. Rep.* **2016**, *6*, 21895. [CrossRef] [PubMed]

19. Dickie, I.A.; Bolstridge, N.; Cooper, J.A.; Peltzer, D.A. Co-invasion by *Pinus* and its mycorrhizal fungi. *New Phytol.* **2010**, *187*, 475–484. [CrossRef] [PubMed]

20. McNeely, J.A. An introduction of human dimensions of invasive alien species. In *The Great Reshuffling: Human Dimensions of Invasive Alien Species*; McNeely, J.A., Ed.; IUCN: Cambridge, UK, 2001; pp. 5–20.

21. Hejda, M.; Pyšek, P.; Jarošík, V. Impact of invasive plants on the species richness, diversity and composition of invaded communities. *J. Ecol.* **2009**, *97*, 393–403. [CrossRef]

22. Olden, J.D.; Rooney, T.P. On defining and quantifying biotic homogenization. *Glob. Ecol. Biogeog.* **2006**, *15*, 113–120. [CrossRef]

23. Ehrenfeld, J.G. Ecosystem consequences of biological invasions. *Annu. Rev. Ecol. Evol. Syst.* **2010**, *41*, 59–80. [CrossRef]

24. Vilà, M.; Espinar, J.L.; Hejda, M.; Hulme, P.E.; Jarošík, V.; Maron, J.L.; Pyšek, P. Ecological impacts of invasive alien plants: A meta-analysis of their effects on species, communities and ecosystems. *Ecol. Lett.* **2011**, *14*, 702–708. [CrossRef] [PubMed]

25. Lazzaro, L.; Mazza, G.; d'Errico, G.; Fabiani, A.; Giuliani, C.; Inghilesi, A.F.; Lagomarsino, A.; Landi, S.; Lastrucci, L.; Pastorelli, R.; et al. How ecosystems change following invasion by *Robinia pseudoacacia*: Insights from soil chemical properties and soil microbial, nematode, microarthropod and plant communities. *Sci. Total Environ.* **2018**, *622*, 1509–1518. [CrossRef] [PubMed]

26. Jeger, M.; Schans, J.; Lövei, G.L.; van Lenteren, J.; Navajas, M.; Makowski, D.B.; Ceglarska, E. Risk assessment in support of plant health. *Efsa J.* **2012**, *10*, s1012. [CrossRef]

27. European Union. Regulation (EU) 2016/2031 of the European Parliament of the Council of 26 October 2016 on protective measures against pests of plants, amending Regulations (EU) No 228/2013,(EU) No 652/2014 and (EU) No 1143/2014 of the European Parliament and of the Council and repealing Council Directives 69/464/EEC, 74/647/EEC, 93/85/EEC, 98/57/EC, 2000/29/EC, 2006/91/EC and 2007/33/EC. *Off. J. Eur. Union* **2016**, *317*, 4–104.

28. European Union. Commission Implementing Regulation (EU) 2018/2019 of 18 December 2018 establishing a provisional list of high risk plants, plant products or other objects, within the meaning of Article 42 of Regulation (EU) 2016/2031 and a list of plants for which phytosanitary certificates are not required for introduction into the Union, within the meaning of Article 73 of that Regulation. *Off. J. Eur. Union* **2018**, *323*, 10–15.

29. Directive, C. Council Directive 2000/29/EC of 8 May 2000 on protective measures against the introduction into the Community of organisms harmful to plants or plant products and against their spread within the Community (Annex II). *Off. J. Eur. Union.* **2000**, *169*, 60.

30. European Food Safety Authority, (EFSA); Dehnen-Schmutz, K.; Jaques Miret, J.A.; Jeger, M.; Potting, R.; Corini, A.; Simone, G.; Kozelska, S.; Munoz Guajardo, I.; Stancanelli, G.; et al. Information required for dossiers to support demands for import of high risk plants, plant products and other objects as foreseen in Article 42 of Regulation (EU) 2016/2031. *EFSA Support Publ* **2018**, *15*, 1492E. [CrossRef]

31. European Food Safety Authority, (EFSA); Panel on Plant Health (PLH); Bragard, C.; Dehnen-Schmutz, K.; Di Serio, F.; Gonthier, P.; Jacques, M.A.; Miret, J.A.J.; Justesen, A.F.; MacLeod., A.; et al. Guidance on commodity risk assessment for the evaluation of high risk plants dossiers. *Efsa J* **2019**, *17*, e05668. [CrossRef]

32. Lorenzo, P.; González, L.; Reigosa, M.J. The genus *Acacia* as invader: The characteristic case of *Acacia dealbata* Link in Europe. *Ann. For. Sci.* **2010**, *67*, 101. [CrossRef]

33. Ratnayake, K.; Joyce, D. Native Australian acacias: Unrealised ornamental potential. *Chron. Horticult.* **2010**, *50*, 19–22.

34. Suryanarayanan, T.S.; Devarajan, P.T.; Girivasan, K.P.; Govindarajulu, M.B.; Kumaresan, V.; Murali, T.S.; Venkatesan, G. The host range of multi-host endophytic fungi. *Curr. Sci.* **2018**, *115*, 1963–1969. [CrossRef]

35. Kaur, T.; Kaur, J.; Kaur, A.; Kaur, S. Larvicidal and growth inhibitory effects of endophytic *Aspergillus niger* on a polyphagous pest, *Spodoptera litura*. *Phytoparasitica* **2016**, *44*, 465–476. [CrossRef]

36. Jiang, M.; Cao, L.; Zhang, R. Effects of *Acacia* (*Acacia auriculaeformis* A. Cunn) associated fungi on mustard (*Brassica juncea* (L.) Coss. var. *foliosa* Bailey) growth in Cd- and Ni-contaminated soils. *Lett. Appl. Microbiol.* **2008**, *47*, 561–565. [CrossRef]

37. Tran, H.B.Q.; Mcrae, J.M.; Lynch, F.; Palombo, E.A. Identification and bioactive properties of endophytic fungi isolated from phyllodes of *Acacia* species. In *Current Research, Technology and Education Topics in Applied Microbiology and Microbial Biotechnology*; Méndez-Vilas, A., Ed.; Formatex: Badajoz, Spain, 2010; Volume 1, pp. 377–382.

38. Li, H.Y.; Li, D.W.; He, C.M.; Zhou, Z.P.; Mei, T.; Xu, H.M. Diversity and heavy metal tolerance of endophytic fungi from six dominant plant species in a Pb-Zn mine wasteland in China. *Fungal Ecol.* **2012**, *5*, 309–315. [CrossRef]

39. Jami, F.; Marincowitz, S.; Slippers, B.; Crous, P.W.; Le Roux, J.J.; Richardson, D.M.; Wingfield, M.J. Botryosphaeriaceae associated with *Acacia heterophylla* (La Réunion) and *Acacia koa* (Hawaii). *Fungal Biol.* **2019**, *123*, 783–790. [CrossRef] [PubMed]

40. González-Teuber, M.; Jiménez-Alemán, G.H.; Boland, W. Foliar endophytic fungi as potential protectors from pathogens in myrmecophytic *Acacia* plants. *Commun. Integr. Biol.* **2014**, *7*, 3–7. [CrossRef] [PubMed]

41. Jami, F.; Slippers, B.; Wingfield, M.J.; Loots, M.T.; Gryzenhout, M. Temporal and spatial variation of Botryosphaeriaceae associated with *Acacia karroo* in South Africa. *Fungal Ecol.* **2015**, *15*, 51–62. [CrossRef]

42. Jami, F.; Slippers, B.; Wingfield, M.J.; Gryzenhout, M. Greater Botryosphaeriaceae diversity in healthy than associated diseased *Acacia karroo* tree tissues. *Australas. Plant Pathol.* **2013**, *42*, 421–430. [CrossRef]

43. El-Sayed, A.S.A.; Moustafa, A.H.; Hussein, H.A.; El-Sheikh, A.A.; El-Shafey, S.N.; Fathy, N.A.M.; Enan, G.A. Potential insecticidal activity of *Sarocladium strictum*, an endophyte of *Cynanchum acutum*, against *Spodoptera littoralis*, a polyphagous insect pest. *Biocat. Agric. Biotechnol.* **2020**, *24*, 101524. [CrossRef]

44. McCabe, E.S. Endophytic Symbionts of *Acacia Victoriae*: Diversity and Potential for Antibiotic Production. Ph.D. Thesis, University of Arizona, Tucson, AZ, USA, 2019.

45. Kokila, K.; Priyadharshini, S.D.; Sujatha, V. Phytopharmacological properties of *Albizia* species: A review. *Int. J. Pharm. Pharm. Sci.* **2013**, *5*, 70–73.

46. CABI. Available online: https://www.cabi.org/isc/datasheet/4008#005AE80D-6260-41E0-A2DA-0B4BD78D5383 (accessed on 20 October 2020).

47. Wulandari, R.S.; Suryantini, R. Growth of *Albizia* in Vitro: Endophytic Fungi as Plant Growth Promote of Albizia. *Int. Sch. Sci. Res. Inn.* **2018**, *12*, 8.

48. Ali, B.Z.; Alfayed, A.A. Endophytic Fungi from Leaves and Twigs of *Albizia lebbeck* and Their Antifungal Activity. *Ibn AL Haitham J. For. Pure Appl. Sci.* **2017**, *27*, 24–33.

49. Gill, D.L. *Fusarium* wilt infection of apparently healthy mimosa tree. *Plant. Dis. Rep.* **1967**, *35*, 111–128.

50. Sharma, N.; Kushwaha, M.; Arora, D.; Jain, S.; Singamaneni, V.; Sharma, S.; Shankar, R.; Bhushan, S.; Gupta, P.; Jaglan, S. New cytochalasin from *Rosellinia sanctae-cruciana*, an endophytic fungus of *Albizia lebbeck*. *J. Appl. Microb.* **2018**, *125*, 111–120. [CrossRef] [PubMed]

51. da Silva, K.L.; Biavatti, M.W.; Leite, S.N.; Yunes, R.A.; Delle Monache, F.; Cechinel Filho, V. Phytochemical and pharmacognostic investigation of *Bauhinia forficata* Link (Leguminosae). *Z. Naturforsch. C.* **2000**, *55*, 478–480. [CrossRef]

52. Silva, K.L.D.; Cechinel Filho, V. Plantas do gênero *Bauhinia*: Composição química e potencial farmacológico. *Quím. Nova* **2002**, *25*, 449–454. [CrossRef]

53. Bezerra, J.D.P.; da Silva, L.F.; de Souza-Motta, C.M. The Explosion of Brazilian Endophytic Fungal Diversity: Taxonomy and Biotechnological Potentials. In *Advancing Frontiers in Mycology & Mycotechnology*; Satyanarayana, T., Deshmukh, S., Deshpande, M., Eds.; Springer: Singapore, 2019. [CrossRef]

54. Hilarino, M.P.A.; de Silveira, F.A.O.; Oki, Y.; Rodrigues, L.; Santos, J.C.; Corrêa, A.; Fernandes, G.W.; Rosa, C.A. Distribuição da comunidade de fungos endofíticos em folhas de *Bauhinia brevipes* (Fabaceae). *Acta Bot. Bras.* **2011**, *25*, 815–821. [CrossRef]

55. Bezerra, J.D.; Nascimento, C.C.; Barbosa, R.D.N.; da Silva, D.C.; Svedese, V.M.; Silva-Nogueira, E.B.; Gomes, B.S.; Paiva, L.M.; Souza-Motta, C.M. Endophytic fungi from medicinal plant *Bauhinia forficata*: Diversity and biotechnological potential. *Braz. J. Microbiol.* **2015**, *46*, 49–57. [CrossRef]

56. Kannan, K.P.; Madhan Kumar, D.; Ramya, P.R.; Madhu Nika, S.; Meenatchi, G.; Sowmya, A.N.; Bhuvaneswari, S. Diversity of endophytic fungi from salt tolerant plants. *Int. J. ChemTech Res.* **2014**, *6*, 4084–4088.

57. Murali, T.S.; Suryanarayanan, T.S.; Venkatesan, G. Fungal endophyte communities in two tropical forests of southern India: Diversity and host affiliation. *Mycol. Progs.* **2007**, *6*, 191–199. [CrossRef]

58. Araujo-Melo, R.; Souza, I.; Oliveira, C.; Araújo, J.; Sena, K.; Coelho, L. Isolation and Identification of Endophyte Microorganisms from *Bauhinia monandra* Leaves, Mainly *Actinobacteria*. *Biotechnol. J. Int.* **2017**, *17*, 1–12. [CrossRef]

59. de Feitosa, A.O.; Dias, A.C.S.; da Ramos, G.C.; Bitencourt, H.R.; Siqueira, J.E.S.; Marinho, P.S.B.; Barison, A.; Ocampos, F.M.M.; do Marinho, A.M.R. Letalidad de citocalasina B y otros compuestos aislados del hongo *Aspergillus* spp. (Trichocomaceae) endófito de *Bauhinia guianensis* (Fabaceae). *Rev. Argent. Microbiol.* **2016**, *48*, 259–263. [CrossRef] [PubMed]

60. Pinheiro, E.A.A.; Carvalho, J.M.; Dos Santos, D.C.P.; Feitosa, A.O.; Marinho, P.S.B.; Guilhon, G.M.S.P.; Santos, L.S.; De Souza, A.L.D.; Marinho, A.M.R. Chemical constituents of *Aspergillus* sp EJC08 isolated as endophyte from *Bauhinia guianensis* and their antimicrobial activity. *An. Acad. Bras. Cien.* **2013**, *85*, 1247–1252. [CrossRef] [PubMed]

61. Bagchi, B. A relative study of endophytic fungi during winter and monsoon in *Bauhinia vahlii* from chilkigarh. *J. Glob. Biosci.* **2019**, *8*, 6257–6265.

62. Pinheiro, E.A.A.; Pina, J.R.S.; Feitosa, A.O.; Carvalho, J.M.; Borges, F.C.; Marinho, P.S.B.; Marinho, A.M.R. Bioprospecting of antimicrobial activity of extracts of endophytic fungi from *Bauhinia guianensis*. *Rev. Argent. Microbiol.* **2017**, *49*, 3–6. [CrossRef]

63. Suryanarayanan, T.S.; Murali, T.S.; Venkatesan, G. Occurrence and distribution of fungal endophytes in tropical forests across a rainfall gradient. *Can. J. Bot.* **2002**, *80*, 818–826. [CrossRef]

64. Ravijara, N.S. Fungal endophytes in five medicinal plant species from Kudremukh Range, Western Ghats of India. *J. Basic Microbiol.* **2005**, *45*, 230–235.

65. Kandasamy, P.; Manogaran, S.; Dhakshinamoorthy, M.; Kannan, K.P. Evaluation of antioxidant and antibacterial activities of endophytic fungi isolated from *Bauhinia racemosa* Lam. and *Phyllanthus amarus* Schum. and Thonn. *J. Chem. Pharm.* **2015**, *7*, 366–379.

66. Pinheiro, E.A.A.; Borges, F.C.; Pina, J.R.S.; Ferreira, L.R.S.; Cordeiro, J.S.; Carvalho, J.M.; Feitosa, A.O.; Campos, F.R.; Barison, A.; Souza, A.D.L.; et al. Annularins I and J: New metabolites isolated from endophytic fungus *Exserohilum rostratum*. *J. Braz. Chem. Soc.* **2016**, *27*, 1432–1436. [CrossRef]

67. Kim, Y.D.; Kim, S.H.; Landrum, L.R. Taxonomic and phytogeographic implications from ITS phylogeny in *Berberis* (Berberidaceae). *J. Plant Res.* **2004**, *117*, 175–182. [CrossRef]

68. Sharma, S.; Gupta, S.; Dhar, M.K.; Kaul, S. Diversity and Bioactive Potential of Culturable Fungal Endophytes of Medicinal Shrub *Berberis aristata* DC: A First Report. *Mycobiology* **2018**, *46*, 370–381. [CrossRef] [PubMed]

69. Mokhber-Dezfuli, N.; Saeidnia, S.; Gohari, A.R.; Kurepaz-Mahmoodabadi, M. Phytochemistry and pharmacology of *Berberis* species. *Pharmacogn. Rev.* **2014**, *8*, 8. [PubMed]

70. Sun, J.; Guo, L.; Zang, W.; Ping, W.; Chi, D. Diversity and ecological distribution of endophytic fungi associated with medicinal plants. *Sci. China Life Sci.* **2008**, *51*, 751–759. [CrossRef] [PubMed]

71. Sati, S.C.; Belwal, M.; Pargaein, N. Diversity of water borne conidial fungi as root endophytes in temperate forest plants of western Himalaya. *Nat. Sci.* **2008**, *6*, 59–65.

72. Erick, T.K.; Kiplimo, J.; Matasyoh, J. Two New Aliphatic Alkenol Geometric Isomers and a Phenolic Derivate from Endophytic Fungus *Diaporthe* sp. Host to *Syzygium cordatum* (Myrtaceae). *Sci. Lett.* **2019**, *7*, 108–118.

73. Yang, R.; Chen, G.; Chen, R. Screening and identification of endophytic fungi with antifungal activities on pathogenic fungi from *Berberis thunbergii* cv. *atropurpurea*. *Acta Bot. Boreali Occident. Sin.* **2013**, *33*, 401–406.

74. Wang, M.N.; Wan, A.M.; Chen, X.M. Barberry as alternate host is important for *Puccinia graminis* f. sp. *tritici* but not for *Puccinia striiformis* f. sp. *tritici* in the US Pacific Northwest. *Plant. Dis.* **2015**, *99*, 1507–1516.

75. Zanin, J.L.B.; De Carvalho, B.A.; Salles Martineli, P.; Dos Santos, M.H.; Lago, J.H.G.; Sartorelli, P.; Viegas, C.; Soares, M.G. The genus *Caesalpinia* L. (Caesalpiniaceae): Phytochemical and pharmacological characteristics. *Molecules* **2012**, *17*, 7887–7902. [CrossRef]

76. Uma Maheswari, N.; Saranya, P. Isolation and identification and phytochemical screening of endophytes from medicinal plants. *Int. J. Biol. Res.* **2018**, *3*, 16–24.

77. Campos, F.F.; Sales Junior, P.A.; Romanha, A.J.; Araújo, M.S.; Siqueira, E.P.; Resende, J.M.; Zani, C.L. Bioactive endophytic fungi isolated from *Caesalpinia echinata* Lam.(Brazilwood) and identification of beauvericin as a trypanocidal metabolite from *Fusarium* sp. *Mem. Inst. Oswaldo Cruz.* **2015**, *110*, 65–74. [CrossRef]

78. Hafsan, H.; Sukmawaty, S.; Masri, M.; Aziz, I.R.; Wulandari, S.L. Antioxidant Activities of Ethyl Acetic Extract of Endophytic Fungi from *Caesalpinia sappan* L. and *Eucheuma* sp. *Int. J. Pharm. Res.* **2018**, *11*, 1–6.

79. de Lima, T.E.F.; Cavalcanti, M.D.S. Endophytes and phylloplane fungi of *Caesalpinia echinata* Lam. of estação ecológica de Tapacurá, PE. *Agrotrópica* **2014**, *26*, 43–50. [CrossRef]

80. Cota, B.B.; Tunes, L.G.; Maia, D.N.B.; Ramos, J.P.; de Oliveira, D.M.; Kohlhoff, M.; Alves, T.M.A.; Souza-Fagundes, E.M.; Campos, F.F.; Zani, C.L. Leishmanicidal compounds of *Nectria pseudotrichia*, an endophytic fungus isolated from the plant *Caesalpinia echinata* (Brazilwood). *Mem. Inst. Oswaldo Cruz.* **2018**, *113*, 102–110. [CrossRef] [PubMed]

81. De Souza, J.T.; Trocoli, R.O.; Monteiro, F.P. Plants from the Caatinga biome harbor endophytic *Trichoderma* species active in the biocontrol of pineapple fusariosis. *Biol. Control* **2016**, *94*, 25–32. [CrossRef]

82. Ali, M.S.; Azhar, I.; Amtul, Z.; Ahmad, V.U.; Usmanghani, K. Antimicrobial screening of some Caesalpiniaceae. *Fitoterapia* **1999**, *70*, 299–304. [CrossRef]

83. Singh, S.; Singh, S.K.; Yadav, A. A review on *Cassia* species: Pharmacological, traditional and medicinal aspects in various countries. *Am. J. Phytomedicine Clin. Ther.* **2013**, *1*, 291–312.

84. Parsons, W.T.; Cuthbertson, E.G. *Noxious Weeds of Australia*, 2nd ed.; Commonwealth Scientific and Industrial Research Organisation (CSIRO): Canberra, Australia, 2001.

85. Jang, D.S.; Lee, G.Y.; Kim, Y.S.; Lee, Y.M.; Kim, C.S.; Yoo, J.L.; Kim, J.S. Anthraquinones from the seeds of *Cassia tora* with inhibitory activity on protein glycation and aldose reductase. *Biol. Pharm. Bull.* **2007**, *30*, 2207–2210. [CrossRef]

86. Hatano, T.; Uebayashi, H.; Ito, H.; Shiota, S.; Tsuchiya, T.; Yoshida, T. Phenolic constituents of *Cassia* seeds and antibacterial effect of some naphthalenes and anthraquinones on methicillin-resistent *Staphylococcus aureus*. *Chem. Pharm. Bull.* **1999**, *47*, 1121–1127. [CrossRef]

87. Ibrahim, D.; Nurhaida, L.S.H. Anti-candidal activity of *Aspergillus flavus* IBRL-C8, an endophytic fungus isolated from *Cassia siamea* Lamk leaf. *J. Appl. Pharm. Sci.* **2018**, *8*, 083–087.

88. Pundir, R.K.; Yadav, D.; Jain, P. Production, optimization and partial purification of l-asparaginase from endophytic fungus *Aspergillus* sp., isolated from *Cassia fistula*. *Appl. Biol. Res* **2020**, *22*, 26–33. [CrossRef]

89. Ruchikachorn, N. Endophytic Fungi of Cassia Fistula L. Ph.D. Thesis, Liverpool John Moores University, Liverpool, UK, 2005.

90. Nigg, J.; Strobel, G.; Knighton, W.B.; Hilmer, J.; Geary, B.; Riyaz-Ul-Hassan, S.; Harper, J.K.; Valenti, D.; Wang, Y. Functionalized para-substituted benzenes as 1, 8-cineole production modulators in an endophytic *Nodulisporium* species. *Microbiology* **2014**, *160*, 1772–1782. [CrossRef] [PubMed]

91. Tapwal, A.; Pandey, P.; Chandra, S. Antimicrobial activity and phytochemical screening of endophytic fungi associated with *Cassia fistula*. *Int. J. Chem. Biol. Sci.* **2015**, *2*, 15–21.

92. Kuriakose, G.C.; Lakshmanan, D.M.; Arathi, B.P.; Kuriakose, G.C.; Kumar, R.S.; Anantha Krishna, A.; Ananthaswamy, K.; Jayabhaskaran, C. Extract of *Penicillium sclerotiorum* an endophytic fungus isolated from *Cassia fistula* L. induces cell cycle arrest leading to apoptosis through mitochondrial membrane depolarization in human cervical cancer cells. *Biomed. Pharm.* **2018**, *105*, 1062–1071. [CrossRef] [PubMed]

93. Call, A.; Sun, Y.X.; Yu, Y.; Pearman, P.B.; Thomas, D.T.; Trigiano, R.N.; Carbone, I.; Xiang, Q.-Y. Genetic structure and post-glacial expansion of Cornus florida L. (Cornaceae): Integrative evidence from phylogeography, population demographic history, and species distribution modeling. *J. Syst. Evol.* **2016**, *54*, 136–151. [CrossRef]

94. Huang, J.; Zhang, Y.; Dong, L.; Gao, Q.; Yin, L.; Quan, H.; Chen, R.; Gu, X.; Lin, D. Ethnopharmacology, phytochemistry, and pharmacology of *Cornus officinalis* Sieb. et Zucc. *J. Ethnopharmacol.* **2018**, *213*, 280–301. [CrossRef] [PubMed]

95. Miller, S.; Masuya, H.; Zhang, J.; Walsh, E.; Zhang, N. Real-Time PCR detection of dogwood anthracnose fungus in historical herbarium specimens from Asia. *PLoS ONE* **2016**, *11*, e0154030. [CrossRef]

96. Osono, T. Endophytic and epiphytic phyllosphere fungi of red-osier dogwood (*Cornus stolonifera*) in British Columbia. *Mycoscience* **2007**, *48*, 47–52. [CrossRef]

97. Zhao, X.; Hu, Z.; Hou, D.; Xu, H.; Song, P. Biodiversity and antifungal potential of endophytic fungi from the medicinal plant *Cornus officinalis*. *Symbiosis* **2020**, *81*, 223–233. [CrossRef]

98. Andrés, M.F.; Diaz, C.E.; Giménez, C.; Cabrera, R.; González-Coloma, A. Endophytic fungi as novel sources of biopesticides: The Macaronesian Laurel forest, a case study. *Phytochem. Rev.* **2017**, *16*, 1009–1022. [CrossRef]

99. Redlin, S.C.; Rossman, A.Y. Cryptodiaporthe corni (Diaporthales), cause of Cryptodiaporthe canker of pagoda dogwood. *Mycologia* **1991**, *83*, 200–209. [CrossRef]

100. Beier, G.L.; Hokanson, S.C.; Bates, S.T.; Blanchette, R.A. *Aurantioporthe corni* gen. et comb. nov., an endophyte and pathogen of *Cornus alternifolia*. *Mycologia* **2015**, *107*, 66–79. [CrossRef] [PubMed]

101. Redlin, S.C. *Discula destructiva* sp. nov., cause of dogwood anthracnose. *Mycologia* **1991**, *83*, 633–642. [CrossRef]

102. Stinzing, A.; Lang, K.J. Dogwood anthracnose. *First detection of Discula destructiva on Cornus florida in Germany*. *Nachr. Des. Dtsch. Pflanzenschutzd.* **2003**, *55*, 1–5.

103. Tantardini, A.; Calvi, M.; Cavagna, B.; Zhang, N.; Geiser, D. First report in Italy of dogwood anthracnose on *Cornus florida* and C. *nuttallii* caused by *Discula destructiva* [Lombardy]. *Inf. Fitopatol.* **2004**, *54*, 44–47.

104. Yun, H.Y.; Lee, Y.W.; Kim, Y.H. Stem canker of giant dogwood (*Cornus controversa*) caused by *Fusarium lateritium* in Korea. *Plant. Dis.* **2013**, *97*, 1378. [CrossRef] [PubMed]

105. Maheshwari, A.; Mmbaga, M.; Quick, Q. *Nigrospora sphaerica* products from the flowering dogwood exhibit antitumorigenic effects via the translational regulator, pS6 ribosomal protein. *Proc. Anticancer. Res.* **2018**, *2*, 8–15. [CrossRef]

106. Hagan, A.; Mullen, J. *Controlling Powdery Mildew on Ornamentals, Alabama Coop. Ext. Sys. Circular ANR-407*; Auburn University: Auburn, AL, USA, 1995.

107. Erdelmeier, C.A.J.; Cinatl, J.; Rabenau, H.; Doerr, H.W.; Biber, A.; Koch, E. Antiviral and antiphlogistic activities of *Hamamelis virginiana* bark. *Planta Med.* **1996**, *62*, 241–245. [CrossRef]

108. Gloor, M.; Reichling, J.; Wasik, B.; Holzgang, H.E. Antiseptic effect of a topical dermatological formulation that contains *Hamamelis* distillate and urea. *J. Complement. Med. Res.* **2002**, *9*, 153–159. [CrossRef]

109. Marcelino, J.A.P.; Gouli, S.; Parker, B.L.; Skinner, M.; Giordano, R. Entomopathogenic activity of a variety of the fungus, *Colletotrichum acutatum*, recovered from the elongate hemlock scale, *Fiorinia externa*. *J. Insect Sci.* **2009**, *9*, 1–11. [CrossRef]

110. Wang, M.; Liu, F.; Crous, P.W.; Cai, L. Phylogenetic reassessment of *Nigrospora*: Ubiquitous endophytes, plant and human pathogens. *Persoonia* **2017**, *39*, 118. [CrossRef]

111. Chen, C.; Verkley, G.J.M.; Sun, G.; Groenewald, J.Z.; Crous, P.W. Redefining common endophytes and plant pathogens in *Neofabraea*, *Pezicula*, and related genera. *Fungal Biol.* **2016**, *120*, 1291–1322. [CrossRef] [PubMed]

112. Abeln, E.C.A.; De Pagter, M.A.; Verkley, G.J.M. Phylogeny of *Pezicula*, *Dermea* and *Neofabraea* inferred from partial sequences of the nuclear ribosomal RNA gene cluster. *Mycologia* **2000**, *92*, 685–693. [CrossRef]

113. Zhou, N.; Chen, Q.; Carroll, G.; Zhang, N.; Shivas, R.G.; Cai, L. Polyphasic characterization of four new plant pathogenic *Phyllosticta* species from China, Japan, and the United States. *Fungal Biol.* **2015**, *119*, 433–446. [CrossRef] [PubMed]

114. Wu, S.P.; Liu, Y.X.; Yuan, J.; Wang, Y.; Hyde, K.D.; Liu, Z.Y. *Phyllosticta* species from banana (*Musa* sp.) in Chongqing and Guizhou Provinces, China. *Phytotaxa* **2014**, *188*, 135–144. [CrossRef]

115. Mahasneh, A.M.; Al-Hussaini, R. Antibacterial and antifungal activity of ethanol extract of different parts of medicinal plants in Jordan. *Jordan J. Pharm. Sci.* **2011**, *108*, 1–26.

116. Abdoul-Latif, F.; Edou, P.; Mohamed, N.; Ali, A.; Djama, S.; Obame, L.C.; Bassolé, I.; Dicko, M. Antimicrobial and antioxidant activities of essential oil and methanol extract of *Jasminum sambac* from Djibouti. *Afr. J. Plant Scie* **2010**, *4*, 038–043.

117. Rambabu, B.; Patnaik, K.S.K. Anti diabetic and anti ulcer activity of ethanolic flower extract of *Jasminum sambac* in rats. *Asian J. Res. Chem.* **2014**, *7*, 580–585.

118. Zhao, G.; Yin, Z.; Dong, J. Antiviral efficacy against hepatitis B virus replication of oleuropein isolated from *Jasminum officinale* L. var. *grandiflorum*. *J. Ethnopharmacol.* **2009**, *125*, 265–268. [CrossRef]

119. Rahman, M.A.; Hasan, M.S.; Hossain, M.A.; Biswas, N.N. Analgesic and cytotoxic activities of *Jasminum sambac* (L.) Aiton. *Pharmacologyonline* **2011**, *1*, 124–131.

120. Agarwal, G.P.; Sahni, V.P. Fungi causing plant diseases at Jabalpur (MP) XI. *Mycopathol. Mycol. Appl.* **1965**, *27*, 136–144. [CrossRef]

121. Wikee, S.; Cai, L.; Pairin, N.; McKenzie, E.H.; Su, Y.Y.; Chukeatirote, E.; Thi, H.N.; Moslem, M.A.; Abdelsalam, K.; Hyde, K.D. *Colletotrichum* species from Jasmine (*Jasminum sambac*). *Fungal Divers.* **2011**, *46*, 171–182. [CrossRef]

122. Starr, F.; Starr, K.; Loope, L. Ligustrum spp.: Privet, Oleaceae. Maui: United States Geological Survey-Biological Resources Division, Haleakala Field Station. 2003. Available online: http://www.hear.org/pier/pdf/pohreports/ligustrum_spp.pdf (accessed on 15 October 2020).

123. Fernandez, R.D.; Ceballos, S.J.; Aragón, R.; Malizia, A.; Montti, L.; Whitworth-Hulse, J.I.; Castro-Diez, P.; Grau, H.R. A Global Review of *Ligustrum Lucidum* (Oleaceae) Invasion. *Bot. Rev.* **2020**, *86*, 93–118. [CrossRef] [PubMed]

124. Che, C.T.; Wong, M.S. *Ligustrum lucidum* and its constituents: A mini-review on the anti-osteoporosis potential. *Nat. Prod. Commun.* **2015**, *10*. [CrossRef]

125. Kim, Y.S.; Lee, S.J.; Hwang, J.W.; Kim, E.H.; Park, P.J.; Jeong, J.H. Anti-inflammatory effects of extracts from *Ligustrum ovalifolium* H. leaves on RAW264. 7 macrophages. *J. Korean Soc. Food Sci. Nutr.* **2012**, *41*, 1205–1210. [CrossRef]

126. Novas, M.V.; Carmarán, C.C. Studies on diversity of foliar fungal endophytes of naturalised trees from Argentina, with a description of *Haplotrichum minutissimum* sp. nov. *Flora: Morphol. Distrib. Funct. Ecol. Plants* **2008**, *203*, 610–616. [CrossRef]

127. De errasti, A.; Carmarán, C.C.; Novas, M.V. Diversity and significance of fungal endophytes from living stems of naturalized trees from Argentina. *Fungal Divers.* **2010**, *41*, 29–40. [CrossRef]

128. Liu, C.; Liu, T.; Yuan, F.; Gu, Y. Isolating endophytic fungi from evergreen plants and determining their antifungal activities. *Afr. J. Microbiol. Res.* **2010**, *4*, 2243–2248.

129. Okane, I.; Lumyong, S.; Nakagiri, A.; Ito, T. Extensive host range of an endophytic fungus, *Guignardia endophyllicola* (anamorph: *Phyllosticta capitalensis*). *Mycoscience* **2003**, *44*, 353–363. [CrossRef]

130. De Errasti, A.; Novas, M.V.; Carmarán, C.C. Plant-fungal association in trees: Insights into changes in ecological strategies of *Peroneutypa scoparia* (Diatrypaceae). *Flora* **2014**, *209*, 704–710. [CrossRef]

131. Xu, F.; Cao, H.; Cui, X.; Guo, H.; Han, C. Optimization of Fermentation Condition for Echinacoside Yield Improvement with *Penicillium* sp. H1, an Endophytic Fungus Isolated from *Ligustrum lucidum* Ait Using Response Surface Methodology. *Molecules* **2018**, *23*, 2586. [CrossRef]

132. Reddy, M.S.; Murali, T.S.; Suryanarayanan, T.S.; Govinda Rajulu, M.B.; Thirunavukkarasu, N. *Pestalotiopsis* species occur as generalist endophytes in trees of Western Ghats forests of southern India. *Fungal Ecol.* **2016**, *24*, 70–75. [CrossRef]

133. Krohn, K.; Farooq, U.; Hussain, H.; Ahmed, I.; Rheinheimer, J.; Draeger, S.; Schulz, B.; van Ree, T. Phomosines H–J, novel highly substituted biaryl ethers, isolated from the endophytic fungus *Phomopsis* sp. from *Ligustrum vulgare*. *Nat. Prod. Commun.* **2011**, *6*, 1907–1912. [PubMed]

134. Schierenbeck, K.A. Japanese honeysuckle (*Lonicera japonica*) as an invasive species; history, ecology, and context. *Crit. Rev. Plant Sci.* **2004**, *23*, 391–400. [CrossRef]

135. Merriam, R.W. The abundance, distribution and edge associations of six non-indigenous, harmful plants across North Carolina. *J. Torrey Bot. Soc.* **2003**, *2003*, 283–291. [CrossRef]

136. Zhang, H.; Sun, X.; Xu, C. Antimicrobial activity of endophytic fungus *Fusarium* sp. isolated from medicinal honeysuckle plant. *Arch. Biol. Sci.* **2016**, *68*, 25–30. [CrossRef]

137. Herrera, J. The reproductive biology of a riparian Mediterranean shrub, *Nerium oleander* L. (Apocynaceae). *Bot. J. Lin. Soc.* **1991**, *106*, 147–172. [CrossRef]

138. Fu, L.; Zhang, S.; Li, N.; Wang, J.; Zhao, M.; Sakai, J.; Haseqawa, T.; Mitsu, T.; Kataoka, T.; Oka, S.; et al. Three new triterpenes from *Nerium oleander* and biological activity of the isolated compounds. *Nat. Prod.* **2005**, *68*, 198–206. [CrossRef]

139. El-Shazly, M.M.; El-Zayat, E.M.; Hermersdörfer, H. Insecticidal activity, mammalian cytotoxicity and mutagenicity of an ethanolic extract from *Nerium oleander* (Apocynaceae). *Ann. Appl. Biol.* **2000**, *136*, 153–157. [CrossRef]

140. Zhao, M.; Zhang, S.; Fu, L.; Li, N.; Bai, J.; Sakai, J.; Kataoka, T.; Oka, S.; Kiuch, M.; Hirose, K.; et al. Taraxasterane-and Ursane-Type Triterpenes from *Nerium oleander* and Their Biological Activities. *J. Nat. Prod.* **2006**, *69*, 1164–1167. [CrossRef]

141. Ramesha, A.; Sunitha, V.H.; Srinivas, C. Antimicrobial activity of secondary metabolites from endophytic fungi isolated from *Nerium oleander* L. *Int. J. Pharma Bio Sci.* **2013**, *4*, B683–B693.

142. Shankar Naik, B.; Shashikala, J.; Krishnamurthy, Y.L. Diversity of fungal endophytes in shrubby medicinal plants of Malnad region, Western Ghats, Southern India. *Fungal Ecol.* **2008**, *1*, 89–93. [CrossRef]

143. Kannan, K.; Madhankumar, D.; Prakash, N.U.; Muthezilan, R.; Jamuna, G.; Parthasarathy, N.; Bhuvaneshwari, S. Fungal endophytes: A preliminary report from marketed flowers. *Int. J. Appl. Biol.* **2011**, *2*, 14–18.

144. Na, R.; Jiajia, L.; Dongliang, Y.; Yingzi, P.; Juan, H.; Xiong, L.; Nana, Z.; Jing, Z.; Yitian, L. Indentification of vincamine indole alkaloids producing endophytic fungi isolated from *Nerium indicum*, Apocynaceae. *Microbiol. Res.* **2016**, *192*, 114–121. [CrossRef]

145. Huang, W.Y.; Cai, Y.Z.; Hyde, K.D.; Corke, H.; Sun, M. Endophytic fungi from *Nerium oleander* L. (Apocynaceae): Main constituents and antioxidant activity. *World J. Microbiol. Biotechnol.* **2007**, *23*, 1253–1263. [CrossRef]

146. Ma, Y.M.; Qiao, K.; Kong, Y.; Li, M.Y.; Guo, L.X.; Miao, Z.; Fan, C. A new isoquinolone alkaloid from an endophytic fungus R22 of *Nerium indicum*. *Nat. Prod. Res.* **2017**, *31*, 951–958. [CrossRef] [PubMed]

147. Vítková, M.; Müllerová, J.; Sádlo, J.; Pergl, J.; Pyšek, P. Black locust (*Robinia pseudoacacia*) beloved and despised: A story of an invasive tree in Central Europe. *For. Ecol. Manag.* **2017**, *384*, 287–302. [CrossRef]

148. European Commission. Commission Regulation No 1143/2014 of 22 October 2014 on the prevention and management of the introduction and spread of invasive alien species. *Off. J. Eur. Union* **2014**, *57*, 35–55.

149. European Commission. Commission Implementing Regulation No 1141/2016 of 13 July 2016 adopting a list of invasive alien species of Union concern pursuant to Regulation (EU) No 1143/2014 of the European Parliament and of the Council. *Off. J. Eur. Union* **2016**, *59*, 4–8.

150. Ferus, P.; Barta, M.; Konôpková, J. Endophytic fungus *Beauveria bassiana* can enhance drought tolerance in red oak seedlings. *Trees Struct. Funct.* **2019**, *33*, 1179–1186. [CrossRef]

151. Vajna, L. *Diaporthe oncostoma* causing stem canker of black locust in Hungary. *Plant. Pathol.* **2002**, *51*, 393. [CrossRef]

152. Krohn, K.; John, M.; Aust, H.J.; Draeger, S.; Schulz, B. Biologically active metabolites from fungi 131 Stemphytriol, a new perylene derivative from *Monodictys fluctuata*. *Nat. Prod. Lett.* **1999**, *14*, 31–34. [CrossRef]

153. Zheng, H.; Zhang, Z.Y.; Han, Y.F.; Chen, W.H.; Liang, Z.Q. Community composition and ecological functional structural analysis of the endophytic fungi in Robinia Pseudoacacia. *Mycosystema* **2018**, *2*, 13.

154. Saikkonen, K.; Faeth, S.H.; Helander, M.; Sullivan, T.J. Fungal endophytes: A continuum of interactions with host plants *Annu. Rev. Ecol. Evol. Syst.* **1998**, *29*, 319–343. [CrossRef]

155. Stone, J.K.; Polishook, J.D.; White, J.F. Endophytic fungi. In *Biodiversity of 25 Fungi. Inventory and Monitoring Methods*; Mueller, G.M., Bills, G.F., Foster, M.S., Eds.; Elsevier Academic Press: San Diego, CA, USA, 2004; pp. 241–270.

156. Busby, P.E.; Ridout, M.; Newcombe, G. Fungal endophytes: Modifiers of plant disease. *Plant Mol. Biol.* **2016**, *90*, 645–655. [CrossRef] [PubMed]

157. Rodriguez, R.J.; White, J.F., Jr.; Arnold, A.E.; Redman, A.R.A. Fungal endophytes: Diversity and functional roles. *New Phytol.* **2009**, *182*, 314–330. [CrossRef]

158. Mucciarelli, M.; Scannerini, S.; Bertea, C.; Maffei, M. In vitro and in vivo peppermint (*Mentha piperita*) growth promotion by nonmycorrhizal fungal colonization. *New Phytol.* **2003**, *158*, 579–591. [CrossRef]

159. Wearn, J.A.; Sutton, B.C.; Morley, N.J.; Gange, A.C. Species and organ specificity of fungal endophytes in herbaceous grassland plants. *J. Ecol.* **2012**, *100*, 1085–1092. [CrossRef]

160. Cheplick, G.P.; Cho, R. Interactive effects of fungal endophyte infection and host genotype on growth and storage in *Lolium perenne*. *New Phytol.* **2003**, *158*, 183–191. [CrossRef]

161. Geisen, S.; Kostenko, O.; Cnossen, M.C.; ten Hooven, F.C.; Vreš, B.; van Der Putten, W.H. Seed and root endophytic fungi in a range expanding and a related plant species. *Front. Microbiol.* **2017**, *8*, 1645. [CrossRef]

162. Arnold, A.E. Understanding the diversity of foliar endophytic fungi: Progress, challenges, and frontiers. *Fungal Biol. Rev.* **2007**, *21*, 51–66. [CrossRef]

163. Hoffman, M.T.; Arnold, A.E. Geographic locality and host identity shape fungal endophyte communities in cupressaceous trees. *Mycol. Res.* **2008**, *112*, 331–344. [CrossRef]

164. Davis, E.C.; Franklin, J.B.; Shaw, A.J.; Vilgalys, R. Endophytic *Xylaria* (Xylariaceae) among liverworts and angiosperms: Phylogenetics, distribution, and symbiosis. *Am. J. Bot.* **2003**, *90*, 1661–1667. [CrossRef] [PubMed]

165. Kowalski, T.; Kehr, R. Fungal endophytes of living branch bases in several European tree species. In *Endophytic Fungi in Grasses an Woody Plants*; Redlin, S.C., Carris, L.M., Eds.; American Phytopathological Society (APS): St Paul, MN, USA, 1996; pp. 67–86.

166. Peláez, F.; Collado, J.; Arenal, F.; Basilio, A.; Cabello, A.; Matas, M.D.; Garcia, J.B.; Del Val, A.G.; González, V.; Gorrochategui, J.; et al. Endophytic fungi from plants living on gypsum soils as a source of secondary metabolites with antimicrobial activity. *Mycol. Res.* **1998**, *102*, 755–761. [CrossRef]

167. Michielse, C.B.; Rep, M. Pathogen profile update: *Fusarium oxysporum*. *Mol. Plant. Pathol.* **2009**, *10*, 311. [CrossRef] [PubMed]

168. Logrieco, A.; Moretti, A.; Solfrizzo, M. *Alternaria* toxins and plant diseases: An overview of origin, occurrence and risks. *World Mycotoxin J.* **2009**, *2*, 129–140. [CrossRef]

169. Samson, A.R.; Hoekstra, E.S.; Frisvad, J.C.; Filtenborg, O. *Introduction to Food and Airborne Fungi*, 6th ed.; The Dutch Centraalbureau Voor Schimmelcultures: Utrecht, The Netherlands, 2000; pp. 64–97.

170. Gautam, A.K.; Sharma, S.; Avasthi, S.; Bhadauria, R. Diversity, Pathogenicity and Toxicology of *A. niger*: An important spoilage fungi. *Res. J. Microbiol.* **2011**, *6*, 270–280. [CrossRef]

171. Phoulivong, S.; Cai, L.; Chen, H.; McKenzie, E.H.C.; Abdelsalam, K.; Chukeatirote, E.; Hyde, K.D. *Colletotrichum gloeosporioides* is not a common pathogen on tropical fruits. *Fungal Divers.* **2010**, *44*, 33–43. [CrossRef]

172. Bisht, S.; Kumar, P.; Purohit, J. In Vitro Management of *Curvularia* Leaf Spot of Maize Using Botanicals, Essential Oils and Bio-Control Agents. *Bioscan* **2013**, *8*, 731–733.

173. Vitale, S.; Santori, A.; Wajnberg, E.; Castagnone-Sereno, P.; Luongo, L.; Belisario, A. Morphological and molecular analysis of *Fusarium lateritium*, the cause of gray necrosis of hazelnut fruit in Italy. *Phytopathology* **2011**, *101*, 679–686. [CrossRef]

174. Yaghmour, M.A.; Bostock, R.M.; Adaskaveg, J.E.; Michailides, T.J. Propiconazole sensitivity in populations of *Geotrichum candidum*, the cause of sour rot of peach and nectarine, in California. *Plant. Dis.* **2012**, *96*, 752–758. [CrossRef]

175. Nam, M.H.; Park, M.S.; Kim, H.S.; Kim, T.I.; Kim, H.G. *Cladosporium cladosporioides* and *C. tenuissimum* cause blossom blight in strawberry in Korea. *Mycobiology* **2015**, *43*, 354–359. [CrossRef] [PubMed]

176. Manca, D.; Bregant, C.; Maddau, L.; Pinna, C.; Montecchio, L.; Linaldeddu, B.T. First report of canker and dieback caused by *Neofusicoccum parvum* and *Diplodia olivarum* on oleaster in Italy. *Ital. J. Mycol.* **2020**, *49*, 85–91. [CrossRef]

177. Sharma, P.; Meena, P.D.; Chauhan, J.S. First Report of *Nigrospora oryzae* (Berk. & Broome) Petch Causing Stem Blight on *Brassica juncea* in India. *J. Phytopathol.* **2013**, *161*, 439–441. [CrossRef]

178. Salvatore, M.M.; Andolfi, A.; Nicoletti, R. The Thin Line between Pathogenicity and Endophytism: The Case of *Lasiodiplodia theobromae*. *Agriculture* **2020**, *10*, 488. [CrossRef]

179. Mantooth, K.; Hadziabdic, D.; Boggess, S.; Windham, M.; Miller, S.; Cai, G.; Spatafora, J.; Zhang, N.; Staton, M.; Ownley, B.; et al. Confirmation of independent introductions of an exotic plant pathogen of *Cornus* species, *Discula destructiva*, on the east and west coasts of North America. *PLoS ONE* **2017**, *12*, 1–26. [CrossRef]

180. Saunders, M.; Glenn, A.E.; Kohn, L.M. Exploring the evolutionary ecology of fungal endophytes in agricultural systems: Using functional traits to reveal mechanisms in community processes. *Evol. Appl.* **2010**, *3*, 525–537. [CrossRef]

181. Maciá-Vicente, J.G.; Ferraro, V.; Burruano, S.; Lopez-Llorca, L.V. Fungal assemblages associated with roots of halophytic and non-halophytic plant species vary differentially along a salinity gradient. *Microb. Ecol.* **2012**, *64*, 668–679. [CrossRef]

182. McKinney, L.V.; Thomsen, I.M.; Kjær, E.D.; Bengtsson, S.B.K.; Nielsen, L.R. Rapid invasion by an aggressive pathogenic fungus (*Hymenoscyphus pseudoalbidus*) replaces a native decomposer (*Hymenoscyphus albidus*): A case of local cryptic extinction? *Fungal Ecol.* **2012**, *5*, 663–669. [CrossRef]

183. Ragazzi, A.; Moricca, S.; Capretti, P.; Dellavalle, I.; Turco, E. Differences in composition of endophytic mycobiota in twigs and leaves of healthy and declining *Quercus* species in Italy. *For. Pathol.* **2003**, *33*, 31–38. [CrossRef]

184. Moricca, S.; Ginetti, B.; Ragazzi, A. Species-and organ-specificity in endophytes colonizing healthy and declining Mediterranean oaks. *Phytopathol. Mediterr.* **2012**, *51*, 587–598.

185. Collado, J.; Platas, G.; González, I.; Peláez, F. Geographical and seasonal influences on the distribution of fungal endophytes in *Quercus ilex*. *New Phytol.* **1999**, *144*, 525–532. [CrossRef]

186. Peršoh, D. Factors shaping community structure of endophytic fungi–evidence from the *Pinus-Viscum*-system. *Fungal Divers.* **2013**, *60*, 55–69. [CrossRef]

187. Martins, F.; Pereira, J.A.; Bota, P.; Bento, A.; Baptista, P. Fungal endophyte communities in above-and belowground olive tree organs and the effect of season and geographic location on their structures. *Fungal Ecol.* **2016**, *20*, 193–201. [CrossRef]

188. Fisher, P.J.; Petrini, O.; Petrini, L.E.; Sutton, B.C. Fungal endophytes from the leaves and twigs of *Quercus ilex* L. from England, Majorca and Switzerland. *New Phytol* **1994**, *127*, 133–137. [CrossRef]

189. Zamora, P.; Martínez-Ruiz, C.; Diez, J.J. Fungi in needles and twigs of pine plantations from northern Spain. *Fungal Divers.* **2008**, *30*, 171–184.

190. Glynou, K.; Ali, T.; Buch, A.K.; Haghi Kia, S.; Ploch, S.; Xia, X.; Ali Çelik, A.; Thines, M.; Maciá-Vicente, J.G. The local environment determines the assembly of root endophytic fungi at a continental scale. *Environ. Microbiol.* **2016**, *18*, 2418–2434. [CrossRef]

191. EPPO Global Database. Available online: https://gd.eppo.int (accessed on 10 October 2020).

192. Tan, M.K.; Timmer, L.W.; Broadbent, P.; Priest, M.; Cain, P. Differentiation by molecular analysis of *Elsinoë* spp. causing scab diseases of citrus and its epidemiological implications. *Phytopathology* **1996**, *86*, 1039–1044. [CrossRef]

193. d'Errico, G.; Carletti, B.; Schröder, T.; Mota, M.; Vieira, P.; Roversi, P.F. An update on the occurrence of nematodes belonging to the genus *Bursaphelenchus* in the Mediterranean area. *Int. J. For. Res.* **2015**, *88*, 509–520.

194. d'Errico, G.; Crescenzi, A.; Landi, S. First report of the southern root-knot nematode *Meloidogyne incognita* on the invasive weed *Araujia sericifera* in Italy. *Plant. Dis.* **2014**, *98*, 1593. [CrossRef]

195. IPPC. *International Plant. Protection Convention*; IPPC & FAO: Rome, Italy, 1997.

196. WTO. *Agreement on the Application of Sanitary and Phytosanitary Measures*; World Trade Organization: Geneva, Switzerland, 1994.

197. Brasier, C.M. Preventing invasive pathogens: Deficiencies in the system. *Plantsman* **2005**, *4*, 54–57.

198. Brasier, C.M. The biosecurity threat to the UK and global environment from international trade in plants. *Plant Pathol.* **2008**, *57*, 792–808. [CrossRef]

199. Sikes, B.A.; Bufford, J.L.; Hulme, P.E.; Cooper, J.A.; Johnston, P.R.; Duncan, R.P. Import volumes and biosecurity interventions shape the arrival rate of fungal pathogens. *PLoS Biol.* **2018**, *16*, e2006025. [CrossRef] [PubMed]

200. Aglietti, C.; Luchi, N.; Pepori, A.L.; Bartolini, P.; Pecori, F.; Raio, A.; Capretti, P.; Santini, A. Real-time loop-mediated isothermal amplification: An early-warning tool for quarantine plant pathogen detection. *AMB Express* **2019**, *9*, 50. [CrossRef] [PubMed]

201. Botella, L.; Bačová, A.; Dvorák, M.; Kudláček, T.; Pepori, A.L.; Santini, A.; Ghelardini, L.; Luchi, N. Detection and quantification of the air inoculum of *Caliciopsis pinea* in a plantation of *Pinus radiata* in Italy. *iForest Biogeosciences For.* **2019**, *12*, 1–193. [CrossRef]

202. Jactel, H.; Desprez-Loustau, M.L.; Battisti, A.; Brockerhoff, E.; Santini, A.; Stenlid, J.; Björkman, C.; Branco, M.; Dehnen-Schmutz, K.; Douma, J.C.; et al. Pathologists and entomologists must join forces against forest pest and pathogen invasions. *NeoBiota* **2020**, *58*, 107. [CrossRef]

Publisher's Note: MDPI stays neutral with regard to jurisdictional claims in published maps and institutional affiliations.

MDPI
St. Alban-Anlage 66
4052 Basel
Switzerland
Tel. +41 61 683 77 34
Fax +41 61 302 89 18
www.mdpi.com

Agriculture Editorial Office
E-mail: agriculture@mdpi.com
www.mdpi.com/journal/agriculture